ECONOMIC ISSUES, PROBLEMS AND PERSPECTIVES

ECONOMIC DEVELOPMENTS AND EMERGING MARKETS OF THE 21ST CENTURY

GLOBAL PRACTICES, STRATEGIES, AND CHALLENGES

ECONOMIC ISSUES, PROBLEMS AND PERSPECTIVES

Additional books in this series can be found on Nova's website under the Series tab.

Additional e-books in this series can be found on Nova's website under the e-book tab.

GLOBAL ECONOMIC STUDIES

Additional books in this series can be found on Nova's website under the Series tab.

Additional e-books in this series can be found on Nova's website under the e-book tab.

ECONOMIC ISSUES, PROBLEMS AND PERSPECTIVES

ECONOMIC DEVELOPMENTS AND EMERGING MARKETS OF THE 21ST CENTURY

GLOBAL PRACTICES, STRATEGIES, AND CHALLENGES

GUY BOSCARINO
AND
DEREK NOTTE
EDITORS

nova publishers
New York

For permission to use material from this book please contact us:
Telephone 631-231-7269; Fax 631-231-8175
Web Site: http://www.novapublishers.com

NOTICE TO THE READER

Library of Congress Cataloging-in-Publication Data

Economic developments and emerging markets of the 21st century : global practices, strategies, and challenges / editors, Guy Boscarino and Derek Notte.
pages cm
Includes index.
ISBN: 978-1-62417-986-0 (hardcover)
1. Economic forecasting--21st century. 2. Technological innovations. I. Boscarino, Guy, editor of compilation. II. Notte, Derek, editor of compilation.
HC59.3.E25 2013
338.9--dc23

2013001535

Published by Nova Science Publishers, Inc. † New York

CONTENTS

PREFACE

In this book, the authors present current research in the study of the global practices, strategies and challenges of global economic development and emerging markets in the 21st century. Topics discussed include the role of social sciences in global economic development; parametric control of economic growth processes and characteristics of short-term cycles; development of the world economy in the system of continental-oceanic polarization; green economies in rural areas through geotourism and geoparks; perspectives on poverty alleviation through community-based enterprise, innovation capabilities and vertical integration mindsets; U-City development for economic competitiveness in an advanced ICT era; EU low-carbon revolution's influence on China and its countermeasures; risk management of floodwater utilization based on water utilities IOT; and poverty, inequality and socio-economic development in Colombia.

Chapter 1 – The need of humans to quench their scientific thirst, to investigate the mysteries of their lives and the national phenomena has created the environment within which renowned scientists were brought up, studied and achieved unprecedented goals. As a prelude to the importance of sciences in formulating our lives, the authors will review a number of scientists who have changed our world with their achievements. In the economic sector, the authors will analyze the strategies of economists and the causes of recessions. They will focus their attention to the factors causing recessions, such as economists, politicians, employers and illiterate persons.

Sciences are studied on the basis of (a) *natural sciences* which explain the phenomena of nature through biology, chemistry, physics, geology and meteorology, (b) *formal sciences* which constitute a system of generating knowledge through mathematics, logic, statistics, information theory and computers, and (c) *social sciences* which trace the origin and development of the human aspects of the world through communication, economics, education, psychology and sociology. This chapter will define and analyze the specific impact of each social branch on economic development in the present century. Practices, strategies and challenges pertaining to international economies, in simple terms are the "main gates" through which economic development can be achieved.

1. Practices: If we accept that economic practices are the mechanisms which constitute the driving force for the economic development then their impact on the relative growth can be negative or positive subject to their effectiveness for various reasons as regards the anticipated target. For example, the population growth in certain countries (1980) blocked the growth of GDP though for demographic reasons this growth was highly desirable. On the

contrary, decline in human fertility in other countries (early 1990) resulted to explosive economic growth. Besides, according to the economists Kelley, A., Schmidt, R., population growth (1980) acted as a brake on economic growth.

2. Strategies: Contemporary economic development strategies such as tax subsidies, training programs, modernization of businesses provide initiatives which are expected to increase cash flow which will contribute to the economic growth.

Quite recently (May 17, 2012) President Obama called on Congress to pass a law according to which nearly two million companies which make new hires or increase wages will receive a tax credits.

3. Challenges: In our world of important changes challenges have been posed related to current economic demands which have to be met by:

- Free trade spirit.
- Strong information structure.
- Supporting SMEs
- Developing Human Recourse Managers.
- Developing Technology and New Businesses.

According to experience what matters is not the concept but the effectiveness of strategies influenced by the prevailing environments.

In the following pages the authors will illustrate the reliability of this notion through researches and authentic global evidences contributing to a science-based sustainable economic development in the 21st Century.

Chapter 2 – In this work the authors briefly consider the components of the parametric control theory and list the main results of its component-wise development. The results of macroeconomic analysis and application of the parametric control theory to solving the problems of economic growth and attenuation of effects of various shocks on the base of a number of mathematical models are presented.

The general representation of computable models of general equilibrium (CGE models) as well as description of the CGE model of the branches of economy is considered. The method of parametric identification of the CGE model of the branches of economy is presented on the base of statistical data on evolution of the national economy of the Kazakhstan Republic.

Within the framework of application of the parametric control theory, the results of estimations of the stability indexes for the estimated CGE model of the branches of economy and examples of macroeconomic analysis along various cuts on the base of this model are given.

The authors choose the instruments of economic policy in the sphere of economic growth on the base of analysis of the sources of economic growth within the framework of the CGE model of the branches of economy and give a solution to one problem of economic growth with application of the parametric control approach based on the considered model.

Also, in this work we consider the problem of choice of the effective coordinated public policy in the sphere of economic growth by means of multiobjective optimization on the base of two CGE models with common economic instruments of optimization.

The results of macroeconomic analysis and parametric control of the processes of economic growth on the base of the Smets-Wouters dynamic stochastic general equilibrium model (DSGE model) estimated by the statistical data of the Kazakhstan Republic are described. In particular, the authors present the estimation of effect of internal shocks on the economic indexes of the Kazakhstan Republic and give a solution to one problem of economic growth on the base of the model under studying.

In the rest of the paper the authors present the results of parametric control of the characteristics of the short-term cycles on the base of the CGE model of the branches of economy and the model of Kondratiev wave, as well as the results of parametric control of effect of shocks on the evolution of national economy on the base of the Smets-Wouters DSGE model estimated by the statistical data of economy of the Kazakhstan Republic.

Chapter 3 – This study examines the main economic mechanisms as reported in scientific literature that are responsible for the performance of the system of "core-periphery" relations in the world economy. It is pointed out that there is a need for a more comprehensive validation of the "core-periphery" model through the use of alternative and complementary methodological approaches.

Such an additional substantiation is suggested in the form of the economic-geographical concept of the continental-oceanic dichotomy to furnish a means of adding immeasurably to the authors' fundamental knowledge of the reasons behind the current structure of the world economy. The aforementioned dichotomy is considered to mean a successive and fundamental divarication of the integral world economy into two opposite and, at the same time, mutually complementing types of economies: continental and oceanic (maritime).

A geoeconomic typology has been performed for a number of key countries of the world by categorizing them as continental and oceanic (maritime), based on assessing the degree of their "transport-geographical continentally". It was possible to unravel the workings of the redistribution mechanism for income (the mass of surplus product) between landlocked and maritime countries in the process of international division of labor. The study highlights the main negative socio-economic implications of increased transport expenditures for landlocked commodity producers and regions.

This study has ascertained a correlation of the degree of transport-geographical continentally of the countries with the level of their socioeconomic development which has been assessed by using the indicators of per capita GNP, per capita export, and life expectancy.

Based on estimating the changes in the volumes of world production and export as regards products of different transportability depending on the remoteness of producers and exporters from the sea, this study revealed the most general regularities of territorial organization of the world economy in the system of continental-oceanic polarization. It is established that the maritime countries and maritime zones play a leading role in the shaping of the world economy and price formation in world commodity markets.

It is suggested that there is a need for a radical change in the international commercial rules in order to eliminate the one-sided economic advantages of the maritime (seaside) countries to the detriment of the continental (landlocked) countries.

Chapter 4 – There are a variety of niche tourism markets which develop depending on tourism supply and demand. In recent decades, geotourism as a new niche market has emerged to introduce geo-heritage sites – geology, geomorphology, geography, and geodiversity – as new tourist attractions and geoparks as new geotourism destinations.

Geoparks are an innovation for the protection of natural, cultural and geological heritage sites and sustainability. The European Geopark Network and the UNESCO Global Geoparks Network have introduced geotourism as one of the main targets for geoparks in parallel with conservation and education. For that purpose, it is crucial to apply the network concept to geotourism activities. Network activity not only provides an opportunity to benefit from the open world market, but also facilitates knowledge transfer. Geoparks, through geotourism and innovative strategies, strive to promote green economy for local communities. This chapter aims at introducing geotourism as a new green economy for rural development and popularization of sustainability and geo-sciences.

Chapter 5 – Recent movements in business research have placed greater emphasis on the "Base of the Pyramid," the billions of people living in extreme poverty. These movements include Transformative Consumer Research, research on Subsistence Marketplaces, and the BoP Protocol, all of which have subtly different goals, but all focus on the improved conditions of consumers living in poverty through a deeper understanding of the individuals and marketplaces. A review of the research in this area reveals a number of articles that have examined these consumers and marketplaces from the bottom-up. On the other hand much of the traditional work in poverty alleviation addresses issues from the national-level, down. The authors propose that a missing, but vital link is the deeper understanding of the marketplaces and systems at the community-level. The authors argue that poverty alleviation efforts focused on empowering individual entrepreneurs, while potentially life-changing for these individuals and their families, tends to result in limited benefits for the community at-large. The authors then discuss the importance of innovation in lifting communities out of from what they describe as "Stage 0," (where communities are exploited for their labor and raw materials). They propose a 4-stage model, ranging from "Stage 0" to "Stage 4" (as communities progress from being exploited for labor/raw materials, to being able to manufacture components/products with older technologies, to being able to manufacture finished systems/products with modern innovations, to finally, being able to innovate independently). Finally, the authors discuss the importance of vertical integration, through a community-based perspective as an enabler of progress, from Stage 0 through the stages. This perspective is unique in that it brings the experience of bottom-up perspectives on poverty alleviation, with the raw potential of innovation as an agent for transformation at the community level.

Chapter 6 – The world economy has undertaken economic restructuring. In a post-industrialization period, information-related industries, associated with urban creativeness, are driving economic growth in the world. Information and Communication Technology (ICT) is critical to make cities competitive in the 21st century. ICT helps to generate high levels of productivity in business sectors and to attract highly educated human capital for new knowledge creation. Governments attempt to provide the advanced ICT infrastructure for achieving a leading role in a global economy since innovative technologies play an important role in the competitiveness of cities. Thus, the primary objective of this paper is to explore the new trend in urban development coupled with ICT infrastructure for enhancing urban competitiveness, using South Korea as case study.

By virtue of advancement in ICT, Korea has initiated to create a Ubiquitous City (U-City). U-City aims to enable any people to access various U-services anywhere and anytime through U-technology. These facilities can strengthen urban competitiveness and improve quality of life attracting highly educated human resources. U-City development is a new approach to improve urban competitiveness in an advanced ICT era.

Chapter 7 – Taking Copenhagen summit on climate change as a symbol, the European countries are focused on the development of low-carbon technologies, and vigorously promote low-carbon revolution to respond to the challenge of climate change, protect secure energy future and capture the economic strategy. The rapid development of low-carbon in Europe, which has a positive impact on China, but at the same time, also constitutes a serious challenge to the status of a new pattern of development in the future world. China should be prepared to develop a low carbon economy strategic planning, strengthen the reserves of long-term carbon strategy and promote the work of the RMB-denominated carbon finance market development, in order to achieve energy conservation and economic development win-win.

Chapter 8 – This chapter uses theoretical analysis combined with case studies to introduce the meaning and conditions of the floodwater utilization in northern Jiangsu, and the key technologies of water utilities IOT architecture and remote sensing equipment, transmission network, water utilities management and feedback mechanism. The chapter explores the function of water utilities IOT technology to strengthen risk management in space, time and capacity, and puts forward suggestions on the formation and improvement of the water utilities IOT, including the establishment of water management information standardization system, the acceleration of the breakthrough of water information sensor core technology, the removal of the industry barriers, regional barriers and departmental barriers.

Chapter 9 – This chapter estimates poverty, inequality and socio-economic trends in Colombia's departments. The authors use the data panel over the period 1993–2007. In this analysis, they suggest a dynamic OLS (DOLS) model to measure poverty, inequality and development trends. Using panel data analysis and cointegration techniques for a dynamic heterogeneous panel the results from DOLS model show variation in the scores across Colombian departments during the sample period. Findings of this analysis should demonstrate that the decrease in poverty and inequality could be achieved through adequate strategies that guarantee development and economic growth with policies concentrated to improve social welfare.

In: Economic Developments and Emerging Markets ...
Editors: Guy Boscarino and Derek Notte

ISBN: 978-1-62417-986-0

Chapter 1

THE ROLE OF SOCIAL SCIENCES IN GLOBAL ECONOMIC DEVELOPMENT

George S. Mouzakitis[*] and Despoina Goutou[†]
Educational Organization e-DEKA,
Aghioi Theodoroi, Almyra, Greece

ABSTRACT

The need of humans to quench their scientific thirst, to investigate the mysteries of their lives and the national phenomena has created the environment within which renowned scientists were brought up, studied and achieved unprecedented goals. As a prelude to the importance of sciences in formulating our lives, we will review a number of scientists who have changed our world with their achievements. In the economic sector, we will analyze the strategies of economists and the causes of recessions. We will focus our attention to the factors causing recessions, such as economists, politicians, employers and illiterate persons.

Sciences are studied on the basis of (a) *natural sciences* which explain the phenomena of nature through biology, chemistry, physics, geology and meteorology, (b) *formal sciences* which constitute a system of generating knowledge through mathematics, logic, statistics, information theory and computers, and (c) *social sciences* which trace the origin and development of the human aspects of the world through communication, economics, education, psychology and sociology. This chapter will define and analyze the specific impact of each social branch on economic development in the present century. Practices, strategies and challenges pertaining to international economies, in simple terms are the "main gates" through which economic development can be achieved.

1. Practices: If we accept that economic practices are the mechanisms which constitute the driving force for the economic development then their impact on the relative growth can be negative or positive subject to their effectiveness for various reasons as regards the anticipated target. For example, the population growth in certain

[*] George S. Mouzakitis: Academic Dean, researcher, Educational Organization e-DEKA.
[†] Despoina Goutou: Assistant Researcher, Educational Organization e-DEKA.

countries (1980) blocked the growth of GDP though for demographic reasons this growth was highly desirable. On the contrary, decline in human fertility in other countries (early 1990) resulted to explosive economic growth. Besides, according to the economists Kelley, A., Schmidt, R., population growth (1980) acted as a brake on economic growth.

2. Strategies: Contemporary economic development strategies such as tax subsidies, training programs, modernization of businesses provide initiatives which are expected to increase cash flow which will contribute to the economic growth.
3. Quite recently (May 17, 2012) President Obama called on Congress to pass a law according to which nearly two million companies which make new hires or increase wages will receive a tax credits.
4. Challenges: In our world of important changes challenges have been posed related to current economic demands which have to be met by:

- Free trade spirit.
- Strong information structure.
- Supporting SMEs
- Developing Human Recourse Managers.
- Developing Technology and New Businesses.

According to experience what matters is not the concept but the effectiveness of strategies influenced by the prevailing environments.

In the following pages we will illustrate the reliability of this notion through researches and authentic global evidences contributing to a science-based sustainable economic development in the 21st Century.

INTRODUCTORY ORIENTATION

During researches, almost around the whole world, we have identified that in spite of the intensive efforts of scientists, which resulted to astonishing developments, practically in all everyday activities in our life span, the intension to achieve the anticipated goals has not been always successful.

In seeking the reason we ascertained the, almost universally accepted, truth: whereas technology develops at accelerated speed, the majority of population is unable to understand and implement innovative and radical practices.

It is our objective to highlight certain cornerstones of scientific outputs connected with data of evolution, important changes, extraordinary achievements in health, space, environmental protection, communication and a host of other sectors including fundamental contributions to cancer research.

According to Bransomb, L. M., (1996) "in all levels of economic activity, the most vulnerable contribution science can make is to inform decisions.

A secondary role is the creation of indigenous technologies or technological improvements. Science-based development implies smart choices of technology made by open-minded innovative and trusting people working in institutional settings that are appropriate to the development conditions at hand" Our presentation of research findings, documented opinions, economic strategies will be based on scientific principles and focus our attenti0n on the role of sciences and scientists contributing to the crucial issues of economic development.

ARGUMENTUM ET SACRIFICUM

Regardless of any religious, philosophical or scientific theories related to the origin and existence of universe, three principles cannot be disputed, the space time, the forms of energy and the physical laws

Yet, for ages there have been controversial opinions between Theism attributing the creation of Universe to an intelligent Designer, whereas Atheists believed that the Universe was the result of undirected process .A Belgian priest George Lemaitre (1920) supported the big bang theory. (Natural Geographic, 2012).

According to this theory, aiming to explain what happened at the very beginning of our Universe (All About Science, 2012), “after its appearance, estimated to about 13.7 million years ago, was inflated (The “Big Bang”) expanded and cooled going from the very small size and extremely hot temperature to the size and temperature of our current universe”.

It is evident that scientists, if left unbiased, work miracles in every sector. For ages, scientists have developed principles through observation and analysis which benefit humanity on condition that they persist in neglecting hereditary beliefs, disregard traditional activities, overlook political influences and religious doctrines.

Scientific researchers have uncovered the secrets of the formation of earth and the evolution after detailed researches, analyses and studies of six important elements, i.e. Geology, Astronomy (recession of the Moon), Earth’s decaying magnetic field, Dinosaur soft tissue, human population and tightly folded rock strata as presented by Burritt, J. (2011), (Table 1)

The roles of scientists in society are to:

(a) conduct various kinds of research to obtain information related to the development of technology,
(b) communicate to society important elements of the world around us,
(c) built a better world,
(d) generate new scientific knowledge, and
(e) mprove means to restore health.

Figure 1. The Origin of Universe.: National Geographic by courtesy of NASA.

Table 1. Evolution Chart

No	Event	Years ago
01	Formation of Earth	4.6 billion
02	Life in bacteria form	3.5 billion
03	Photosynthetic bacteria appearance	1.4 billion
04	Multi-celled plants and animals appearance	1.0 billion
05	Crustaceans appearance	650 million
06	Fish appearance	510 million
07	Land plants appearance	420 million
08	Insects appearance	380 million
09	Seed plants and amphibians appearance	360 million
10	Reptiles appearance	340 million
11	Mammals appearance	213 million
12	Birds appearance	150 million
13	Flowering plants appearance	140 million
14	Humans appearance	2 million
15	Christian religion appearance	2012 years
16	Islamic religion appearance	1402 years

Source: Ontario Consultants on Religious Tolerance, 20.

Figure 2. Hiroshima after bombing.

They quite often risk their lives either because they act against conservatism or testing the results of their experimental outputs. Through their example, they have demonstrated how problems can be solved if systematic mechanisms are implemented.

Quite often the anticipated benefits are not attained for various reasons, mainly due to the high percentages of illiteracy in undeveloped and under development countries or due to political bias to "scientists" who turn the beneficial implementation to catastrophic results.

Einstein's atomic fission which led to the Hiroshima disaster by, the atomic bomb on August 6, 1945 (figure 2) followed by the total disaster of Nagasaki, on September 12, 1945 (figure 3).

Figure 3. Mushroom cloud.

Library of Congress.

Figure 4. Karl Scheele.

Nobel's invention of dynamite was the cause of a disastrous explosion which killed 8 and injured 200 more in Jellico (Tennessee) on September 21, 1906.

1. Karl Scheele (December 9, 1742 – May 21, 1786). Died from testing his discoveries.

Scheele (9 September 1742 – 21 May, 1786, figure 4) was a brilliant German-Swedish pharmaceutical chemist who discovered many chemical elements – the most notable of which were oxygen, which he called "fire air" because it supported combustion manganese (1774) molybdenum (1778) tungsten (1781), and chlorine, as well as chemical compounds such as citric acid, glycerol, hydrogen cyanide. He, also, investigated that air was a mixture of two gases, which he named "fire air" and "foul air". One more important scientific discovery, quire revolutionary, was the identification of lime, silica and iron, in a specimen of purolusite.

Scheele had the habit of testing his discoveries. His luck was to die of symptoms strongly resembling mercury poisoning.

2. Elizabeth Ascheim date of birth probably August 3, 1905. Killed by X-Rays.

The American Elizabeth Fleischman Ascheim (figure) married her doctor Dr Woolf after her mother died. Because of his medical position, Woolf was very interested in the new discovery of Wilhelm Conrad Röntgen – x-rays. His new wife became equally interested and undertook studies in electrical science.

Figure 5. Elizabeth Ascheim.

Figure 6. Marie Curie.

Eventually she bought an x-ray machine which was the first x-ray lab in San Francisco. She and her husband spent some years experimenting with the machine – using themselves as subjects. Elizabeth died of an extremely widespread and violent cancer.

3. Marie Curie I (November 7, 1867 – July 4, 1934). Died of radiation exposure.

In 1898, the French-Polish Curie (figure 60 and her husband, Pierre, discovered radium. She spent the remainder of her life performing radiation research and studying radiation therapy. Her constant exposure to radiation led to her contracting leukemia and she died in 1934. Curie is the first and only person to receive two Nobel prizes in in two different fields: chemistry and physics. She was also the first female professor at the University of Paris. She enjoyed a high esteem and admiration by scientists worldwide and shared with her husband the Nobel Prize in Physics (1963) and this time alone the Nobel Prize in Chemistry (1919)

4. Galileo Galilei (February 15, 1564 – January 8, 1642). He blinded himself.

The Italian Galileo's work on the refinement of telescope opened up the possibilities of generations, but it also ruined his eyesight. He was fascinated with the sun and spent many hours staring at it – leading to extreme damage to his retinas. This was the most likely cause of his near blindness in the last four years of his life. Because of his life's work Galileo (figure 7) is sometimes referred to as the "father of modern physics".

5. Antoine Levoisier (August 26, 1743 – May 8, 1794). Physicist sentenced to death.

Levoisier (figure 8) is considered the father of modern chemistry. He gave the name of oxygen (1778) and hydrogen (1783) and constructed the metric system, put together the first extensive list of elements, and helped to reform chemical nomenclature. He was also the first to establish that sulfur was an element (1777) rather than a compound. He discovered that, although matter may change its form or shape, its mass always remains the same.

During the French Revolution, on May 8 was sentenced to death and was guillotined the same day.

- Science discovers fundamental information about how the universe, and the natural world work.
- Technology is the practical application of that information. It embraces the attempt of human beings to improve existing means or tools, it creates new mechanisms in an effort to improve living conditions.
- In economic terms, science and technology are essential constituents of effective investments.

Figure 7. Galileo Galilei.

Library of Congress.

Figure 8. Antoine Levoisier.

Nearly all businesses need science and technology. Just picture a business environment without computers, mobile phones, developed transportation means, and describe the prevailing work conditions and their impact on economic development. (Brudloss, I. P., (2003) commenting on the contribution of physics to the Information Age supports that “nearly everyone would agree that the computer, the transistor and the world wide web are among the greatest inventions of the 20th century Economists know that to-day’s entire world economy is linked to these technologies.

The American Physical Society (2010) supports that the contribution of physics in the economic development of countries is important due to the provision of means to enhance the:

- information technology,
- green eco-friendly energy production,
- provision of solutions to difficult transport problems,
- health care standards, sanitary provisions, clear water supply,
- modern knowledge-based agricultural and food production,
- predicting and communicating ecological, meteorological events (tsunamis, earth quakes and other disasters).

Most importantly, Muranzi, R. (2005) the Rwanda’s minister of education, science, technology and scientific research describes the role of physics in the economic development emphasizing the traits of this contribution:

- Physics is the basic sciences and its concepts and techniques underpin the progress of all other branches of science.
- It is a cross-cutting discipline that has applications in many sectors of economic development.
- The role of physics in sustainable development was recorded at the United
- Nations Millenium Summit in 2000.
- In October 2005, politicians, educators and physicists from many countries met in Durban, South Africa, to consider the role of physics in creating a sustainable future for developing countries.
- With the help of medical physics and information technology, Rwanda is making some progress in the area of access to the right equipment for the diagnosis of diseases and the efficient communication of medical data.
- The Institute of Physics invested in the future of Rwanda,by sponsoring a team of physics teachers from the United kingdom to train teachers in the use of physics apparatus.

The evidences provided so far match with our opinion that ineffective outputs are not due to false concepts but to erroneous interpretations of the characteristics of professional and academic environments.

Analyzing the various economic systems in relation to specific strategies and the anticipated goals, we do not intend to criticize the economic principles of renowned economists, but to trace the reasons which led to recessions globally.

The fact that the US have witnessed 30 years of recession from 1902 to 2009 and UK 11-year-duration-recessions from 1919-2012 does not mean that economists, including the 23 Nobel prize winners (2000 –2011) are to blame for these catastrophic crises.

Nor can we support that the seven Nobel economists (six from US and one from Cyprus (Table 2) are responsible for the crisis.

It is not disputed that all these years of recessions have caused drawbacks in all the dimensions of our lifestyles. Yet, we cannot accept that the economic principles of outstanding economists during these economically disastrous periods can be attributed to inadequate, ineffective strategies.

In conformation, we will review the biographical sketches of some of them, relate their active periods with recessions and depressions (in economic terms) and finally we will analyze the real reasons on the basis of our findings of researches.

In Scientia Opportunitas

Economists are scientists who study the concepts of a very important branch of social sciences, i.e. economy, and devise methods to measure and fuel economic prosperity (Timimi, K., 2010).

The overall studies, researches and observation of economists have resulted to strategies to meet economic issues, to explain how economic models function and how to provide opportunities for economic growth, which will benefit people and release them from poverty. Besides, among the responsibilities of the economists is the evaluation of the costs and benefits of public policies (Ross, R., 2008).

Additionally, economists conduct researches to draft policies and propose solutions for problems caused by economic recession. Most specifically, the duties of economists are decided according to their position in the business environment.

It is supported that scientific behavior of economists is different according to their position. When employed by an investment bank they are expected to provide information for effective investments. As faculty members they will be asked to conduct researches, as a governmental officials they will be responsible to plan economic strategies, prepare budgets etc. (Wolf, M., (2012)

1. Adam Smith (1723-1790), (Keth Tribe and Mizuta Hirashi (2002).

Adam Smith, (figure 9) a Scottish philosopher, is considered the father of modern economies and pioneer of political economy.

Table 2. Nobel winners in the last four years

2011	Christopher A. Pissaridis	Cyprus
	Thomas J. Sargent	US
2010	Peter A. Diamond	US
	Dale T. Mortensen	US
2009	Elinor Ostrom	US
	Oliver Williamson	US
2008	Paul Krugman	US

In his "Wealth of Nations" he explained the benefits of a free market economy which he through self-interest will improve economy as a whole. He supported that in case individuals care to maximize their revenues, they contribute more to the progress of societies. Moreover, he was against the formation of monopolies. "A monopoly granted either to an individual or to a trading company has the same effect as a secret in trade or manufactures.

The monopolists, keeping the market constantly under stocked, never fully supplying the demand, sell their commodities much above the natural price, and raise their expenses, whether they consist of wages or profit, greatly above their natural rate. The price of monopoly is the highest which can be got. The natural price, or the price of free competition, on the contrary, is the lowest which can be taken, not upon every occasion.

2. Karl Marx (1818-1883).

Karl Marx, (Figure 10) a German philosopher, sociologist and political economist is considered the founder of communism. Marx has had a powerful influence upon mankind. His work "Das Kapital", has been considered a masterpiece. It was almost completed between 1844 and 1883, in a period of democratic nationalism, trade unionism and revolution.

Figure 9. Adam Smith.

Figure 10. Karl Marx.

Indeed, within Marx's lifetime, a new revolutionary tradition was born associating his name with that tradition (Kreis, S., 2008)

The three volumes of 2.500 pages "Das Capital" reveal his belief that economics was the driving force behind human society. He saw two important factors as determining how society functioned; the means and the relations of production. The means of production could be described as all those things that are necessary to produce goods such as labor, materials, and technology. The relations of production refer to the way humans interact with these means of production.

Marx saw capitalism as the current mode of production but like other modes of production it had gone before long. This was because it had an inherent flaw due to the fact that it required one group of people to exploit another group. (Yang, Y. (2010).

3. John Maynard Keynes (1883-1946.)

He is considered the most important economist of the twentieth century due to his book "General Theory of Employment. Interest and Money".

He advocated the policy of governmental intervention to lessen the negative effects which result from depressions and recessions. In a Chapter entitled 'The General Theory" he clarified that: "I have called this book the *General Theory of Employment, Interest and Money",* emphasizing the prefix *general.*

The object of such a title is to contrast the character of my arguments and conclusions with those of the *classical* theory of the subject, upon which I was brought up and which dominates the economic thought, both practical and theoretical, of the governing and academic classes of this generation, as it has for 100 years past"

Politically, Keynes (figure 11) is a lifelong member of the liberal party, one of the two main political parties in the UK until the 1920s (Shidelski, R., 2003). He supported the eugenics movement having served as director of the British Eugenics Society from 1937 to 1944. In 1946 he declared that: " eugenics are the most important, significant and genuine branch of sociology which exists". (Keynes, J. M., (1946)

4. Friedrich von Hayek (1899-1992).

He is undoubtly the most eminent of the modern Austrian economists who belongedto the classical liberal school and was best known for its defense. He further advocated free-market capitalism. Hayek (figure 12) shared the Nobel prize in Economic Sciences (with his political rival Gunnar Myrdal) for his *" pioneering work in the theory of money and economic fluctuations and penetrating analysis of the interdependence of economic, social and institutional phenomena".*

Figure 11. John Maynard Keynes.

His contribution to the Functional Efficiency of Economic Systems is profound and original. His scientific books and articles in the twenties and thirties aroused widespread debate.

5. Milton Friedman (1912-2006).

An American economist who advocated the free market. He challenged the policies of Keynes which, he argued, were responsible for causing worldwide inflation.

Therefore, he was against governmental intervention in a market economy. Keynesianism was a great reformation of economic thought. It was followed, inevitably, by a counter reformation.

A number of economists played important roles in the great revival of classical economics between 1950 and 2000, but none has been as influential as Milton Friedman. (figure 13) Another important achievement was that he succeeded in reviving the "Quantity Theory of Money" (QTM). According to QTM if the amount of money in an economy doubles, price levels also double, causing inflation (the percentage rate at which the level of prices is rising in an economy). The consumer therefore pays twice as much for the same amount of the goods or services (Heakel, R., 2010)

Figure 12. Friedrich von Hayek.

Figure 13. Milton Friedman.

In the simplest form, the theory is expressed as:

$MV = PT$ (the Fisher Equation)

Each variable denotes the following: M = Money Supply V = Velocity of Circulation (the number of times money changes hands, P= Average Price Level, T= Volume of transactions of goods and services.

Money supply multiplied by the number of times money changes hands equals the average priced level multiplied by the volume of transactions of goods and services.

6. Franco Modigliani (1918-2003).

The Italian Franco Modigliani (figure 14) is a Nobel prize winner (1985) *"for his pioneering analysis of savings and of financial markets"*

Together with Merton Miller he formulated the *Modigliani-Milton Theorems* in corporate finance, demonstrating that under certain assumptions, the value of a firm is not affected by whether it is financed by equity (selling shares) or debt (borrowing money)

He is, also, the originator of the *cycle hypothesis*, an attempt to explain the level of saving in the economy proposing that consumers would aim for a stable level of consumption throughout their lifetime, saving during their working period and spending during retirement.

The life-cycle hypothesis is a purely microeconomic theory. However, Modigliani has shown in a number of later works that the hypothesis has a number of macroeconomic applications (Scheth, A. 2012)

Figure 14. Franco Modigliani.

7. Joseph Eugene Stiglitz (1943-).

American economist (figure 15) winner of Nobel prize (2001) shared with George A. Akerlof (US) and A. Michael Spence (US) for their *"analyses of markets with asymmetric information."*

Most microeconomic models simply assumed that wages and prices were fixed. But, of course, during the great depression wages and prices had fallen considerably. The problem was not that they were absolutely fixed, but with the dynamics of adjustment. (from his autobiography).

Figure 15. Joseph Eugene Stiglitz.

The economists reviewed in the preceding pages are examples of outstanding scientists with brilliant minds, inventive thoughts, capable of formulating economic strategies of permanent value. Nevertheless, they have witnessed recessions (Table 3)

It is not disputed that all these years of recessions have caused drawbacks in all the dimensions and forms of our lifestyle.

As to what causes recession, there are contrasting opinions and controversial beliefs not only among scientists but, also, economic analysts, journalists. Borade, G., (2012), supports that: "Periods of recession have followed dramatic increases in the price of oil.

Table 3. Recessions during the active lifespan of the cited economists

ECONOMISTS and RECESSIONS	1920 1921	1923 1924	1926 1927	1929 1933	1937 1938	1945	1948 1949	1953 1954	1957 1958
Duration months	18	14	13	55	13	8	11	10	8
John Maynard Kenes	X	X	X	X	X				
Friedrich von Hayek					X	X	X	X	X
Milton Friedman							X	X	X
ECONOMISTS and RECESSIONS	1960 1961	1969 1970	1973 1975	1980	1981 1982	1990 1991	2001	2007 2009	2009 2012
Duration months	10	11	16	6	16	8	8	48	
Milton Friedman	X	X	X	X	X	X			
Franco Modigliani	X	X	X	X	X	X			
Joseph Eugene Stiglitz				X	X	X	X	X	X

1. Larry Elliott (2012), (figure 16) British journalist, economics editor of the Guardian (educated at St. Albans School and Fitzwilliam College of the University of Cambridge) in a rather offensive way claims that "many big names are in frame as to who to blame for the great recession.

Figure 16. Larry Elliott.

Figure 17. Business News.

Every financial crisis has its totemic moment; a decision that even at the time seems to defy logic and in retrospect is seen as an act of gross stupidity". His list of those who considers responsible consists of economists, US presidents, politicians chief executives of Goldman Sachs and Standard and Proofs.

2. The international best-selling author and innovator, ***Canadian Ernie Zelinski***, with over 700.000 copies sold worldwide, in his article "Causes of Economic Recession" published in the "BUSINESS NEWS" in 2010 (figure 17) compares 10 causes of economic recession expressed by economists with ten causes he considers they lead to the economic crisis. From the comparison of these two columnists, it is evident that whereas economists refer to the causes or recession in purely economic terms, Lewinski (figure 18) attributes 40% of the causes to human interference It is reported that "he deeply believes in the powers of creativity and well planned international actions which are the most important elements revealing how ordinary individuals can attain goals for personal prosperity and freedom.

For cross-checking reasons, we thought it would be advisable to carry out an empirical research with a sample of 1036 participants, randomly selected among business people (SMEs, CAO of industrial units), faculty members and individuals of varied professionals from three areas in Greece (Athens, Korinthos and Thessaloniki) one from Bucharest (Romania) and one from North Cyprus.

Figure 18. E. Zalinski.

Table 4. Comparison of two different opinions about recession causes

No	Causes according to economists	No	Causes attributed by the author
01	Excessive interest rates	01	Economists
02	Underconsumption	02	Individuals following advice of economists
03	Overproduction	03	Greed-including average citizens
04	Currency crises	04	Governments
05	Lack of consumer confidence	05	Money
06	Inflation	06	Stock Markets
07	Speculation and economic bubbles	07	Groupthink
08	National debt	08	Corporations
09	National recession cycle	09	Credit and Debit
10	Inadequate government interference	10	Definition of recession itself

The period of the research: 20 July to 29 August, 2012,mixed method of telephone calls, e-mails, personal interviews. The problem statement was:

> "Of the following five categories, which two do you consider responsible for the economic recession?"

Their responses appear in Table 5.

A deep analysis of these findings prove that:

[1] Faculty members believe that economists have a small contribution to the creation of economic crises
[2] Employers share almost equal percentage in the sense that it is not their responsibility to dictate economic policies but rather they included to those damaged.
[3] Politicians rank second in the list of responsibility because the general opinion is that either they are not capable of implementing economic strategies or their concern is mainly concentrated on vote hunting!
[4] The factor of illiteracy is recognized to be the main reason which causes economic recession in as much as it is linked with those who intentionally break the laws.

Table 5. Findings of research

No.	Categories	Persons	%
01.	Economists	70	06.76
02.	Politicians	414	39.96
03	Employers	78	07.53
04	Illiterate People	436	42.08
05.	Failure to obey the rules and laws	38	03.67
	Total	1.036	100.00

Deus ex Machina

In the ancient Greek tragedy the phrase "god from the machine" meant a plot device involved in an unsolvable problem which suddenly was solved thanks to an unexpected intervention of some new event or a mighty force, say god. The Latin poet Horace translated this phrase to Latin and has been used in Roman drama in similar situations.

In view of the hard times, people are experiencing due to the economic recession, like Greece, Spain, Portugal, Italy, ordinary people do not see any sign of recovery. In order to find long-lasting solutions they are about to wish for a "deus ex machine" solution.

However, in order to establish economic stability we have to investigate the real reasons which led to the economic recession. Back to the Elliott's references to faulty decisions and practices on the part of the administration, the majority of opinions, accepts that they have contributed to the economic recession which has expanded like a domino effect.

Economists' Faults

[a] A decision of the Federal Reserve to cut interest rates and leave them at rock bottom levels for two years resulted to cheap borrowing costs which encouraged Americans to load on debt to buy homes, even with no income.
[b] The price assets of the financial markets should be based on all the publicly available information.

Financial Institutions

[a] Goldman Sachs are blamed as credit rating agencies because they implemented rating policies the wrong way.
[b] Standard and Poor's are considered to have acted in rating top AAA to erroneous entities.
[c] Additionally, Shah, A., (2010) commented that: " the global financial crisis, brewing for a while, really started to show its effects in the middle of 2007 and into 2008."

Politicians-Parliamentarians

[a] Elliott (op.cit) considers a mistake the removal "of constraints on bankers, making finance more important at the expense of manufacturing and the reduction of union power in the UK of 80s." He criticized "the loss of a million manufacturing jobs and an ever-widening trade deficit".
[b] Krugman, P., (2012) admitted that: "In the case of Greece, matters were made far worse than necessary by the way Europe' s leaders imposed their policies.
(d) Experience has shown that in many countries MPs have not been as responsible as required for contributing to their nations endeavors to restore conditions for economic growth and stability. The reasons can be attributed to:
- Limited awareness of the real dimensions of specific cognitive areas.
- Vote-hunting syndrome.
- Complicated activities unrelated to their parliamentary duties.
- Sometimes hidden interests confronting governmental policies.

(e) Finally, according to Yavlinsky, G., (2012) "academic researchers and governmental decision makers should not lose sight of the fact that even comparatively sophisticated ways of responding to the crises, such as imposing taxes, will not resolve fundamental problems, which are not simply economic.

Employers

Employers constitute part of the problem in a recession period. The catastrophic experience from recession is gained through a chain development like this:

Loss of jobs==➔increase of unemployment==➔drop of income=➔consumption expenses reduced=➔slowdown of production==➔limited turnover

Davis M., (2008) describes a life cycle of a typical operation scheme within a recession environment consisting of six steps which we present in schematic form as follows (Figure 17) (a) During recession revenues and profits decline (# 1), (b) Hiring of new employees is postponed (# 2), (c) Salaries of existing staff are dramatically reduced (# 3) (d)The company stops buying equipment and material. (# 4) (e) In turn, this results to deferred new production (# 5)

- And finally, the market share of the company (# 6) is reduced with equal sharp shrinkage of cash flow.

Illiteracy

The United Nations define illiteracy as the inability to read and write a simple message in any language. Based on surveys, the illiteracy rate was at least 44% of the world's population in 1950 and dropped to 16% in 1998.

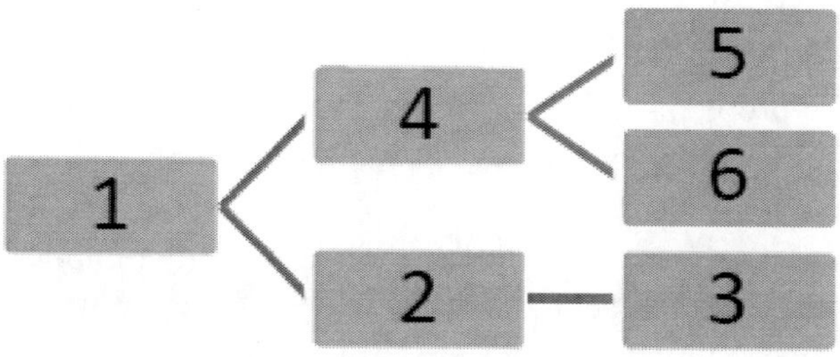

Figure 19. Typical company's operation during recession periods.

Table 6. Primary enrolment ratio within relevant age

No	Region	Male %	Female %
01	Sub-Saharan Africa	74	69
02	South Africa	95	88
03	Middle East and North Africa	94	89
04	Latin America and Caribbean	96	96
05	East Asia and Pacific	97	96
06	Central Eastern Europe (CEE)	92	91
07	Commonwealth of Independent States (CIS)	92	91
08	Developed Regions	97	97

Source: UN Statistics Database Division.

By geographical area, the highest illiteracy rate is found in the less developed nations of Africa, Asia and South America ranging from 60.1% (Mouritania) to 84.3% (Niger) By 1970, the US Office of Education considered at least six years of schooling to be the minimum criterion for functional literacy. In 1990 over 5% of the adult population living in the US did not meet that criterion. Mainly for study purposes we distinguish two types of illiteracy

(a) Complete illiteracy: The lack of the most basic reading and writing skills (U.N. definition)
(b) Functional illiteracy: The lack of a minimum of capabilities needed to function in society. (US Office of Education)
(c) In the countries of Northern Europe the illiteracy rate ranges from 0.2% (Estonia) and 0.3% (Lithuania) to 1.0% (Denmark, Ireland, Norway, Sweden and UK).

(d) In the countries of Southern Europe the relative rate ranges from 1.1% (Italy) to 5.1% (Portugal).
(e) In the African countries the rate is from 60.1% (Mouritania) to 84.36 (Niger),
(f) It is anticipated that in the twenty first century the world rate of illiteracy will increase because only 25% children worldwide attend school. The primary level enrolment ratio (percentage of total population is as follows (Table 6)

Educationalists believe that illiteracy tends to be one of the most important factors which cause economic recession. The effects of illiteracy in society can be;

(a) Increase of unemployment,
(b) Low standard of living,
(c) Suffer of health problems due to inability to read medical directions
(d) Adult illiteracy costs $240 billion in lost industrial productivity

The figures of illiterate persons globally, are indicative of the importance this category of persons may play in the development of national economies. With a percentage of 20% illiteracy worldwide, it is a situation which certainly should be given special attention and concern since this figure tends to result to tremendous problems In our era, there are quite a number of causes of illiteracy which can be classified in three causes:

Social Reasons

(a) Poverty: Parents with low financial means consider education as a waste of time.
(b) Drop outs: Lack of interest on the part of children, limited means.

Family Reasons

(a) Children are grown up in an environment which does not support learning
(b) Preference of work to school attendance. Parents prefer to place children to work than to send them to school

Psychological Reasons

(a) Disabilities make children feel unable to attend.
(b) Lack of motivation due to poor performance at school (Figure 20)

The rate of illiterate persons in the various geographical regions range from 8.90% in South Africa to 34.88% in East Asia and Pacific (Table 7).

It is evident that the basic concern should be the design of remedial measures aiming at reducing the rates of illiteracy. The most effective means is to organize and run educational and training courses both for academic and professional purposes. The management of such project requires the budget of national governments to include increased percentages of Gross Domestic Product (GDP).

Basically, there is a close relationship between illiteracy rates and percentage of GDP for educational expenditures.

[a] In countries with illiteracy 1.0% the budget for educational expenditures ranges between 4.6 (Germany) to 8.5 (Denmark).

[b] Additionally, it has been noted that in countries white per capita income above $50.000 the literacy rates 90% or more.

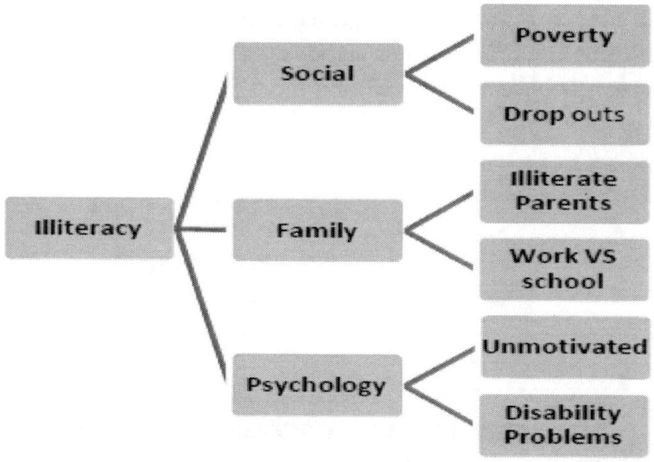

Figure 20. Causes of illiteracy.

Table 7. Illiteracy data globally

Geographic Region	Total population (in millions)	Illiteracy data	Year	Source
Sub-Saharan Africa	853.6 M. 425.9 F. 427.0	M. 110.7 F. 132.6	2010	World Bank
South Africa	50.6 M. 24.5 F. 26.1	M. 1.3 F. 3.2	2011	South Africa Statistics
Middle East and North Africa	331.3 M. 166.6 F 164.6	M 10.0 F. 10.1	2011	World Bank
Latin American and Caribbean	582.5 M. 287.8 F 294.7	M. 11.5 F. 11.9	2011	World Bank
East Asia and Pacific	1.961.6 M 1.004.3 F.957.2	M. 301.4 F. 382.9	2011	World Bank
Central Eastern Europe (CEE)	733.6 M. 366.0 F. 367.6	M 29.2 F. 33.1	2011	UNESCO Statistics
Commonwealth of Independent States (CIS)	277.8 M. 139.2 F. 138.6	M. 11.1 F. 12.5	2011	UN Statistics Division
Developed Regions	275.2 M. 135.4 F. 139.8	M 4.1 F. 4.2	2011	OECD Statistics
Total	5.066.2	M 479.3		
		F 590.5		

According to Premi, M. (2002) "with almost two thirds of India's population aged 7 years of age and above, literacy rates indicate a very significant increase for both males and females"

- In the case of Arab countries, "there is a great awareness at governmental and organizational levels of the seriousness of illiteracy.
- Several measures were taken which set up policies and strategies to coordinate the efforts to combat illiteracy with direct technical support from the UNESCO regional office in Beirut, Lebanon. (Hammond, H.(2005)
- The results have been impressive (Table 8)

We feel that the recourse to educational and/or training courses delivery as a means of reducing the illiteracy rates globally is the dues ex machine.

Table 8. Combat of women's and men's illiteracy in the Arab Countries

Category	1980	1990	2000
Illiterate women 15-24(1)	44.9%	29.9%	19.4%
Illiterate women over 15 (2)	64.9%	51.9%	40.2%
Illiterate men 15-24 (1)	20.4%	16.3%	12.7%
Illiterate women over 15	14.2%	9.4%	6.1%

Source: Compiled from Non-formal education for men and women, UNESCO.

Table 8A. Illiterate persons per age-group

94 million in Africa	34 million in America
24.3 million in Asia	9 million in Europe

In quite a lot of countries organizations like UNESCO, and EU Commission are involved in remedial mechanisms related to illiteracy rates in an effort to combat this problem which is among the most influential factors of economic recession.

1. European Commission Education and Training (2011)

(a) "One in five 15-year-olds in Europe, as well as many adults, lack basic reading and writing skills, which makes it harder to find a job and puts them at risk of social exclusion. The European Commission has set up an independent group of experts to identify ways of raising literacy levels".

(b) "Androulla Vassiliou, the European Commissioner for Education, Culture, Multilingualism and Youth, said: "*The launch of this group marks the start of a Europe-wide campaign to address the root causes of low literacy levels. Too many Europeans find themselves locked out of the job market and find it hard to contribute fully to society because they lack basic reading and writing skills.*"

(c) "The latest results of the Program for International Student Assessment (PISA), a worldwide survey on the performance of 15-year-olds in reading, maths and science, show that one in five has low reading skills.

According to the OECD, Europe has slightly improved since 2000. but greater efforts will be needed if EU countries are to achieve their target of fewer than 15% of low performers in reading, maths and science by 2020".

2. UNESCO, Literacy as a factor in development (1965)

(a) "We must find ways of mastering programs in all forms of technological progress and diffusion of constantly renewed ideas, in order to turn them into a means of literacy. The full utilization of human resources is the determining factor in development. Illiteracy is the outcome of a complex series of economic and social factors.
(b) "The illiteracy figures for each continent show very clearly that an immense effort remains to be done before the goal of universal literacy can be reached. The number of illiterate adults in the particularly active age-group of 15 to 44 years is: (Table 8A)

Therefore, it seems essential to adopt immediate measures at national and international levels.

(c) Illiteracy is mainly caused by factors like these:
- Underdevelopment.
- Instruction given in an official language which is different to the mother tongue of the majority of the children.
- Special diverse circumstances which hinder the development of school education.
- Resistance on the part of the population.

(d) Besides, literacy plays an important role in the overall development, since it
- Creates new attitudes,
- Introduces new forms of behavior related to economic and social development.
- Constitutes an investment producing quick returns.

3. Hasan, A., (2012), Pakistan: "Ways to combat illiteracy"

(a) Hasan, A., in his article in a weekly magazine, attributes the high rates of illiteracy to two major reasons:
- Very low governmental expenditure for education,
- Due to high poverty levels, children, instead of going to schools, find jobs in order to contribute to the family expenses.

(b) To combat illiteracy he proposes:
- Operation of evening schools.
- Keeping people aware of the benefits of education towards a brighter future.
- Combat of corruption in schools, which destroys the future of the nation.
- Change of the mentalities of parents regarding education, for example the opinion of parents that school attendance of their daughters is worthless.

4. New York Times: National Policy Urged to Combat Adult Illiteracy (1988)

(a) The journalist Fiske, E., in his article in the New York Times describes the inability of millions of Americans to read and write as "a root cause of a wide range of social problems,"
(b) On the other hand, Arthur Sulzberger, publisher of the New York Times reveals that " up to 60 million Americans - one-third of the adult population cannot read their

local newspaper", which means that such people are "cut off from a flow of vital information that is essential to involved and productive lives."

5. Major plans to combat illiteracy in South Africa (2008)

(a) According to Magubane, K (2006) Prof. Kader Asmal, former minister of education said that "Literacy is a strong weapon through which Africa could claim its rightful place among the continents. It is therefore imperative that every child in Africa must learn to read and write. African leaders and experts must commit themselves to making this happen."
(b) Illiteracy is a priority problem in South Africa. The latest government statistics indicate that 4.7 million people are totally illiterate (never attended school) and their 4.9 million adults are functionally illiterate (left school before Grade 7).

6. Rwanda: Government continues to combat illiteracy (2012)

- Gahiji Innocent, chief political columnist of "News of Rwanda" describes some effective governmental initiatives in their attempt to combat illiteracy: Of the 73 schools for adults which had been set up since June 2011, the number has extended to 122 schools, as announced by David Kera; in charge of education in Nyabihu district.
- From the estimated 3600 people who would take the lessons in 73 classes, there are 122 classrooms with 5440 people learning how to read and write, reveals David Kera
- The reason behind this increase is attributed to the illiterate persons desire to learn reading and writing, which they did in big numbers.
- Facilities and allowances were given to the teachers.

7. Volunteer teachers combat illiteracy in Burkina Faso

The United Nations Development Program (UNDP) in the annual report for 2010-2011 provides data for Burkina Faso (Africa) to combat illiteracy Burkina Faso has one of the highest illiteracy rates in the world. The school rate of enrollment is only 44.6 %t. Adults account for more than 70 % of illiterate people.

According to the UNO program volunteer teachers support the efforts of the country to combat illiteracy. (Figure 21)

Class sizes are large, sometimes more than 120 pupils. To solve these issues, more than 80 volunteers have been recruited after being attended training courses at the National School for Primary School Teachers.

- The National Volunteers Program of Burkina Faso (PNVB) is jointly financed by the United Nations Development Program (UNDP), the United Nations Volunteers Program (UNVP) and the Government of Burkina Faso

8. A Braille Writing Tutor to combat illiteracy

According to Kalra, N., et al. (2011) surveys reveal that more than 90% of the world's 145 million blind and 16 million visually impaired live in developing communities. Of these

145 million less than 3% are literate. This is partly due (a) to the lack of trained teachers and (b) to their inability to work on the traditional slate and stylus.

Despite the advantages of the use of Braille system in education and its impact on literacy, there are a number of barriers.

In the case of India, the main barrier is attributed to limited opportunities for education, since parents of blind children are reluctant to recognize the value of education for their children. Besides, children do not always receive sufficient training for using the Braille system. To meet these problems, the team of the Carnegie Mellon University constructed the Braille Writing Tutor. The system was tested in real classroom conditions thanks to the co-operation of the Mathru School for the Blind, Bangalore, India. Blind users were asked, at the end of the pilot use of the system, to respond to the following statements.

Figure 21. A United Nations Volunteer (UNV) teacher conducting a small class in Burkina Faso. (Photo UNDP).

Table 9. Test responses

No.	Statements	Response %
01.	I found the Braille Writing Tutor difficult to use	20.7
02.	It took me a long time to learn how to use the tutor	27.3
03.	I cannot understand the voice used by the tutor	3.2
04.	I found it difficult to distinguish between the buttons in the E-slate	48.8

Table 10. Response of users of Braille Writing Tutor (BWT)

Statements related to Braille Writing Tutor use
I find the tutor useful 83.3%
I dislike using the tutor 0.0%
I prefer writing on a regular slate to writing on E-slate 83.4%
I want to continue using the Braille Writing Tutor 100.0%
I think the tutor will help me in learning to write Braille 100.0%

In his way their reaction as well as the effectiveness of the system. would be tested. All characteristics of the system which might create problems have been almost readily accepted by the majority of users and only the recognition of the buttons seems to create a problem. Furthermore, a second test battery was tested: (Table 10)

In order to identify the reliability and applicability of our belief that the increase of literacy levels constitutes the main driving force towards the restoration of economic stability and the much-needed relief of the disastrous effects of recession, we investigated the current status related to the mechanisms of combating recession in a representative sampling. The US have started a systematic effort to combat adult illiteracy.

Pakistan, South Africa, Rwanda, Burkina Faso and the team of Carnegie Melton University, Robotics Institute in co-operation with the Mathru School for the Blind, Bargalore, India have implemented effective projects to reduce illiteracy rates not only for ordinary persons but, also, for persons with disabilities.

- Educationalists believe that all these initiatives sooner or later will attain the anticipated results.
- Others with a high degree of pessimism claim that only miracles can produce such results' What is the truth? Youth, F. (2008) reporter of the Christian Post revealed that nearly 9 of 10 Americans believe miracles have occurred in the past and can occur today, according to a new national survey

The new survey also found a majority (56 percent) of Americans claiming to have seen themselves situations and circumstances like this. Which are miraculous" or "unexplainable by science. On the contrary, Hume, D., (1748) a Scottish philosopher, economist and essayist in his essay: "An enquiry concerning Human Understanding" (Section X, Of Miracles) supports that (Figure 23)

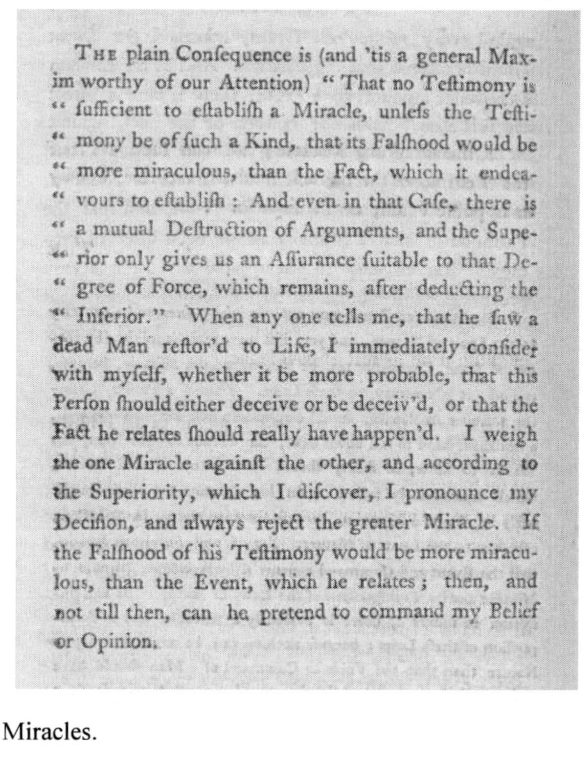
THE plain Conſequence is (and 'tis a general Maxim worthy of our Attention) "That no Teſtimony is "ſufficient to eſtabliſh a Miracle, unleſs the Teſtimony be of ſuch a Kind, that its Falſhood would be "more miraculous, than the Fact, which it endeavours to eſtabliſh: And even in that Caſe, there is "a mutual Deſtruction of Arguments, and the Superior only gives us an Aſſurance ſuitable to that Degree of Force, which remains, after deducting the "Inferior." When any one tells me, that he ſaw a dead Man reſtor'd to Life, I immediately conſider with myſelf, whether it be more probable, that this Perſon ſhould either deceive or be deceiv'd, or that the Fact he relates ſhould really have happen'd. I weigh the one Miracle againſt the other, and according to the Superiority, which I diſcover, I pronounce my Deciſion, and always reject the greater Miracle. If the Falſhood of his Teſtimony would be more miraculous, than the Event, which he relates; then, and not till then, can he pretend to command my Belief or Opinion.

Figurer 23. Hume's Of Miracles.

"The plain consequence is (and it is a general maxim worth of our attention. No testimony is sufficient to establish a miracle, unless the testimony be of such a kind that its falsehood would be more miraculous than the fact which it endeavors to establish.

And even in that case, there is a mutual destruction of arguments, and the Superior only gives us an Assurance suitable to that degree of force., which remains after deducting the "inferior". I weigh the one Miracle against the other, and according to the Superiority, which I discover, I pronounce my decision

And always reject the great Miracle. If the falsehood of his Testimony would be more miraculous, then the event, which he relates, then and not till then can he predent to command my Belief or Opinion."

We will not dispute with the religious beliefs or Hume's philosophic arguments about "Miracles". Instead we will quote the criticism of the renowned Greek Philosopher Ioannis N. Theodoracopoulos (1966) about THE CAPITAL of Karl Marx: "The principles and strategies of his "The Capital" failed to succeed simply because they were addressed to people educationally unable to comprehend his socialistic principles". Therefore, instead of "Miracle" we will accept the intervention of the "Deus ex machine" or else the educational and training courses delivery.

However, we mean a systematic, structured instruction process and not a random selection of opinions which are groundless, ineffective and do not produce any outcome. To this end a group of expertise consisting of policy-makers, sociologists, economists, educators, researchers and educationalists should be involved in the education and/or training courses on the basis of identified characteristics related to economic means on individual and national settings, family structures, social care and market formation and demands.

To this end, six scientists of separate cognitive areas must be involved: Sociologists, (b) Researchers, (c) Policy Makers, (d) Economists, Educationalists and (f) Educators should be involved in the design and implementation of the curriculum. The sequence of activities should be based on the following pattern: (Table 11)

In case this procedure , fully documented as to its effectiveness, is followed, it is anticipated that the aims of the educational process will be attained. A useful tool for deciding about the content of educational/training courses, is this formula

$$A= B+C$$

where A denotes the needed qualifications, B represents the available ones and C stands for the content of the complementary education and training courses in order to satisfy the prevailing demands.

In view of the foregoing documentation, practices and strategies which have been implemented in the past and resulted to the disastrous financial consequences and led to the contemporary economic recession will be pinpointed.

The remedial process which is scheduled to follow will aim to practices and strategies with innovative and effective orientation.

Table 11. Sequence of activities

1. SOCIOLOGISTS Study and explain the needs of societies
2. RESEARCHERS Define key problems in all aspects of societies
3. POLICY MAKERS Formulate overall activities of societies
4. ECONOMISTS Develop economic policies and strategies
5. EDUCATIONALISTS Specialists in theory and methods of education
6. EDUCATORS Instructors of particular subjects

Interim Recap

According to Schwab, K., and Greenhill, R., (2012) "the report of the forum contributes to the understanding of the key factors determining economic growth, helps to explain why some countries are more successful than others in raising income levels and opportunities for their respective populations, and offers policy makers and business leaders an important tool in the formulation of important economic policies"

Practices

It is generally recognized that economic practices play a most crucial role in the sustainable economic development by (a) identifying the strong business companies in a specific region, (b) assisting existing businesses to expand, (c) strengthening businesses which produce products and services for export, (d) promoting the revitalization of low- and moderate-income communities. These practices have failed to accomplish their objectives by systematically supporting a sustainable economic development globally, due to multilateral systems operating by deliberately faulty actions, unethically structured interests, unbalanced national policies, as will be highlighted below:

1. European 27 Members- Countries

(a) A most crucial problem in the countries of the EU which are situated in the border of the Schengen Area is the illegal undocumented immigrants. Coming from five major areas (Table 12) constitute a major economic, sociological and political issue. The problems include: (i) Unemployment, (ii) Health, (iii) Accommodation (iv) Cultural deviations, (v) Behavioral discrepancies and (vi) Educational issues The term "undocumented immigrants" refers to third-countries nationals without the required permit authorizing them to regularly stay in the EU member-states. Their impact on the recession is decisive. Numbers of immigrants are shown on the following table 13, though the authenticity of the figures, mainly in Greece is disputed (Table 13)

The situation in Greece is unbelievably awkward. The demographic data, as of 2011, reveal an economic and social issue which cannot be restored easily (Table 14) Greek Media claim that citizens are afraid to walk around in fear of robberies , thefts or murders by undocumented immigrants.

Table 12. Countries of origin of immigrants

No.	Geographical areas	immigrants
01	Eastern Europe	53.0 %
02	Asia and Ocean	12.5 %
03	Northern Africa	17.1 %
04	Sub-Saharan Africa	8.7 %
05	Latin America	8.7 %

Table 13. Number of illegal undocumented immigrants

No	Country	Illegal (000) Undocumented immigrants	Year
01.	Austria	100.0	2005
02.	France	32.9	2011
03	Germany	1.000.0	2009
04	Greece	2.500.0 (?)	2011
05	Italy	349.0	2007
06	Netherlands	225.0	2006
07	Portugal	90.0	2010
08	Spain	18.5	2010
09	United kingdom	947.0	2009

Table 14. Hellenic demographic Data

Population	Million	%	Source
Total	10.79	100.00	Hellenic Statistical Authority, 2011
Active	4.95	45.87	Hellenic Statistical Authority,2011
Unemployed	1.91	17.70	International Labor Office, 2011
Illegal immigrants	0.91-2.50	8.43 23.16	Hellenic Ministry of Civil Protection Estimate.

The Government has failed to find a solution. Repatriation of the illegal immigrants is not possible because most of them are of unknown origin. Besides, the issue of travelling documents might break the Schengen agreement on the part of Greece.

(b) The European countries are divided into five categories in terms of economic condition which causes irregularities and discrepancies according to the GDP per capita (Purchasing Power Parity /PPP) estimates of International Monetary Fund), as of 2012

- From US$ 7.284 (Montenegro) to US$ 24.702 (Cyprus)
- From US$ 25.417 (Georgia) to US$ 44.744 (Luxembourg)

- From US$ 60.557 (Slovenia) to US$ 180.987 (Ireland)
- From US$ 201.441 (Finland) to US$ 419.065 (Belgium)
- From US$ 714.704 (The Netherlands) to US$ 3.154.800 (Germany)

With the advent of globalization, the mobility of people from one country to another has been facilitated. In spite of the Schengen agreement, this flow tends to develop to illegal immigration. As a consequence individuals of low-income and literacy move from the poorer countries to the wealthy ones, in search of a better fortune. When this movement increases, the workforce of the countries of origin decreases dramatically and the respective workforce of the host country increases at uncontrolled rates. However, there are different but closely related features of the same concept. This is the case with the EU 27 and the economic practices. As reported by Tonkin, S., (2010), the World Economic Forum at its Global Competitiveness Report for 2011-2012 disclosed the competitiveness rankings according to which the EU27 countries are positioned in places between the second (Sweden) with score 5.56 (against 5.63 of the first Switzerland) and Greece placed ninetieth out of a total 142, with score just 3.99.

Sweden (3rd)

Sweden "has been placing significant emphasis on creating the conditions for innovation-led growth. The quality of its public institutions is first-rate, with a very high degree of efficiency, trust, and transparency. Private institutions also receive excellent marks (2nd), demonstrating the highest ethical behavior (3rd), supported by strong auditing and reporting standards (2nd) and well-functioning corporate boards (1st). Goods and financial markets are very efficient, although the labor market could be more flexible (25th). With a strong focus on education over the years they rank 2nd for higher education / training) and 2nd in technological adoption."

Finland (4th)

It boasts well-functioning and highly transparent public institutions (4th) It occupies the top position in the higher education and training pillar(1st.) and has provided the workforce with the skills needed to adapt rapidly to a changing environment and has laid the groundwork for high levels of technological adoption (12th) Health and Primary Education (1st), Technology readiness (12th) and Financial Market (9th).

Germany (6th)

Since the previous assessment, the quality of its public institutions and the efficiency of its goods markets have deteriorated slightly. The country is ranked an excellent 2nd for the quality of its infrastructure, due to the first-rate facilities across all modes of transport. The goods market is quite efficient, characterized by intense local competition (9th) and low market dominance by large companies (3rd). Germany's business sector is highly sophisticated, especially in production and distribution channels. German companies are among the most innovative in the world, spending heavily on Research and Development (5th).

The United Kingdom (10th)

It gains lost ground rising by two more places .It improves its performance by benefitting from the strengths of its labor market (7th). The United Kingdom continues to have sophisticated (6th) and innovative (13th) businesses harnessing the latest technologies for productivity improvements and operating in a very large market (6th for market size). contributing to productivity enhancements. On the other hand, the country's macroeconomic environment (85th) represents the greatest problem on its competitiveness, with a double-digit fiscal deficit in 2010 (138 position) .The situation is made worse by the mounting public debt (77 % of GDP in 2010, 120th).

France (18th)

The country's infrastructure is among the best in the world (4th), with outstanding transport links, energy infrastructure, and communications. The health of the workforce and the quality and quantity of education are other clear strengths (ranked 16th for health and primary education and 20th for higher education and training). These elements are the basis for a business sector that is aggressive in adopting new technologies for productivity enhancements The sophistication of the country's business culture and its leadership in the area of innovation supported by a highly developed financial market (16th) and a large market size(7th), are important attributes to boost the country's growth potential."

Belgium (15th)

It has "outstanding health indicators and a primary education system that is among the best in the world (2nd). also an exceptional higher education and training system, top-notch management schools, and a strong on-the-job training Its goods market is characterized by high levels of competition and an environment that facilitates new business creation. Business operations are distinguished by high levels of sophistication and professional management. However, there are concerns about government inefficiency (56th), and the economy is burdened by persistent deficit spending and high public debt."

Ireland (29th)

After having declined in rank over the past two decades it continues to benefit from a number of strengths, including its excellent health and primary education, its strong higher education/training (22nd) and its well-functioning goods and labor markets. These attributes have fostered a sophisticated and innovative business culture (ranked 22nd for business sophistication and 23rd for innovation). The country's macroeconomic environment raises significant concern (118th), with its budget deficit of more than 32 percent in 2010, following the government's bailout of the banking sector, placing Ireland last 142."

Spain (36th)

After two years of sharp decline that led it to fall from 29th place in 2008–09 to 42nd place last year this year's progress can be attributed to slight improvements in several areas measured by the Index, as well as deterioration in the performance of other economies that previously ranked ahead of Spain.

Despite a sluggish economic recovery and an important weakening of its macroeconomic stability (falling from 66th to 84th position), the country has managed to improve its performance thanks to a greater use of Information Communication Technology (ICT) (28h) and its resilience in terms of R and D investment and innovation capacity. Further improvement of these factors will be crucial for its future recovery and a much-needed economic transformation."

Portugal (45th)

The country "improves its competitiveness performance slightly despite its critical financial situation, which led to a recovery plan earlier in the year—and notwithstanding the negative economic forecasts for the next two years as the consolidation plans start to reduce public spending.

This positive development is largely led by an increase in ICT use throughout the economy (18th) and an improvement in the quality of its overall infrastructure (23rd), especially of roads (5th). Despite this slight progress, the country still holds one of the poorest competitive positions among advanced economies and suffers from serious weaknesses.

In addition to the well-documented macroeconomic difficulties of a national savings rate below 10 percent (128th), a high deficit (122nd), and high public debt (128th) that hinder the availability of financial resources from local companies, the economy suffers from rigidities in its labor market (112th) and a disconnect between salaries and productivity (112th) that have hampered Portugal's capacity to remain internationally competitive."

Grèce (90th)

Greece falls another seven places in the rankings, remaining the lowest-ranked country of the European Union. In the context of the ongoing sovereign debt crisis, Greece continues to fall in the macroeconomic environment pillar, dropping to 140th position this year. Greece's financial markets are assessed more poorly than in the past, at 110th, showing particularly low confidence on the part of investors, and goods market 107

The evaluation of public institutions (e.g., government efficiency, corruption, undue influence) continues to suffer and is ranked a low 89th overall. Another major area of concern is the country's inefficient labor market (126th), which constrains Greece's ability to emerge from the crisis, demonstrating the importance of recent efforts to increase the retirement age and increase labor market flexibility."

The key indicators of the above cited 10 countries have been based on sources as follows:

1. Population: U.N. Population Fund, State of World Populations, 2010 based on national resources.
2. GDP in US$ per reviewed country
3. GDP per capita in US$
4. Share of local GDP in world total.
5. Monetary Fund World Economic Outlook Database (2010) from national resources (Table 15)

Apparatus Criticus

This title, in the English version meaning critical matter, is meant to designate supplementary information intended to facilitate the understanding of the data provided in the following tables of competitiveness rates of representative countries from all geographical areas.

Table 15. Comparative economic data: EU countries

Country name	Population In millions	GDP in US dollars	Per capita US dollars	Share to GDP In World total
Sweden	9.3	455.8	48.875	0.48%
Finland	5.3	239.2	44.489	0.25%
Germany	81.1	3.315.6	40.631	3.96%
UK	61.9	2.247.5	36.120	2.94%
Belgium	10.7	465.7	42.630	0.53%
France	62.6	2.582.5	41.019	2.96%
Spain	45.3	1.409.9	30.639	1.84%
Portugal	10.7	229.3	21.559	0.33%
Ireland	4.6	204.3	45.689	0.23%
Greece	11.2	305.4	27.307	0.41%

For more than three decades, the World Economic Forum's annual *Global Competitiveness Reports* have studied and recorded the many factors underpinning national competitiveness Since 2005, the World Economic Forum has based its competitiveness Index, which measures the micro- and macroeconomic foundations of national competitiveness.

Competitiveness is defined as the set of institutions ,policies, and factors that determine the level of productivity of a country.

The links between competitiveness and economic growth takes the form of a chain sequence:

Competitiveness==→level of productivity.===→level of prosperity =→ rates of return===→ fundamental drivers of growth

Xavier Sala-I-Martin (2012) supporting the opinion of the Austrian-Hungarian-American economist and political scientist Schumpeter, J (1942) states that for ages economists specify theories ranging from Adam Smith's theory to contemporary economic principles highlighting the importance of investments in physical capital and infrastructure. This theory, though developed 70 years ago, is still valid.

He, also, mentioned that since 2005, when Global Competitiveness index (GCI) has been used by an increasing number of countries and institutions in assessing national competitiveness the strengths and weaknesses of companies are identified in an undisputable way. The relevant data are quite useful for policy reforms.

Following this principle, the World Economic Forum evaluates the competitiveness rates of countries based on 12 pillars:

- Pillar ONE- Institutions: This pillar refers to the environment based on the legal and administration framework within which individuals, firms and governments try to generate wealth.
- Pillar TWO-Infrastructure covers the sectors of transportation and communication for the access of less-developed communities in order to benefit from the economic activities and services.
- Pillar THREE-Macroeconomic Environment is of greatest importance for business since by means of its stability constitutes one of the most crucial factors which increase the productivity of a notion.
- Pillar FOUR: Health and Primary Education: The links between health and education, denote that healthy workforce plays important role in the countries competitiveness and productivity, whereas the quality of basic education increases the efficiency of its workers.
- Pillar FIVE: Higher Education and Training: The quality of higher education is an important contribution to the economies which "want to move up the value chain" (Becker, G. S., (1993) Besides, globalized economy requires well-educated workforce.
- Pillar SIX: Goods Market Efficiency provides the right mix of products and services through supply-and-demand conditions. "The recent economic crisis has highlighted the degree of interdependence of economies worldwide and the degree to which growth depends on open markets". (Almeida, R., and Cameiro, P., (2009)
- Pillar SEVEN: Labor Market Efficiency strengthens the possibility and flexibility to shift workers from one activity to another rapidly and at low cost.
- Pillar EIGHT: Financial Market development can make capital available for private sector investment from sound sources like loans from banking sector, well-regulated securities, exchange venture capital and other.
- Pillar NINE: Technological Readiness is essential for firms to compete and meet the market demands. ICT contributes to the increased efficiency and competitiveness of countries. (Xavier Sala-I-Martin, 2003)
- Pillar TEN: Market Size affects productivity since globalization has rendered international markets a substitute for domestic markets, mainly for small companies.
- Pillar ELEVEN: Business Sophistication creates the environment necessary for the higher efficiency in the production of goods and services. It concerns the quality of a country's overall network and the operations and strategies of individual firms.
- Pillar TWELVE: Innovation is important for economies as they attempt to enhance knowledge and the possibility to integrate technologies in economy.

These pillars are divided into three categories:

(a) Basic requirements: pillars 1-4, denoting faction driven economies through Institutions, Infrastructure, Microeconomics Environment, Health and Primary Education
(b) Efficiency enhances economies: pillars 5-10, denoting efficiency driven economies through Higher Education/Training, Efficiencies of Goods and Labor markets, Markets size.
(c) Innovation and sophistication factors: pillars 11-12.

Review of Economic Data 2011-2012 of European Countries

Table 16. Comparison data among 5 European countries

No	Specific Pillars	SE	FI	DE	UK	BE
	Overall Position	3	4	6	10	15
	Overall Score	5.61	5.47	5.41	5.39	5.20
01	Institutions:	88	4-5.98	19-5.27	15-5.34	27-5.03
02	Infrastructure	13 – 5-74	19-5.62	2-6.35	6-6.09	17-5.65
03	Macroeconomics	13- 6.08	20-5.21	30-5.43	85-4.54	60-4.90
04	Education: Primary and Health	18 – 6.35	1-6.76	23-6.27	14-6.42	2-6.75
05	Higher Education	2 - 5.81	1-6.09	7-5.73	16-5.47	5-5.75
06	Goods Market	7 - 5.21	21-4.89	26-4.79	19-4.97	14-5.06
07	Labor Market	25 – 4.82	15-4.94	64-4.41	7-5.36	44-4.81
08	Financial Market	11 – 5.24	9-5.34	39-4.54	20-4.94	28-4.76
09	Technological Readiness	2 - 6.29	12-5.75	14-5.61	8-6.08	11-5.80
10	Market Size	31- 4.59	54-4.15	5-6.00	6-5.77	26-4.78
11	Sophisticated Business Culture	2 - 5.83	63-4.00	4-5.66	6-5.41	11-5.30
12	Innovation	2 –5.78	90-2.91	7-5.39	13-4.94	15-4.83

Table 17. Comparison data among 5 European countries

No	Specific Pillars	FR	IE	ES	PT	EL
	Overall Position	18	29	36	45	90
	Overall Score	5.14	4.77	4.54	4.40	3.92
01	Institutions:	23-5.57	23-5-19	88-3.58	51-4.20	96-3.52
02	Infrastructure	4-6.30	14-5.70	78-3.24	23-5.44	45-4.52
03	Macroeconomics	86-4.60	13-3.78	72-4.21	111-4.14	140-3.29
04	Education: Primary and Health	16-8.37	5-6.59	88-5.44	34-6.12	37-6.09
05	Higher Education	20-5.24	22-5.15	32-4.90	35-4.82	46-4.65
06	Goods Market	38-4.56	13-5.10	66-4.23	62-4.27	107-3.88
07	Labor Market	68-4.38	17-5.34	119-3.84	122-3.79	126-3.88
08	Financial Market	16-5.00	115-3.44	64.4.14	78.3.98	110-3.52
09	Technological Readiness	13-5.63	17-5.34	28-4.95	19-5.31	42-4.21
10	Market Size	7-5.74	58-4.12	13-5.44	45-4.35	42-4.43
11	Sophisticated Business Culture	14-5.14	22-4.33	34-4.51	50-4.19	77-3.79
12	Innovation	17-4.71	23-4.37	39-3.55	32-3.71	88-2.48

2. North America

(a) The United States, according to the report of the forum: "continues the decline that began three years ago, falling one more position to 5th place. While many structural features continue to make its economy extremely productive, a number of escalating weaknesses have lowered the US ranking in recent years.

- US companies are highly sophisticated and innovative, supported by an excellent university system that collaborates admirably with the business sector in R and D.
- However there are some weaknesses in particular areas that have deepened since past assessments.

- The business community continues to be critical toward public and private institutions (39th). In particular, its trust in politicians is not strong (50th). In comparison with last year, policymaking is assessed as less transparent (50th) and regulation as more burdensome (58th).
- A lack of macroeconomic stability continuous to be the United States greatest area of weakness (90th).
- Over the past decade, the country has been running repeated fiscal deficits, leading to burgeoning levels of public indebtedness that are likely to weigh heavily on the country's future growth.
- On a more positive note, after having declined for two years in a row, measures of financial market development are showing a hesitant recovery improving from 31st last year to 22nd overall this year in that pillar"
- The US economy is characterized by a moderate unemployment rate and high levels of research and capital investment. (Mintz, S., (2012))
- As of 2012, the country remains the world's largest manufacturing country, representing a fifth of the global manufacturing output (Vargo, F., 2012
- Of the world's 500 largest companies, 133 are headquarted in the US, twice the total of any other country (CNN, 2010)
- The total employed individuals in US are estimated to 154.4 millions, of which 22 million are governmental employees (McFeatters Dale, 2010)
  - 53% of workers are hired by small businesses and 38% work in large businesses. (Pope, C., 2012)
  - Unemployment rate is estimated at 9.4% as of May,2009 whereas part-time employed represented the 15.9% due to the financial crisis 2007-2010), (Bureau of Labor Statistics, US Government, 2009)
  - The overall financial position of US (2009) amounts to 50.7 trillion of debt Owed by households, businesses and governments representing more than 3.5 times the overall GDP of US (Components of US debt)

(b) Canada is accepted by the forum that it continues to benefit from highly efficient markets (with its goods, labor, and financial markets ranked 12th, 5th, and 13th, respectively), well-functioning and transparent institutions and excellent infrastructure In addition, the country has been successful in nurturing its human resources: it is ranked 6th for health and primary education and 12th for higher education and training. As we have noted in recent years, improving the sophistication and innovative potential of the private sector, with greater R and D spending and producing goods and services higher on the value chain, would enhance Canada's competitiveness and productive potential going into the future

3. Africa

According to statistical information and demographic data Africa is the world's largest and most populous continent with a total area 30.368.609 km2 representing the 6% of the total earth's area and the 20.4% of the total land area. (Sayre April Pulley, 1999)

Population has been estimated by the UN Population fund to 1.022.234.000 as of 2009, published, also, in the African News (2009)

The continent's nominal GDP ranges from US$ 0.196 billion (Sao Tome and Principe) to US$ 357,259 billion (South Africa) out of a total of US$ 1,730,072 billion for Africa and US$ 62.909.274 billion globally.

(a) Five African countries are included in the Global Competitiveness Index, namely: South Africa (rank 54, score 4.32), Algeria (rank 86, score 3.96), Libya)
(b) During a forum which was sponsored and organised in Burkina Faso by the Government of Burkina Faso, the African Development Bank, the United Nations Economic Commission for Africa (UNECA) and software company, Microsoft it was discussed the role of technology both in governmental and private sectors in the ICT.

According to the United Nations Economic Commission for Africa (UNECA, 2007) the purpose was reported to be:

- to share specific experiences of technology solutions in education, agriculture, healthcare and governmental operations.
- to provide an opportunity for African countries to learn.
- to combat the general shortage of skills and experience in the region.
- to encourage an overall increase in technology investment.

Additionally, the President of Burkina Faso expressed his ambition "This Forum will be an occasion for African government, their advisory institutions and ICT experts to share their experiences, so that we can build an inclusive information society to serve the sustainable development of our countries."

Besides, the African Development Bank discusses partnership arrangements with Microsoft on meeting poverty problems through education, technology access and training initiatives.

Over the last three decades, the Bank has spent over US $ 1 billion to finance projects on telecommunication, US$ 500 million in master plan studies and US$ 200 million in institutional development.

South Africa

South Africa is described as "the highest-ranked country in Sub-Saharan Africa. It benefits from the large size of its economy, particularly by regional standards (ranked 25th in the market size pillar). It does well on measures of the quality of institutions such as intellectual property protection (30th), property rights (30th), the accountability of its private institutions (3rd). Particularly impressive is the financial market development (4th), at a time when trust is returning only slowly in many other parts of the world. South Africa does reasonably well in more complex areas such as business sophistication (38th) and innovation (41st), benefiting from good scientific research institutions (30th) and strong collaboration between universities and the business sector in innovation (26th). Thus, South Africa is the most competitive economy in the region. However, in order to further enhance its competitiveness the country will need to address some weaknesses

Ghana

Ghana is mentioned in the 114th as last year. The country continues to display strong public institutions and governance indicators with relatively high government efficiency. Some aspects of its infrastructure are good by regional standards, particularly the state of its ports (66th). Financial markets are relatively well developed (61st). On the other hand, education levels continue to lag behind international standards at all levels, labor markets continue to be characterized by inefficiencies, and the country is not harnessing new technologies for productivity enhancements Finally, Ghana suffers from macroeconomic instability (it is ranked a low 139th in this pillar), with the government running high fiscal deficits and building up significant debt.

An analysis related to the issue of Higher Education and Economic development in Africa (Bloom, D., et al.-2006) concluded that:

- "Increasing tertiary education may be important in promoting faster technological catch-up and improving a country's ability to maximize its economic output".
- "Technological catch-up still plays an important role in movement toward the production possibility frontier".
- "Tertiary education may improve technological catch-up and help to maximize Africa's potential to achieve its greatest possible economic growth"
- Higher education will not make a difference in Africa if other barriers play negative role. Higher education creates the potentials, but governments and private actions must seize the opportunities"

4. Asia

The continent's nominal GDP ranges from US$ 616 (East Timor to US$ 5,878,257 (China) In terms of GDP in million the top 5 countries are:

- China US$ 5,878,257
- Japan US$ 5.458.872
- India US$ 1,537,966
- Russia US$ 1,465,079
- Korea US$ 1,007,080

On the other hand, the lowest 5 countries are:

- Tajikistan US$ 5.578
- Kyrzystan US$ 4.444
- Malvides US$ 1.433
- Bhuton US$ 1.397
- East Town US$ 616

The area of the continent (first in the world) is subject to dispute, probably due to the use of different measuring systems. The total area is estimated at something between 43.608.000

km2 (New York Times Atlas, 1992) and 44.390.000 km2 (Concise Columbia Encyclopedia, 1989).

Population is 3.879.000.000 (Sayre April Pulley (op.cit)

As concerns the competitiveness ranking for 2010-2011 out of the 139 countries 23 countries are from the region of Asia, among which eight are included in the top thirty, namely China ranking 27 with score 4.84 and Japan ranking t6 with score 5.37 The United Nations Statistics Division have distinctly defined regions of Asia by rather arbitrary assessments of political circumstances. The major sub regions are as follows: Central Asia, East Asia, North Asia, Southeast Asia, West Asia, and South Asia From the economic point of view, it is important that countries from all regions are included in the Global Competitiveness index of the World Economic Forum.

a. Central Asia

Kazakhstan, as revealed by the forum report, "due to its economic growth, combined with earlier tax and financial sector reforms, has dramatically improved government finance from the 1999 budget deficitt level of 3.5% of GDP to a deficit of 1.2% of GDP in 2003. Government revenues grew from 19.8% of GDP in 1999 to 22.6% of GDP in 2001, but decreased to 16.2% of GDP in 2003.

On November 29, 2003 the Law on Changes to Tax Code which reduced tax rates as adopted. The value added tax fell from 16% to 15%, the social tax from 21% to 20% and the personal income tax, from 30% to 20%. On July 7, 2006, the personal income tax was reduced even further to a flat rate of 5% for personal income in the form of dividends and 10% for other personal income.

Energy is the leading economic sector. Production of crude oil and concentrate from the oil and gas basins of Kazakhstan amounted to 51.2 million tons in 2003. Kazakhstan raised oil and gas condensate exports to 44.3 million tons in 2003. Gas production in Kazakhstan in 2003 amounted to 13.9 billion cubic meters, up 22.7% compared to 2002, including natural gas production of 7.3 billion cubic meters. Kazakhstan holds about 4 billion tons of proven recoverable oil reserves and 2,000 cubic kilometers

According to industry analysis, expansion of oil production and the development of new fields will enable the country to produce as much as 3 million barrels (480,000 m^3) per day by 2015, and Kazakhstan would be among the top 10 oil-producing nations in the world. Kazakhstan's oil exports in 2003 were valued at more than $7 billion, representing 65% of overall exports and 24% of the GDP. Kazakhstan instituted an ambitious pension reform program in 1998. As of January 2005, the pension assets were about $4.1 billion.

The banking system of Kazakhstan is developing rapidly and the system's capitalization now exceeds $1 billion. The National Bank has introduced deposit insurance in its campaign to strengthen the banking sector. Several major foreign banks have branches in Kazakhstan, including Citibank, and HSBC.

Despite the strength of Kazakhstan's economy for most of the first decade of the 21st century, the global financial crisis of 2008–2009 has exposed some central weaknesses in the country's economy. The year on year growth of Kazakhstan's GDP dropped 19.81% in 2008. Four of the major banks were rescued by the government at the end of 2008 and real estate prices have sharply dropped. According to the 2010-2011 World Economic Forum in Global Competitiveness Report Kazakhstan is ranked 72nd in the world in economic competitiveness."

b. East Asia

China, as per the forum report "continues its steady progression in the rankings, rising by one rank to 26th. Indeed, it has improved its score and rank each year since 2005. The country continues to lead the Brazil, Russia, India, China, South Africa (BRICS) economies by a significant margin, with South Africa— China's performance improves in most pillars of the Global Competitiveness Index (GCI).

Its macroeconomic situation is very favorable (10th), despite a prolonged episode of high inflation. China is one of the world's least indebted countries, with a savings rate of 53 percent of GDP, and runs moderate budget deficits. These factors, combined with good economic prospects, contribute to an improvement of the quality of its sovereign debt far greater than that of the other BRICS. China also achieves relatively high standards in terms of health and basic education (32nd), with positive trends in health indicators and nearly universal access to primary education, which is well assessed."

Japan is described as having "fallen three places to rank 9th, with a performance similar to last year's. 23 The country continues to enjoy a major competitive edge in business sophistication and innovation, ranking 1st and 4th, respectively, in these two pillars. Company spending on R and D remains high and Japan benefits from the availability of many scientists and engineers, but supporting a strong capacity for innovation. Indeed, in terms of innovation output, this pays off with the second-highest number of patents per capita. Further, companies operate at the highest end of the value chain, producing high-value-added goods and services. The country's overall competitive performance, however, continues to be dragged down by severe macroeconomic weaknesses (113th), with high budget deficits over several years (135th), which have led to the highest public debt levels in the entire sample by far (over 220 percent of GDP in 2010).

c. Southeast Asia

Malaysia is reported by the forum " to have reached the 21st position, registering improvements across the board. The country's progress is particularly noteworthy in the institutions and macroeconomic environment pillars, as well as in several measures of market efficiency. Among the prominent advantages of this strong and consistent performance are its efficient and sound financial sector— which places among the world's most developed, just behind Singapore and Hong Kong—and its highly efficient goods market, ranked 15th. In addition, its macroeconomic situation has improved markedly over the past year to reach 29th place, even though the country continues to run a budget deficit of about 5 percent of GDP. Malaysia needs to improve its performance in education and technological readiness. In terms of higher education and training (38th), improving access remains a priority in light of low enrollment rates of 69 percent (101st) and 36 percent (66th) for secondary and tertiary education, respectively."

Vietnam's competitiveness, as mentioned by the forum, "declines dropping six places to 65th. The country loses ground in 10 of the 12 pillars of the GCI and only a significant improvement in the macroeconomic environment limits its fall in the rankings. The 2010 budget deficit was still too large, at 6 percent of GDP, and inflation moved back to near double-digit levels after having briefly receded the year before. Vietnam will have to build on its strengths while addressing the economy's numerous challenges. Among its competitive strengths are its fairly efficient labor market (46th) and its innovation potential (66th).

However, the challenges going forward are numerous and significant. Infrastructure, strained by rapid economic growth, remains a major challenge for the country despite some improvement in recent years, with particular concerns about the quality of roads (123rd) and ports (111th). And although education appears to be satisfactory in terms of quality, enrollment rates at all levels remain low (64th, 103rd, and 110th for primary, secondary, and tertiary enrollments, respectively). In order to further improve its competitiveness Vietnam must continue to strengthen its institutional environment."

d. West Asia and Southwest Asia

Qatar: According to the forum report "It keeps its position as the most competitive economy in the region due to improvements in its macroeconomic environment, business sophistication, and innovation. Its strong performance in terms of competitiveness rests on solid foundations made up of a high-quality institutional framework, a stable macroeconomic environment and an efficient goods market. Low levels of corruption and undue influence on government decisions, high efficiency of government institutions, and high levels of security are the cornerstones of the country's very solid institutional framework. Despite efforts to strengthen the financial sector, its trustworthiness and confidence and the soundness of banks are assessed as low and under-protected by the business community and legal rights of borrowers and lenders. Given its high wage level, diversification into other sectors will require the country to raise productivity by continuing to promote a greater use of the latest technologies and by fostering more openness to foreign competition"

Saudi Arabia as stated in the forum report "maintains the second-best place in the region. The country has seen a number of improvements to its competitiveness in recent years, which have resulted in a solid institutional framework, efficient markets, and sophisticated businesses.

The macroeconomic environment benefitted from rising energy prices, which resulted the budget balance into surplus in 2010, although it still remained significantly below pre-crisis levels. Additionally, the country's largest stimulus package among the G-20 contributed to improving infrastructure in the country. which faces important challenges going forward. Health and education do not reach the standards of other countries at similar income levels.

As a result, the country continues to occupy low ranks in the health and primary education pillar. More efficient use of talent will increase in importance as global talent shortages loom on the horizon and the country attempts to diversify its economy, which will require a more skilled and educated workforce. The use of the latest technologies can be enhanced as this is an area where Saudi Arabia continues to lag behind other Gulf economies."

e. South Asia

Pakistan (118th) as disclosed by the forum report: " partially comes back from last year's significant drop in rank. Yet, in several categories, it remains one of the poorest performers of the developing Asia region, and indeed of the entire sample of economies. It is particularly worrisome that Pakistan earns its lowest marks, with no sign of improvement, in the most basic areas of competitiveness, namely institutions (107th), infrastructure (115th), health and primary education (121st), and the macroeconomic environment (138th). In order to benefit from the scale advantages associated with its significant market size (30th), Pakistan will have to decrease regulatory rigidities in the labor market (now ranked 136th) and reduce barriers to domestic and foreign competition in order to render the markets for goods and services more efficient (93rd). By boosting the technological adoption of firms and the public would allow for considerable productivity increases in the country.

India is reported "to rank 56th in this year's assessment. The country demonstrates only minor changes in its competitiveness performance since last year. Its gap with China is widening: the score difference between the two economies has increased sixfold between 2006 and today. The country's supply of transport, ICT, and energy infrastructure remains largely insufficient and ill-adapted to the needs business (89th). Indeed, the Indian business community continues to cite infrastructure as the single biggest hindrance to doing business in the country. It must be noted, however, that the situation has been slowly improving since 2006, although this does not translate into a higher ranking because other countries have been improving faster. The picture is similar in the health and basic education pillar (101st). Despite improvements across the board over the past few years, public health and education quality remain a prime cause of concern. Corruption (99th) fuels this discontent. The macroeconomic environment (105th) continues to be characterized by large and repeated public deficits".

5. Latin America and Caribbean

Venezuela (124th) is revealed to "continue to fall in the rankings, despite a slight improvement in its overall score. The poor quality of the country's public institutions is ranked the worst in the sample at 142nd place. Severe weaknesses in its markets efficiency—especially for its goods and labor markets, where it is the worst performer—and a deterioration in the macroeconomic stability have led Venezuela to feature at the bottom of the region and among the least competitive countries in the world. Despite its tertiary education enrollment rate (8th), the overall quality of the educational system is weak (121st). This, added to a lack of sophisticated businesses (124th) and poor innovation potential (126th), critically constrain the competitiveness performance of the country"

Brazil (rank 53rd overall) is reported to " benefit from several competitive strengths, including one of the world's largest internal markets (10th) and a sophisticated business environment (31st), thus allowing for important economies of scale and scope. Moreover, the country has one of the most efficient financial markets (40th) and one of the highest rates of technological adoption (47th) and innovation (44th) in the region. On a less positive note, Brazil still suffers from weaknesses that hinder its capacity to fulfill its tremendous competitive potential. The lagging quality of its overall infrastructure (104th) its macroeconomic imbalances (115th), the poor overall quality of its educational system (115th), the rigidities in its labor market (121st), and insufficient progress to boost competition (132nd) are areas of increasing concern".

Argentina is accepted that "remains fairly stable this year at 85th, rising two positions in the rankings. The extraordinary competitive potential of the country that benefits from a large domestic market size (22nd) and a population that has a fairly high level of education, with one of the highest tertiary education enrollment rates in the region (21st), remains unfulfilled because of both a lack of trust in its institutions (134th) and the large inefficiencies in its allocation of goods (137th), as well as labor (131st) and financial (126th) resources. More precisely, the country's low public trust of politicians (138th), the uncertainties and favoritism in the decisions of government officials (139th), and the poor assessment of government efficiency (138th) contribute to weakening the foundations of the country's institutional framework. High barriers to trade (142nd) bring about an inefficient allocation of resources in the goods market.

Moreover, a lack of confidence in the financial system (116th) hinders the economy's capacity to mobilize and channel financial resources in an efficient manner. The result is that businesses in the country face difficulties in accessing equity through local markets (114th), loans (134th), or venture capital (129th) to finance their investment decisions"

Peru, at 67th place is considered to "continue to move upward in the rankings. Its improvements in macroeconomic stability (52nd) thanks to a better control of inflation and reduction of the government deficit and debt, coupled with a friendlier environment for entrepreneurship with fewer procedures (34th) and less time (91st) needed to start a business, have contributed to strengthening the country's competitive edge. Moreover, the consolidation of the efficiency gains in both the labor (43rd) and financial (38th) markets, the relatively large size of the domestic market (43rd), and the country's openness to international trade have contributed to sustaining Peru's competitiveness progress in the past four years.

Yet, Peru's economy still faces a number of important challenges that hamper its competitiveness potential. A relatively weak public institutional environment (103rd), an insufficiently developed transport infrastructure network (93rd), an educational system in need of higher quality (128th), and the very low level of innovation (113th) are areas for further effort."

6. Australia

Australia: According to the forum report "the country's most notable advantages are its efficient financial system (6th), supported by a banking sector that counts among the most stable and sound in the world, ranked 4th. Also noteworthy is its very good—and improving—performance in education: Australia ranks 11th in both the health and primary education subpillar and the higher education and training pillar. Australia's macroeconomic situation is satisfactory in the current context (26th), especially when considering the difficulties many other economies face in this area. Despite repeated budget deficits in recent years, its government debt, at 22.3% of GDP, is the second lowest among the advanced economies behind Luxembourg.

Australia's public and private institutions are transparent and efficient ranked 17th and 8th respectively. Australia still lags behind the top performers of the GCI when it comes to innovation (22nd) and business sophistication (29th), two critical drivers of competitiveness for advanced economies. Finally, because of intensifying trade in commodities, the country's transport infrastructure, particularly seaports, has been increasingly improved in recent years and it lags behind the world's best transportation means"

The key indicators referring to the above mentioned countries have been based on sources as follows:

1. Population: UN Population Fund, State of World Population.
2. GDP in US$, International Monetary Funds (I.M.F.)

3. GDP per capita I.M.F.
4. Share of GDP in world total: International Monetary Funds Outlook Database (2010) based on national sources

The data provided by the above reliable sources (Tables 18 to 21) represent important tools which assist any analysis related to demographic and economic elements for representative countries from the geographical areas of North America, Africa, Asia, Latin America and Caribbean, Australia.

Table 18. Comparison data among representative countries

NO	Countries by Continent	Population in millions	GDP in US $ million	GDP per Capita in US $	GDP world Share %
NORTH AMERICA					
01	US	317.6	14.657.8	47.284	19.74
02	Canada	33.9	1.574.1	46.215	1.80
AFRICA					
03	South Africa	50.5	357.3	7.158	0.71
04	Ghana	24.3	31.1	1.312	0.08
ASIA					
05	Kazakhstan	15.8	138.4	8.883	0.26
06	China	1.354.9	5.878.3	4.382	13.61
07	Japan	127.0	5.458.9	42.820	5.82
08	Malaysia	27.9	238.0	8.423	0.56
09	Vietnam	89.0	103.6	1.174	0.37
10	Qatar	1.5	129.5	76.168	0.20
11	Saudi Arabia	26.2	443.7	16.996	0.84
12	Pakistan	184.8	174.9	1.050	0.63
13	India	1.214.5	1.538.0	1.265	5.40
LATIN AMERICA and CARIBBEAN					
14	Venezuela	29.9	290.7	9.960	0.47
15	Brazil	195.4	2.090.3	10.816	2.94
16	Peru	29.5	152.8	5.172	0.37
17	Argentina	40.7	370.3	9.138	0.85
18	AUSTRALIA	21.5	1.235.5	55.590	1.19

Review of Economic Data of the following Countries for 2011-2012 (Table 19) including rank and score of the Forum overall evaluation

(a) In North America (US and Canada)
(b) Africa (South Africa, Ghana)
(c) Asia (Nine countries including China, Japan, Saudi Arabia, India)
(d) Latin America and Caribbean (4 countries including Venezuela and Argentina
(e) Australia

Pillar Codes:

1. Institution
2. Infrastructure
3. Macroeconomic Environment
4. Health and Primary Education

Review of Economic Data of Countries in North America (US, Canada) Africa (South Africa, Ghana) Asia (nine countries) Latin America and Caribbean (four countries) Australia as stated in Table number 19 for 2011-2012 with the overall evaluation of the Forum:

Table 19. Comparative data from World Forum on Competitiveness

N0	Countries by Continent	Rank and Score	1	2	3	4
NORTH AMERICA						
01	US	5- 5.43	39- 4.64	16- 5.68	90- 4.49	42- 6.05
02	Canada	12- 5.33	11- 5.07	11- 5.86	49- 5.06	6-6.58
	AFRICA	vvvvvvvv	vvvvvvv	vvvvvvv	vvvvvvv	vvvvvvv
03	South Africa	50- 4.34	50- 4.23	60- 4.13	116-4.08	45-6.00
04	Ghana	114 3.65	61- 3.96	110-2.84	139-3.49	124 4.28
	ASIA					vvvvvvvv
05	Kazakhstan	72 4.18	94- 3.54	82-3.74	18-5.86	85-5.46
06	China	26 4.90	48- 4.32	44-4.65	10-5.22	32- 6.16
07	Japan	9 5.40	24- 5.18	15-5.69	113-4.20	9-6.00
08	Malaysia	21 5.08	30- 4.94	20-5.22	29-5.50	38-6.14
09	Vietnam	65 4.24	87- 3.63	90-3.59	65-4.78	73-5.66
10	Qatar	14 5.24	14- 5.34	27- 5.17	5-6.40	22-6.28
11	Saudi Arabia	17 5.17	12- 5.47	25- 5.31	12-6.03	61-5.78
12	Pakistan	118 3.58	107- 3.36	115-2.77	138-3.62	121-4.36
13	India	56 4.30	69- 3.84	89-3.60	105-4.30	101-5.25
LATIN AMERICA And CARIBBEAN						
14	Venezuela	124 3.51	142-2.42	117-2.72	128-3.85	84-5.48
15	Brazil	53 4.32	77.3.72	64-3.99	115-4.16	87-5.45
16	Peru	67 4.21	95-3.54	88-3.62	52-5.02	97-5.36
17	Argentina	85 3.99	134-2.93	81-3.70	62-4.88	56-5.80
18	AUSTRALIA	20 5.11	13-5.29	26-5.62	10-6.51	10-6.51

Pillar Codes:

5. Higher Education and Training,
6. Goods Market Efficiency
7. Labor Market Efficiency,
8. Financial Market Development (Table 20)

Review of Economic Data of Countries in North America, Africa, Asia, Latin America and Caribbean, Australia for 2011-2012 (Overall evaluation included)

Table 20. Comparative data from World Forum on Competitiveness

No	Countries by Continent	Rank and Score	5	6	7	8
	NORTH AMERICA	vvvvvvvvv		VVVVVVV	VVVVVVV	
01	US	5 – 5.43	13 – 5.57	24-4.80	4- 5.57	22-4.87
02	Canada	12 – 5.33	12 – 5.59	12- 5.12	5 - 5.42	13-5.20
AFRICA						
03	South Africa	50 – 4.34	73 – 4.09	20- 4.66	95 – 4.06	4–5.48
04	Ghana	114 – 3.65	104- 3.35	72- 4.20	79- 4.25	61-4.16
ASIA						
05	Kazakhstan	72 - 4.18	65 – 4.13	87- 4.07	21- 4.58	48- 4.42
06	China	26 – 4.96	58- 4.34	45 4.48	36- 4.88	48- 4.48
07	Japan	9 - 5.40	19 5.27	18- 4.98	12- 5.04	32- 4.54
08	Malaysia	21 – 5.08	38- 4.76	15- 5.06	20- 4.87	3- 5.53
09	Vietnam	65 - 4.34	103- 4.37	75- 4.16	45- 4.60	73- 4.00
10	Qatar	14 – 5.24	50- 4.62	17- 5.04	22- 4.86	18- 4.86
11	Saudi Arabia	17 - 5.17	36- 4.81	4- 5.25	50-4.57	16- 5.06
12	Pakistan	119 - 3.53	122- 3.01	93-4.06	136-3.47	70-4.03
13	India	56 - 4.31	87- 3.88	70- 4.21	81- 4.20	21- 4.91
LATIN AMERICA and CARIBBEAN						
14	Venezuela	124 – 3.51	67- 4.17	142-.2.89	148-2.88	132-3.11
15	Brazil	63 – 4.32	57. 4.35	113-3.81	83.4.19	43.4.47
16	Peru	67 – 4.21	77- 4.00	50- 4.37	43- 4.62	38-4.54
17	Argentina	85 – 3.94	54- 4.48	137-3.23	131-.3.52	126-3.26
18	AUSTRALIA	20 – 5.11	11-5.62	22- 4.84	13-5.04	6-5.38

Table 21. Comparative data from World Forum on competitiveness

	Countries by Continent	Rank and Score	9	10	11	12
North America						
01	US	5- 5.43	20- 5.26	1-6.92	10- 5.35	5- 5.37
02	CANADA	12-5.31	16- 5.46	10-5.44	24- 4.51	10-5.67
	AFRICA	vvvvvvvvv	vvvvvvvvv		vvvvvvvvv	vvvvvvvvv
03	South Africa	50- 4.34	76- 3.60	25- 4.81	38- 4.32	41- 3.53
04	Ghana	114-3.65	116-2.99	81- 3.42	99- 3.51	99- 2.84
	ASIA					vvvvvvvvvv
05	Kazakhstan	72- 4.18	87- 3.44	55- 4.12	109- 3.42	110-2.67
06	China	26- 4.96	72- 3.52	6.77	37-4.27	29-3.92
07	Japan	9- 5.40	25- 5.06	4- 6.12	11- 5.91	9- 5.59
08	Malaysia	21- 5.08	44-4.20	29- 4.75	20- 4.98	24- 4.37
09	Vietnam	65- 4.24	79- 3.51	33- 4.59	87- 3.72	66- 3.16
10	Qatar	14- 5.24	33- 4.74	59- 3.86	12- 5.27	18- 4.69
11	Saudi Arabia	17- 5.17	43- 4.33	23- 4.92	17- 5.11	26- 4.16
12	Pakistan	119- 3.53	115-2.94	30- 4.67	76- 3.80	75- 3.10
13	India	56- 4.31	93-3.31	3- 6.16	43- 4.27	33-3.58
LATIN AMERICA and CARIBBEAN						
14	Venezuela	124- 3.51	92- 3.36	41- 4.46	124- 4.46	126- 2.50
15	Brazil	63- 4.32	54- 5.98	10- 5.61	31- 4.54	44- 3.50
16	Peru	67- 4.21	69- 3.65	48- 4.34	65- 3.93	113- 2.72
17	Argentina	85- 3.94	64- 3.71	24- 4.86	79- 3.78	78- 3.08
18	AUSTRALIA	20-5.11	22- 5.11	19- 5.10	29-4.87	22- 4.48

Pillar Codes:

9. Technological Readiness
10 Market size
11 Business sophistication
12 Innovation (Table 21)

The detailed analyses and evaluation of collected data related to countries of various levels in all the continents disclose that the degree of response to the 12 pillars is closely linked to the degree of competitiveness.

However, whereas, as stated by Dube, C., (2008), competition policy is generally accepted that it contributes to economic development, it is disputable whether countries which have adopted competition practices perform better than those which have not adopted such practices.

We support the contrary opinion based on the World Economic Forum which provides detailed data evidencing that there is a quite close relationship between the evaluated general scores of the countries and the successful response to the requirements of the 12 pillars.

The following tables will prove with undisputed data the importance of the compliance of operations of governments, institutions and individuals which formulate the competitiveness scores to the most important pillars. Indeed, through structured practices, competitiveness contributes to the economic development of countries through (a) appropriate use of energy in order to lower costs and increase productivity, (b) management of renewable resources and (c) a healthy workforce. (Scwab, K., and Greenhill, R. op.cit).

For comparison reasons, we have compared data related to the response of countries to the 12 pillars which influence their competitiveness This table refers to:

- The top ten EU27 countries-members according to the Global Index of Competitiveness Index (GCI).
- The bottom 5 countries-members of EU27
- Ten selected countries from all continents with high average general score
- The bottom 5 countries from the participants in the world forum according to their general score.

Among the economic cycles, the importance of competitiveness in the formation of economic strategies is accepted worldwide. To this end, the role of the World Economic Forum has been important. Specifically, Schwab, K., and Greenhill, R., (2012) in their preface in the publication of the 2011-2012 index support that: "The World Economic Forum has, for more than three decades, played a facilitating role in this process by providing detailed assessments of the productive potential of nations worldwide.

The *Report* contributes to the understanding of the key factors determining economic growth, helps to explain why some countries are more successful than others in raising income levels and opportunities for their respective populations, and offers policymakers and business leaders an important tool in the formulation of improved economic policies and institutional reforms"

On the other hand, three faculty members of Harvard University, Porter, M. E., Sachs, J. D., and McArthur, J. W., (2012) accept that:

- Global competitiveness means "the set of institutions and economic policies in supportive of high rates of economic growth in the medium term."
- The Current Competitive Index (CCI) uses microeconomic indicators to measure the "set of institutions, market structures, and economic policies supportive of high current levels of prosperity."
- "A nation's prosperity depends on its competitiveness, which is based on the productivity with which it produces goods and services. An understanding of the microeconomic foundations of competitiveness is fundamental to national economic policy."

Importantly, by studying the tables presented in the preceding pages, we can specify some concluding thoughts in order to proceed to the effective planning of strategies to meet the prevailing economic problems.

Table 22. Comparison between responses to the 12 pillars and competitiveness

No	Countries name	Score	Average	No	Countries Name	Score	Average
X1	EU 27 Top Ten			X3	Selected Top 10		
01	Sweden	5.61	5.61	01	Switzerland	5.74	5.79
02	Finland	5.47	5.99	02	Singapore	5.63	5.77
03	Germany	5.41	5.49	03	Hong Kong	5.36	5.57
04	The Netherlands	5.41	5.62	04	Canada	5.53	5.40
05	Denmark	5.40	5.50	05	US	5.48	5.39
07	Belgium	5.20	4.97	07	Japan	5.40	5.30
08	France	5.14	5.04	08	Qatar	5.24	5.09
09	Austria	5.14	4.85	09	Malaysia	5.08	4.63
10	Luxemburg	5.03	5.18	10	Saudi Arabia	4.34	4.58
X2	EU 27 Bottom five from 142			X4	Bottom 5 from 142		
01	Latvia	4.24	4.24	01	Yemen	3.06	3.09
02	Slovac Republic	4.19	4.32	02	Angola	2.98	2.76
03	Bulgaria	4.16	4.10	03	Burundi	2.95	2.74
04	Romania	4.08	4.00	04	Haiti	2.90	2.80
05	Greece	3.94	3.94	05	Chad	2.87	3.73

STRATEGIES

Because of the reliability of the Global Competitive Index (GCI) the Advisory Board of Sustainability and Competitiveness has advised the World Economic Forum to " integrate the concept of sustainability into the GCI.

In order to distinguish a short-to-medium-term view while the Sustainable Competitiveness Index (SCI) will be used for longer-term view, say to 20 years ahead.

The SCI will be organized in five sub-indexes (Schwab, K., and Greenhill, R., (op.cit).

Namely:

(a) Sub-index 1: Human Capital which will provide healthy and well-educated workers who will cope with the globalization requirements
(b) Sub-index 2: Market Conditions, that is goods markets which will ensure healthy competition for products and services, labor markets which will contribute to the economy development through the best of their workers abilities, financial capital which will ensure that capital is invested in the most efficient way, and market size.
(c) Sub-index 3:Technology and innovation are critical constituents of the possibility of local and/or national economies to adjust to the rapidly changing economies, globally,
(d) Sub-index 4: Policy environment and enabling conditions related to relevant pillars of the GCI,
(e) Sub-index 5: Physical environment plays crucial role in the sustainable competitiveness. (Table 23, Schwab, K., editor, 2012).

Table 23. Sustainable Competitiveness Index data

No	SCI Framework	Pillar	No	SCI Framework	Pillar
	Human Capital			Technology-Innovation	
01	Health and Primary Education	4	7	Technological Readiness	9
02	Higher Education and Training	5	8	Business Sophistication	11
03	Social Cohesion	NEW	9	Innovation	12
	Market Conditions			Environmental Policy	
04	Labor Market Efficiencies	7	10	Institutions	1
05	Financial Market Development	8	11	Macroeconomics Environment	3
06	Market Size	10	12	Infrastructure	2

For a long time in the past, policy-makers and economists have formulated strategies in order to combat recession or develop economies in a host of countries.

Yet, the majority of these attempts have failed to attain the anticipated goals.

The problems of failure are attributed to various reasons. They can be classified to three sectors according to the individuals or groups attitude:

1. Mechanisms of Decisions Making

(a) The ways companies or governments make decisions and choose what tools to use as well as when and how to use them (Merchant, N.(2008)
(b) Strategies based on theory rather than on real needs identification.
(c) Lack of correct identification of the structures of markets.
(d) Lack of specific planning

2. Policy Makers

(a) Lack of competent teams to get involved with the problems.
(b) Insufficient data available to policy makers to secure an appropriate policy.
(c) Lack of understandable measures.

3. End Users

(a) The low educational level of the consumers.
(b) The reaction of consumers to new practices.
(c) Inability to adopt policies.
(d Lack of motivation

Additionally, Johnson, P., (2012) attributes the failure of strategies to communication deficits, leadership misguided judgments, no plan behind the idea, passive management and defective motivation.

To formulate a strategy for the development of local, national or international economies, certain steps should be taken into consideration related to data analysis, acquisition of knowledge and skills development and strengths and weaknesses identification. (Table 24)

A proposal of Leadbester, C., and Meadway, J., (2008) is an example of how recession problems can be solved in the UK:

It follows that the UK needs a Total Innovation strategy that draws together public and private, social and commercial innovation and entrepreneurship.

The recession will create a new platform of growth of business entities and industrial units. The following points constitute a highlight of his opinions aiming to combat recession:

- Recessions are equally social, political and economic challenges. The causes lie in malfunctioning and poorly regulated financial networks.
- Whether strategies respond to recession depends on how the challenges are framed by policy-makers and politicians.
- Leadbester (op.cit) quoting the opinion of Joseph Schumpeter reveals that "recessions could provide a platform for innovation and economic growth by unleashing a process of creative destruction".
- The challenge for government policy is to create opportunity out of adversity.
- UK should use the recession to develop new dynamic growth sectors.

  - High-potential innovative new firms, establish economic growth and social transformation for the rest of the economy.
  - UK should be better able to make creative use of the communication revolution in business and culture.
  - Finally, UK need a human capital strategy to encompass young people staying on at school and college with a focus on entrepreneurship and business skills.

On the other hand, education is a considered contributor to economic prosperity. Developed countries enjoy the benefits of wealthy economies. Developing countries , due to low educational level, have to work hard in their effort to achieve the same privileges.

For decades, economic theorists have proposed ways of managing recessions, which for various reasons have not been followed as is the case with the Keynesian theory.

Keynes supported that a circular flow of money in the economy is based on the practice "my spending becomes part of your earnings, and your spending becomes part of my earnings". This doctrine can be quite effective subject to appropriate economic policies.

In the case of Greece, the situation seems to be irreversible unless correct measures are decided by the newly elected government.

Some indicative data are self-explicit:

- GDP (2012) Euros 263.59 billion
- Public debt (2012) Euros 437.60 billion or 166.1% of GDP
- Imports (2012) Euros 40.81 million
- Exports (2012) Euros 17.44 million
- Government Revenue Euros 85.96 million
- Government Expenditure Euros 100.80 million
- Horizontal cut of public and private employees' salaries by 30-35%
- Horizontal cut of public and private pensions by 22-32%
- VAT increase for most commodities and services to 23%

Horizontal cut of salaries/pensions==➔ weakens consumption power which ==➔reduces businesses income and obliges owners==➔to fire employees who as a result==➔lessen even more the buying and selling standards of markets and therefore==➔decrease the amount of VAT returns and payments of income taxes and=➔minimize the planning and implementation of governmental projects for social care.

These measures are quite ineffective and instead of combating recession they probably lead to bankruptcy.

Almost the same opinion is supported in a review of the Greek case by Andre Cabbanes (2012)

"To understand the problem of Greece, we ought to distinguish monetary flows within the country from those across borders. They entail different accounting records, liabilities and problems. Within the country, a very special economic agent is the government, the revenues of which normally come from taxes.

It is a common sense to understand that:

- Within the country: the Greek government kept its economic policy which consisted in spending more than it received from taxes and minor state entrepreneurial activities, and, in the past, printing drachmas to make up for its budget deficits.
- Across Greek borders The Greek government began to borrow euros mostly from foreign investors. Big European lenders (Société Générale, Deutsche Bank, etc.) were happy to lend to Greece, because the interest rate was high and at the same time they knew that Europe would refund them should Greece fail.
- Greece's trade went into deep imbalance. Greece paid its imports in part with exports, in part with promises, and in part with euros, which left the country. The

Greek State is nearing defaulting on its foreign liabilities, and it is running out of euros for domestic expenses. The eurozone, concerned, hesitates on what to do.

- Meanwhile, the Greek government began to reduce the salaries of civil servants, pensions of retirees and public spending. As a result, a growing number of Greeks don't have enough money for their everyday expenditures, and therefore shops and local producers suffer as well.
- Against an ineffective policy, other economists advocate that recessions, in general, can be solved through mechanisms which will support economic growth by a number of initiatives promoting investments.

In order to secure financial stability, increased profitability and competitiveness, companies plan strategies to expand their business activities internationally.

A company, whether a SME or an industrial unit, in order to succeed in expanding its activities, has to take into consideration a series of important issues related to important factors.

Table 24. Steps for successful strategies

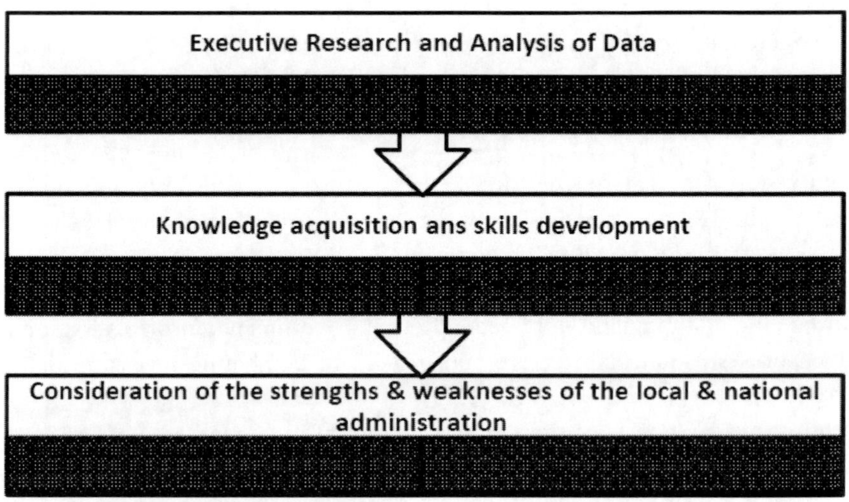

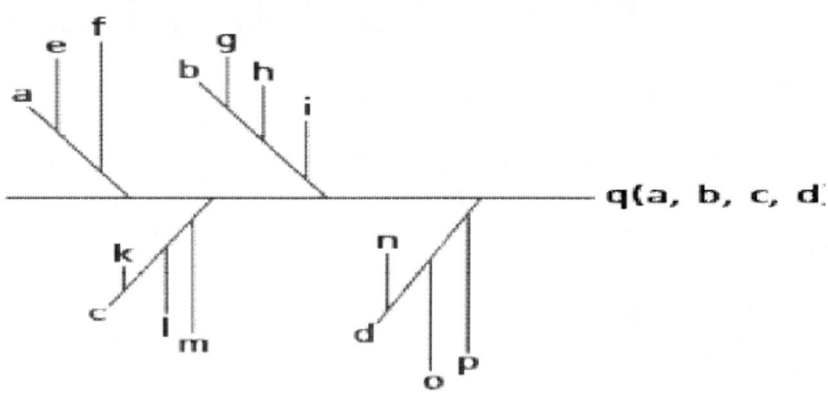

Figure 23. Business Factors, part I.

Of the 12 pillars of the World Economic Forum the following six considered as key factors.

- Innovation
- Health and Primary Education
- Technological Development
- Infrastructure
- Education
- Market Structure

[a] INNOVATION-I (Pillar 12)
  (e) Production process sophistication,
  (f) Extent of marketing.
[b] INNOVATION-II
  (g) quality of research,
  (h) University-Industry links,
  (f) Availabiliy of scientists.
[c] HEALTH and PRIMARY EDUCATION (Pillar 3)
  (k) Quality of health care,
  (l) Primary education enhansement,
  (iii) General Welfare measures.
(d) TECHNOLOGICAL DEVELOPMENT (Pillar 9)
  (n) Availability of latest technology,
  (o) Internet,
  (p) e-commerce.

Part-I

[a] INFRASTRUCTURE (Pillar 2)
  (e) Transportation: Roads, Railways Network, Port Facilities, Airports,
  (f) Energy: Electrical Power Supply, Fixed-Mobile Telephone Lines.
[b] EDUCATION
  (g) Quality of educational system, internet access to schools,
  (h) Secondary and/or Vocational Higher Education, on-the-job -training
  (i) Educational research
[c] MARKET STRUCTURE-_I
  (k) Goods – Competition, Anti-monopoly Policy, Taxes Rates,
  (l) Labor: Wage Determination.
  (m) Hiring and Firing Process, Female participation
[d] MARKET STRUCTURE-II
  (n) Financial: Availability of Services,
  (o) Access to Loans,
  (p) Investments.

Apart from these eight pillars which contribute to the necessary economic growth, there are parameters which hinder any possibility of success and make investors reluctant to embark on the investment effort and undertake the eminent risk of final failure.

According to a recent research conducted by the World Economic Forum (2012) respondents were asked to select the five most problematic factors of doing business, ranking {1} for the most problematic and [5] for the least danger.

The most important responses are shown below by majority of percentage:

- Corruption of individuals and/or services 17.8%
- Inadequately educated workforce 16.1%
- Governmental Bureaucratic procedures 15.1%
- Inadequate Infrastructure 10.5%
- Restrictive labor regulations 09.0%

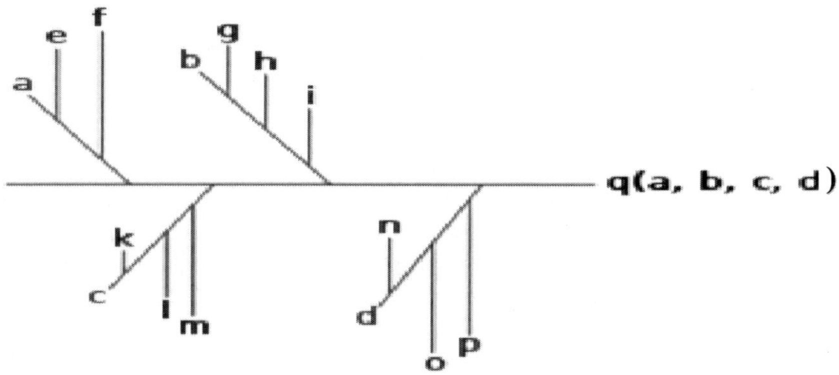

Figure 24. Business factors – part II.

The following table reveals a comparison data among the sample countries we have studied so far in the above five areas of importance, for the establishment of new companies or expansion of existing ones (Table 25)

Codes: [1] Corruption, [2] Inadequately Educated Workforce, [3] Governmental Bureaucracy, [4] Inadequate Infrastructure, [5] Restrictive Labor Regulations, [6] Documents required to start a business, [7] Days required to get a license.

By studying the above data important conclusions can be inferred related to the most problematic factors which affect business development:

[1] Corruption

Among the sample countries assessed, it is shown that, with the exception of Greece (13.6), seven of the listed countries range from 0.4 (Finland) to 4.1 (Portugal) As concerns the other geographical areas, among the top five counties are : a) two from Latin America (Peru, 16.5 and Argentine, 12.7), (b) two Asian countries (Kazakhstan, 17.3, India 16.7) (c) one from Africa (South Africa, 11.5) philosophical terms, it denotes abuse of power by public officers for own benefit

[2] Inadequately Educated Workforce

It is generally accepted that educated persons constitute the driving force in the promotion of businesses, due to the qualifications and capabilities they have acquired through schooling. Inadequately educated people cannot cope with innovative approaches in business, familiarize themselves with technological advances and develop easy contacts with the modern business environment.

Table 25. Comparison of factors influencing the establishment and operation of businesses

Countries by region	1	2	3	4	5	DOC	DAYS
EU 27							
SWEDEN	0.5	7.2	9.3	6.3	24.1	3	16
FINLAND	0.4	2.2	1.8	1.4	23.3	3	14
GERMANY	2.0	11.0	11.2	3.0	20.7	9	15
UK	0.9	6.8	13.5	5.0	8.4	6	13
FRANCE	0.9	4.0	9.1	2.3	23.8	5	7
BELGIUM	0.7	8.1	14.8	2.6	11.5	3	4
IRELAND	1.2	2.6	17.9	8.8	9.8	4	13
SPAIN	3.1	6.3	14.6	2.6	18.3	10	42
PORTUGAL	4.1	5.4	16.4	1.0	16.5	6	6
GREECE	13.1	0.9	22.1	3.6	8.2	15	19
NORTH AMERICA							
US	2.1	6.3	3.2	3.2	5.4	6	6
CANADA	0.6	8.5	17.7	11.4	7.2	1	6
ASIA							
KAZAKHSTAN	17.3	12.3	10.8	5.3	2.1	6	19
CHINA	8.5	7.3	10.9	7.8	4.5	14	18
JAPAN	0.1	1.1	13.9	2.4	7.4	8	21
MALAYSIA	9.6	8.7	12.8	6.4	10.4	9	17
VIETNAM	5.7	7.4	4.2	7.8	1.8	9	44
QATAR	0.0	10.6	6.2	8.2	27.7	18	12
SAUDI ARABIA	2.5	17.0	10.2	8.8	22.0	4	51
PAKISTAN	11.6	4.9	8.3	8.7	3.3	10	21
INDIA	16.7	3.3	13.5	17.0	2.3	12	29
LATIN AMERICA and CARIBBEAN							
VENEZUELA	10.4	0.3	12.1	2.7	11.9	17	141
BRAZIL	6.4	6.4	10.5	15.1	12.2	16	121
PERU	16.8	7.5	16.0	10.4	6.4	6	27
ARENTINA	12.1	2.8	7.3	3.6	2.8	14	26
AFRICA							
SOUTH AFRICA	11.5	16.8	19.6	8.0	16.1	6	22
GHANA	10.8	5.7	7.6	10.2	1.4	7	12
AUSTRALIA	0.0	7.7	12.3	10.3	16.6	2	2

The EU27 member-countries and countries of North America do not seem to face education and/or training problems in contrast to the countries of Asia (Saudi Arabia 17.0) and Africa (South Africa 16.8) where the situation tends to be problematic.

[3] Bureaucracy

This phenomenon has to do with all public services and organizations which formulate policies, implement regulations and laws. The paradox is that of the 26 sample countries assessed 19 are bureaucratic, among which there are countries characterized by progressive business philosophy, like US (13.7), Canada (17.7), A possible explanation might be that these countries behave in a bureaucratic way in order to discourage labor immigrants to take up residence in these countries in seek of jobs!

[4] Infrastructure

It is a very crucial contributor to economic growth since it is connected with all means of transport of goods and people and provides facilities for effective and easy communication. With only six countries with two-digit score, from Ghana (10.2) to India (17.05), it is documented that nearly most countries in the world invest on the quick and effective communication.

[5] Restrictive Labor Regulations

They present an increased score probably in order to prevent a large scale labor immigration, usually from poor Asian and African countries. On the other hand, restrictions aim at protecting employees from employers misconducts and, also, to supervise and regulate decision young children's employment. Besides, International Labor Office (ILO) during the convention number 138 related to the Rights of the Child addressed the minimum age for employment, hours and conditions of work in order to protect them from the various forms of exploitation.

[6] The Number of Procedures

The number of procedures required for starting a business constitute a purely bureaucratic procedure to prevent undesirable business activities, and, hence, discouraging investments. This requirement starts from one document (Canada and New Zealand) to eighteen (Uganda). For 142 countries, a percentage of 74.73% of the total 190 countries globally the survey has shown the following data (Table 26) As to the days required to start a business, the responses are embarrassing since they range from one day (New Zealand) to 694!!! (Suriname). The following data (Table 27) reveals the days required in the countries –members of the economic forum. Economists and policy makers are striving to formulate strategies to combat recession, at first stage, and then attempt to achieve the growth of economy in as many countries globally as possible. For this very purpose The World Bank, the UNO, the EU27 and even the OECD have launched strategic responses to the economic crises.

It is accepted that our era is characterized by accelerated development in technology in all sectors of our daily activities, at a pace which the majority of people globally cannot follow.

One reason is the reluctance of people to adapt to contemporary norms which contradict inherited behaviors of conservatism.

Table 26. Procedures required to start a business

No	Documents	Countries
1	1-4	22
2	5-6	42
3	7-8	29
4	9-11	25
5	12-18	22

Table 27. List of the days required to finalize the procedures for the establishment of a business

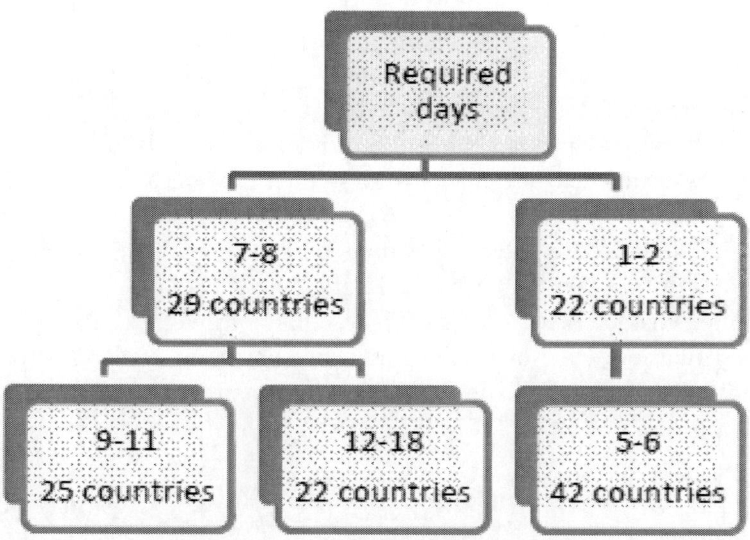

Table 28. Capacity for Innovation

No	Value Scale	Number of countries	%
1	5.8	2	1.40
2	5.2 - 5.7	5	3.51
3	5.0 – 5.1	4	2.82
4	4.0 – 4.8	16	11.27
5	3.0 – 3.9	49	34.50
6	1.5 – 2.9	66	46.50

It is not a coincidence that in the survey of the World Economic Forum as to the way the companies of all countries-members obtain technology, 19.02% responded by conducting formal researches and 80.92% from licensing or imitating foreign companies (Table 28).

Motivation and willingness are not the only factors of successful implementation of economic strategies.

The orientation towards contemporary concepts and principles requires adequately training workforce in order to be able to comprehend and apply the technological developments, which constitute part of innovation.

It is supported by scientists that technology plays a leading role in innovation and can be considered the foundation basis for it.

According to Robertson, J., Sorbello T., and Unsworth, K., (2008), of School of Management, Queensland of Technology, Brisbane, Australia: "The global business environment is now exceedingly competitive, with a high rate of innovation failure and the need to find ways to support organizations to successfully implement innovation is greater than ever. Technology diffusion agencies have emerged for this reason, and can be a powerful tool,".

As for the use of internet, Blind., K., of the Tecchnical University of Berlin,supports that: The information and communication technologies ICT and especially the Internet has not only changed the economy, but also the society including the political system worldwide.

Consequently, the Internet is not only the consequence of innovation at ICT, but also enabler for various types of innovation not only in the private, but also in the public sector.

Nevertheless, the importance of the regional dimension and the role of innovation remained strong or are even strengthened due to the role of knowledge and trust between the partners which builds on close personal face-to-face interactions over time".

Finally, a survey conducted in Canada, by ESI International, Inc. (2011) highlights the importance of training workforce in order to collaborate, which is fundamental to organizations trying to improve their innovation performance. Besides, it is proved that educated workforce perform their duties more effectively.

Comparing three characteristic sectors of three pillars of the GCI, which are interrelated, we identify that countries with lower categories of scores in availability of technology, internet users and staff training are mainly countries from undeveloped or under development areas. (Table 29)

[1] Availability of latest technologies: To what extent are the latest technologies available in your countries? (a) 1= not available, 7=widely available

[2] Internet Users: Percentage of individuals using the internet

[3] Extent of staff training: To what extent do companies in your country invest in training and employee development? (a) 1- hardly at all, 7= to a great extent.

PROMISING STRATEGIES

While recession expands globally as a domino effect, individual countries and international organizations as well design mechanisms to strengthen national economies and improve the quality of life.

An overview of such mechanisms will be observed and discussed as regards five large geographical areas, EU27, North America, Latin America, Asia and Africa.

Table 29. Data of position of 142 countries in three crucial pillars

No	Technology	Count	%	Internet	Count	%	Train.	Count.	%
1	3.1-.9	18	12.7	4.2-15.0	44	31.0	2.3- 2.9	11	7.8
2	4.0 .9.	53	36.6	16- 39.9	36	25.4	3.0- 3.9	61	43.0
3	5.0-.9	41	29.6	40- 69.9	37	26.0	4.0- 4.9	56	39.4
4	6.0-.9	30	21.1	70- 95	25	17.6	5.0- 5.6	14	9.8

[1] EU 27

Policy at Union Members Level

The President of the European Council Herman van Romuy (2012) in an exclusive article to Public Service Europe urged for an anti-recession strategy to be followed by council as a whole but, also, by individual partners:

[1] "Restoring confidence in the eurozone is key to bringing the whole of Europe back onto a path of economic growth. The euro area has taken decisive action on all fronts to rise to the challenge posed by this crisis with initiatives to push forward the necessary fiscal consolidation and structural reforms in our member states."

[2] "It is now important to implement the two programmes agreed by the European Council in December - the fiscal compact treaty and the crisis mechanisms. Furthermore, our crisis mechanisms will be strengthened".

[3] "We are working with our international partners to increase the resources of the International Monetary Fund and single currency members have already announced a contribution of €150bn. Restoring financial stability is key. But a reform agenda based on fiscal austerity alone is not enough".

[4] "We need to focus also on growth and job creation. Which are what we need. We must urgently put in place an anti-recession strategy, mobilizing means and efforts at the EU level and at member state level. The single market is a key driver to Europe's economic growth. We should fully use its potential across all sectors, products, services, energy, the digital economy".

Additionally, President Jose Manuel Barroso, in a speech held before the European Parliament (2012) summarized the main elements of the EU's response to the crisis and for future growth by 2020.

- Combined public and private investment levels to reach 3% of EU's GDP.
- 75% employment rate for women and men aged 20-64 by 2020.Better educational attainment reducing school drop outs rates below 10%, at least 40% 0f 30-34 year-olds with third level education
- Employment: 75% of the 20-64year-olds to be employed.

- R and D: 3% of the EU's GDP to be invested in R and D.
- Climate change energy: (a) greenhouse gas emissions 20% lower than 1990 (b) 20% energy from renewable sources (c) 20% increase of energy efficiency.
- Poverty/social inclusion: At least 20 million fewer people in or at risk of poverty and social exclusion. The above features give an overall view of the targets of EU policy for 2020. They are meant to be used by national countries, and support educational improvements which will reduce unemployment and combat poverty, increase innovation practices in economy and R and D procedures and finally invest in technologies which will create new businesses and provide increased job opportunities.

[2] North America

United States

The United States are considered the leading power worldwide and the political, economic and social conditions prevailing in this country contribute to a great degree to the solution problems in countries not only in the respective geographical area but in all other continents as well. Towards the next decades of the 21st century the US are gearing their policies

President Barack Obama in his address on November 17, 2010 declared that: "For in a global economy, the key to our prosperity will never be to compete by paying our workers less or building cheaper, lower-quality products. That's not our advantage. The key to our success – as it has always been – will be to compete by developing new products, by generating new industries, by maintaining our role as the world's engine of scientific discovery and technological innovation. It's absolutely essential to our future." In line with this target, the Administration proceeded to a series of strategic decisions and priorities in an effort either to maintain or increase prospects of welfare for the nation through " economic growth and competitiveness on many critical dimensions"

New Initiatives

- "The Administration proposed Wireless Initiative which will help businesses reach 98% of Americans with high-speed wireless access within five years and also facilitate the creation of a nationwide public safety network. The initiative will support advances in security, reliability, other critical wireless features; in health, education, transportation, and other application areas".
- "The patent reform agenda is essential to reducing the enormous backlog of patent applications.
- The delivery of innovative goods and services to market, contributes to economic growth and the creation of high-paying jobs".
- "To accelerate the development of clean energy technologies, the President has proposed a Clean Energy Standard to help us reach a goal of delivering to help us reach a goal of delivering 80% of the nation's electricity from clean courses by 2035.

(1) The Startup America initiative will facilitate entrepreneurship across the country, increasing the success of high-growth startups that create broad economic growth quality and competitiveness.

- Invest in the Building Blocks of American Innovation which will dive America's future economic growth and competitiveness
- Educate the next generation with 21st century skills and create a world-class workforce. President Obama is taking continuous steps to improve our educational system from early education through college and graduate school, and to promote student achievement and careers.
- Broaden American leadership in fundamental research. The commercial innovations that drive economic progress often depend on breakthroughs in fundamental science. President Obama has implemented the largest increase in federally-funded research in history and is making continuous investments to double funding for three key basic research agencies".
- Build a leading 21st century infrastructure. denotes "a renewed commitment to investing in the roads, rails, and runways that America's businesses need to be efficient and innovative".
- Develop an advanced information technology ecosystem. President Obama has developed a comprehensive strategy to create the IT ecosystem needed for 21st century innovation which encompasses the critical information, computing, and networking platforms that increasingly supports the national economy".

(2) "Promote Market-Based Innovation. American businesses are our engines of innovation. They bring American ingenuity to the marketplace, where new ideas are proven, commercialized, and diffused. It is imperative to promote a national environment for innovation and entrepreneurship".

(a) Support innovative entrepreneurs. President Obama has expanded lending support and tax credits for small businesses while supporting well-functioning capital markets for businesses of all sizes".
(b) Catalyze innovation hubs and encourage development of entrepreneurial ecosystems. President Obama continues to emphasize the potential of "innovation hubs," looking for new opportunities to bring talented scientists and entrepreneurs together to support innovation.
(c) Catalyze Breakthroughs for National Priorities where innovation is critical but market failures impede progress, government can help spur technological advances. Priorities include developing alternative energy sources, reducing costs and improving care with health

(3) Unleash a clean energy revolution. New and improved energy technologies will play central roles in the 21st century global economy, and provide economic growth and create jobs of the future while confronting environmental challenges and enhancing our energy security.

(a) Accelerate biotechnology, nanotechnology, and advanced manufacturing. The President is committed to investments in innovation that promise to drive better health, future economic growth, and quality jobs in America.
(b) Develop breakthrough space capabilities and applications. Space capabilities play critical roles in global communications, navigation, and Communications.

[3] Asia

Malaysia

Analysts consider Malaysia a multiethnic, upper-middle-income country heavily based financially on the income from its natural resources (rubber, tin, palm oil, Petroleum) and manufacture. Malaysia is member of Association of Southeast Asian Nations (ASEAN) together with Brunei Darussalam, Cambodia, Indonesia, Laos, Myanmar, Philippines, Singapore, Thailand and Vietnam ASEAN basic indicators for Malaysia are reported as follows (Table 30)

Following the independence from the UK (1957), Malaysia enjoyed relative prosperity which , among others, resulted to a total income rising at 6-7% each year from 1970 to 2000 (Yusof, Z. A., and Bhattasali, D., (2011)

Manufacturing receives special attention, with separate sections on electronics and electrical products, the palm oil products, the automotive industry, rubber-based products.

Other key indicators reveal that:

- In Malaysia, the pace of growth of the economy in the first quarter of 2012 moderated to 4.7 per cent against 5.7% in third quarter and 5.2% in the fourth quarter.
- On the demand side, public consumption grew by 5.9% year-on-year, whereas. net exports dropped by 20.8% due to the economic crisis.
- The Auto Index finally settled to 120 points, the Retail Trade Index registered to 119.4 points. On the contrary, the Residential Property Index fell by 5.2 points to 118.5 and the Tourist Market Index fell by 7.9 points to settle at 117.3 However, prospects are very promising as a result of the policies implemented:
- The supply side will expand in all sectors as well as manufacturing. As an overall result, all major economic sectors are projected to register growth in 2012, whereas the services sector is expected to continue playing an important role in driving the country's economy. (Malaysia Institute of Economic Research (2012).

[4] Latin America

Peru

After the transition to representative democracy (2001), the three priorities of the government of Peru policy are (a) increasing the quality and quantity of employment; (b) fighting against poverty by means of support of education and health; (a) consolidating democracy and the rule of law. The main donors, including the EC, have for many years supported the development of Peru, with irregular results, by funding various sectors. Peru was actually one of the largest main recipients of EC aid in Latin America during the 1990s, especially directed towards social development (health and education), food security and rural development.

Table 30. Basic indicators of Malaysia

No	Sector	Amount	Year
01	GDP current price	US$ 193.107.7 million	2009
02	GDP per/capita	12.353.3	2009
03	Exports	US$ 156.890.9 million	2009
04	Imports	US$ 123.330.5 million	2009
05	Population	28.86 million	2012
06	Poverty	2.2%	2012

The co-operation between EU and Latin America according to Article 177 of the Treaty includes:

a) the fostering of sustainable economic and social development.
b) the gradual integration of the developing countries into the world economy
c) the fight against poverty. Together with democratic reconstruction, and the need for reforms the fight of the administration against poverty and social exclusion is at the top of the Government's agenda, through

- The promotion of Human Rights
- The strengthening of State institutions
- The concentration of efforts on the fight against poverty, by means of short-term employment and income generating measures related to social infrastructure, roads, housing, strengthening of local governments, etc. (total budget US$ 614 million).
- The reform of health and education systems aiming at a qualitative and quantitative improvement.
- The structural reform agenda, based on: fiscal reform, continuation of privatizations, strengthening of financial supervision,, support to local governments.

The government aims to reform the education system to provide it with flexibility and institutional capacity: intending to link secondary education campaigns), The economic reform process has exposed Peru to a number of challenges. The international community generally considers President Toledo's reform agenda to be sound and successful due to the following tasks:

- Consolidating democracy and the rule of law.
- Socio-economic development.
- Fighting against poverty through education and health.

Between 1990 and 2001, co-operation amounted to more than 276 million in terms of payments11, which places Peru among the principal beneficiaries of Community cooperation in Latin America

The Member States of the EU have maintained close bilateral relations with Peru for several decades, for the economic, social and political development.

EU co-operation funds received by Peru between 1990 and 1998 amounted to US$1,768 million, which makes it the third largest recipient of European aid in Latin American. Based on the experience of co-operation in Peru, the following principles should be respected to ensure greater effectiveness of EC donor activity:

- EC contributions should ideally concentrate on key sectors.
- While co-operation with the authorities has generally worked well, there may still be a need to better articulate this relationship.
- The EC co-operation strategy will pay close attention to environmental and disaster issues. At the WTO ministerial Conference it was agreed to provide trade technical assistance as a central component of the negotiations under that agenda.

[5] Africa

Ghana

The Senior Minister and Chairman of the National Development Planning Commission of Ghana introducing the Country Report of the International Monetary Fund (2006) presented an overview of the economic situation in Ghana. Selected sections of his preface are quoted below:

[1] Poverty Reduction Strategy Papers (PRSPs) are prepared by member countries in broad consultation with stakeholders and development partners, including the staffs of the World Bank and the IMF. With annual progress reports, they describe the country's macroeconomic, structural, and social policies in support of growth and poverty reduction

[2] The central goal of the new policy is to accelerate the growth of the economy so that Ghana can achieve middle-income status within a measurable planning period.

[3] Side by side with this 20-year transformation of Ghana's social and economic infrastructure, an explosive population boom had got underway, fuelled by the classic rural-urban drift of younger people. These costs of social change were paid out of the profits from the cocoa industry and a newly emerging timber export business.

[4] The need for a rapid and radical transformation of the structure of Ghana's internal production and foreign trade is one of the most widely-shared beliefs.

[5] The restoration of competitive democratic politics in the 1990s seemed paradoxically to work against economic stabilization

[6] The internationally accepted definition of a middle-income country is highly elastic, ranging from the lower limits of an average income of $750 per capita per

[7] In an educational program focused on poverty reduction, the standard way to meet this target was to improve and equalize access to a six-year basic education for all

children up to the age of 12 and school attendance was made obligatory for all children for 11 years -- from 4 to 15 --including two years of Kindergarten, and three years of Junior High School with a genuine secondary school content.

The extensive report on Ghana deals with key issues pertaining to the overall strategy related to the economy of the country (International Monetary Fund, 2006):

[1] Through macroeconomic stability and modest economic growth Ghana's current socio-economic development provides a middle income status with a per capita income of at least US$1,000 by the year 2015.
[2] The vulnerability of the economy due to the reliance on the export earnings and the social structure constitute constraints to achieving accelerated wealth creation and poverty reduction.
[3] The capacity of the private sector will be strengthened to perform as the engine of growth and poverty reduction by improving Ghana's access to global and regional markets.
[4] Accelerating the provision of irrigation infrastructure, promoting selective crop development, modernizing livestock development , improving access to mechanized agriculture are among the priorities of the agricultural sector.
[5] Support services which will develop improved productivity in agriculture include transportation, energy, science and technology.
[6] Employment creation will ensure that the benefits of accelerated growth will increase job opportunities equal for men, women, youth and people with disabilities.
[7] The main goal of Human Resource Development under GPRS II is to ensure the development of a knowledgeable , well-trained personnel and well-trained workforce.
[8] Priority policy interventions will increase participation in education and training, produce skilled workforce for industry and enactment of the Disability Bill.
[9] Health projects will provide quality health services, financing support of the poor, efficiency in service delivery.
[10]The estimated total resources required for these sectors and also those not included in this review will be US$ 8,06 billion (2006-2009)

CHALLENGES

The term challenge denotes the difference between an existing situation and the specifically desired outcome.

The President of the Confederation of Asia Pacific Chambers of Commerce and Industry (CACCI) in his address to the Fourth Annual Convention of Ceylon Chamber of Commerce recognized that: "trade liberalization and market efficiency and competitiveness by the new economic order, have called for adjustments in economic policies.

In a highly competitive world, the role of data, information and knowledge has become more important than ever before. The speed with which information is generated and exchanged will decide the fate of business in this new global order.

These developments have led to the emergence of the eight challenges/Issues six of which are discussed below:

Figure 25. The signature of J. H. Mensah at the end of his preface to the IMF Country Report No. 06/225.

1. Keeping the Free Trade Spirit Alive

"There is a need to sustain the region's commitment to openness and liberation. Advantages of the system:

(a) Increased production due to expanded market size,
(b) Consumer satisfaction is, almost, guaranteed, as free trade expands to global markets which provides opportunities for decisions as to the quality and cost of consumables.

2. Building Up a Strong Information Structure

"Continuing improvements in the information structure capable of linking into a global network are considered worthwhile". As suggested by Oshikoya, T. W., and Hussain, M. N., (2012) of the African Development Bank: "Since man-made comparative advantage can only be acquired by knowledge and brainpower, the newly emerging knowledge-based economic structures have far reaching implications with regard to labor markets and the roles of technical education, human capital formation and research and development in the economic growth".

3. Bringing SMEs into the Mainstream of Economic Development

"The trend towards globalization has made it imperative for SMEs to remain competitive by improving their products, recognizing marketing strategy and developing new technologies". The importance of SMEs in the economic development was emphasized during a seminar organized by The European Foundation for the Improvement of Living and Working Conditions in Berlin (2006)

4. Human Resource Development

"The development of human resources is crucial in improving enterprise efficiency, introducing market innovation and increasing opportunities for market expansion".

5. Technology Development

"To survive the keen competition in the global market, firms must continually upgrade their technology developing schemes of cooperation in technology transfer and increased technology-based investments" The possibility of applicability of specific technological advances may not be fit under any conditions. According to, Kebede, K., et al. 2008): "A technology that has proven to be appropriate and sound in a given area or under specific cultural conditions may not be appropriate under different conditions emerging from local culture, economic needs of countries and geographic circumstances."

6. Exploiting Opportunities for New Businesses

"The rise of new fields of industry and business would do much to revitalize economies and industries and create new job positions." Depending on the country characteristics, demographic data, even geographical location, formalities for establishing a new business vary.

Apart from the legal requirements, factors which should be considered for providing opportunities for new businesses include cultural environment of the area identification of market characteristics, marketing strategies business identity, communicative means and competitiveness.

Review of Challenges in Selected Areas

1. EU27 Countries

In a paper on Demographic Challenge, Gesano G., and Heins, F., in collaboration with Naldini A., support that "Population is a key factor in respect to several economic and social issues. The size of population and its rate of change affects the GDP per capita.

An adequate availability of labor supply, in quantity and quality, for the different needs of labor markets is a basic condition for the economic development and maintenance of any region and country Sectors in which challenges play a very important role are *Population ageing, Disparities mainly in ethical, health and public policies sectors, Globalization and Demographic.*

The prospects of the population fluctuation in the EU27 members-countries and the UN development regions are shown in the following table 37.

Sectors directly affected by the Demographic Challenges include:

Table 37. Total population in the EU27 and UN development regions

Regions	Year	Medium '000	High '000	Low '000
EU27	2005	489.866	489.866	489.866
	2020	505.316	515.298	495.250
	2030	505.619	529.408	481.538
More developed countries	2005	1.216.550	1.216.550	1.216.550
	2020	1.268.343	1.294.280	1.242.195
	2030	1.281.628	1.343.902	1.218.619
Less developed countries	2005	5.295.726	5.295.726	5.295.726
	2020	6.406.489	6.556.370	6.256.626
	2030	7.027.267	7.418.272	6.637.156
Least developed countries	2005	761.846	761.846	761.846
	2020	1.059.484	1.082.675	1.036.294
	2030	1.271.634	1.338.741	1.204.697

Source: UN World population Prospects.

Labor Markets

The EUs working age population, as stated by the 4th Cohesion Report, will be shrinking by 2020 to an extent that can no longer be compensated to increased participation rates (EC 2007).

Education

According to business analysts labor productivity depends to a large degree on the education level of the labor force. The Regions 2020 projection (EC, 2009), predicts a slight drop in the share of people lacking a complete secondary education from 29% in 2005 to 25% in 2020.

Indicators show that the growth of tertiary education levels will not be sufficient to catch up with other developed countries such as the US and that the large disparities between EU regions will remain. In Higher Education, the EU27 scores are only 22% of people aged 25-64 as compared to 45% in Canada, 39% in the US and 30% in Australia and Korea.

Globalization

Globalization is claimed to increase integration of world markets due to reduced transaction costs and limited barriers to the exchange of goods, services, capital, people, ideas, information and knowledge. Yet, its benefits are enjoyed by those regions which are in the most favorable position as compared with other regions.

As prosperity of regions depends heavily on exports, participation in global trade favors enterprises that hold a strong market position through branding, product innovation and process innovation. High education and extensive international connections provide a significant competitive advantage. On the other hand, disparities in personal income create another severe challenge which hinders any prospect of development. (Table 38)

Table 38. Classification of EU27 countries according to per capita income

No	Income per capita	Country	No	Income per capita	Country
	< 27.000 Euro		15		Luxemburg
01		Sweden	16		Malta
02		Netherlands	17		Poland
	Euros 20.000-27.000		18 19		Slovakia Slovenia
03		Austria	20		UK
04		Belgium		Around E.16.000	
05		Cyprus	21		Czech Rep.
06		Denmark			Spain
07		Estonia		E.12.000-13.000	
08		Finland	22		Greece
09		France	23		Hungary
10		Germany	24		Poland
11		Ireland	25		Portugal
12		Italy		E.4.000-7.500	
13		Latvia	26		Bulgaria
14		Lithuania	27		Rumania

2. Countries Candidates for Joining the EU27

Candidate countries are striving to be admitted to the EU27 However, it is not only the challenge of these countries to meet a number of criteria in the sector of political and economic reforms. Political reasons and refusals on the part of the EU members have frozen the relative procedures: Croatia's admittance has been delayed due to Slovenia's blocking until September 2009 for lack of full compliance of Croatia to the International Criminal Tribunal for Yugoslavia. (CICTY)

- FYROM: The procedure has been complicated by the long-standing disagreement of Greece over the official name of the country.
- ICELAND has resisted for decades to join the EU in spite of the extensive ties with members of the Union.
- MONTENEGRO: Admittance is delayed until the country complies with laws, fundamental rights and the fight against corruption and organized crime.
- SERBIA: Its admittance has been confronted by EU members because they consider the country responsible for the violent dissolution of Yugoslavia in 1990 and the conflict over Kosovo in 1999.
- TURKEY faces problems because of the EU disputes having to do with the case of the divided island of Cyprus, being a EU27 member.

Table 39. Challenges faced by the six countries candidates for EU27

CROATIA	MONTENEGRO
Privatization of Enterprises	Need for Strategic Projects
Restructure of Public Administration	Energy Problems
Fiscal Consolidation	Infrastructure, Transportation
Flexibility of Labor Markets	SERBIA
Business Environment	High Unemployment Rates
FYROM	Stagnant Household Incomes
Judicial Reforms	Large Trade Deficit
Institutional Framework	Unemployment of 15-24 years old (43%)
Administrative Reforms	TURKEY
ICELAND	Fundamental Structure
National Currency drop by 90%	Poor Anatolia District
Market Capitalization	Increased Competition
Collapse of 3 Major Banks	Strong Currency Problems

Table 40. Economic and demographic data of candidate countries, I

No	ECONOMIC and DEMO-GRAPHIC DATA	CROATIA	FYROM	ICELAND
01	Population in million	4.42	2.06	0.33
02	GDP in US$ million	67.94	10.78	14.58
03	Employed in million	1.37	NA	0.15
04	Unemployment as % of labor force	12.34	32.18	6.47
05	Governmental revenue as % of GDP	35.44	31.50	41.72
06	Governmental expenditure In % of GDP	41.54	33.70	42.37
07	Investments in % of GDP	25.31	NA	17.05
08	Exports in million	2.23	12.96	2.33
09	Imports in million	2.21	10.73	4.71
10	External Debt as GDP %	90	60	35

The challenges of the above candidate countries show a political rather than economic situation which needs effective strategies based on political consent and not groundless economic theories (Table 39) Analyzing the economic and demographic indicators of the candidate countries (Tables 40 and 41) we will draw some conclusions in our attempt to highlight the real reasons of economic crises which lead the countries to the threshold of bankruptcy.

Table 41. Economic and demographic data of candidate countries – II

No	ECONOMIC and DEMOGRAPHIC DATA	MONTENEGRO	SERBIA	TURKEY
01	Population in million	0.63	7.42	72.97
02	GDP in US$ million	0.44	48.40	876.58
03	Employed in million	NA	1.75	48.50
04	Unemployment as % of labor force	NA	19.75	11.01
05	Governmental revenue as % of GDP	42.25	38.02	33.02
06	Governmental expenditure In % of GDP	44.78	40.71	34.55
07	Investments in % of GDP	22.50	21.65	22.18
08	Exports in % of GDP	5.34	10.22	6.13
09	Imports in % of GDP	0,21	8.53	7.79
10	External debt as GDP %	N.A.	80	36

An in depth critique of the presented data pinpoints certain crucial listings of income and expenses which clearly explain the reasons of economic crises. The role of the following budget entries are decisive in the economic growth of any country: [1] Governmental revenue, [2] Governmental expenditure, [3] Income from exports, [4] Expenses for imports, [5] Unemployment rates and [6] Debt as percentage of GDP.

[1] Revenues are money generating mechanisms mainly by means of direct taxes (income taxes, inheritance and other) or indirect taxes (VAT) and exploitation of public property.

[2] Expenditure refers to payment salaries and pensions, general military forces, health and education policies. In theory, all expenses related to administration should be covered totally from the revenue since this sort of expenditure is the driving force for a well-structured country. Ideally, the formula to be applied should be

$a>b$ or $a=b$

where [a] represents the revenue and [b] stands for the expenditure. Most countries globally have been trying to cover this deficit by implementing measures which are quite inappropriate and ineffective, like horizontal cuts of salaries and pensions, increase of taxation, firing civil servants. All these hinder the cash flow and worsen the situation.

On the other hand, governments borrow at high interest rates without any possibility of repayment

[3] In the balance of trade sector we traced that two countries experience financial problems. Iceland a deficit of US$ 2.38 million and Turkey a deficit of US$ 1.66

[4] Unemployment rates are, also, contributors of economic recessions In FYROM the percentage of unemployment is 32.18% of labor force, in Serbia 19.75%, in Croatia 12.34% and in Turkey 11.09%

[5] Heavy borrowing is a catastrophic means of attempting to face financial problems. The relative figures are quite disappointing and recovery, under these conditions as shown below resembles a daydream episode. (Table 42)

3. Global Economic Associations

For decades, countries characterized by the strength of their economies and the prospects of emerging economies, have established associations to combat economic crises and promote rules of understanding between themselves.

The most important associations, globally, range from 1962 (G10) to 2005 (BRICS)

G10

G10 was established in 1962 by members of the International Monetary Funds (IMF) Belgium, Canada, France, Italy, Japan, The Netherlands, UK and US, which were subsequently joined by the Central Banks of Germany and Sweden. Later Switzerland has joined the group with a minor role.

Of these ten members, seven are also members of EU27 (Belgium, France, Italy, The Netherlands, UK, Germany and Sweden) The Governors of G10 meet every second month at the Bank on International Settlements. In these meetings, they consult and co-operate on economic, monetary and financial matters.

The involvement of the group in monetary issues led them to sign the Smithsonian Agreement (meeting of December 1971) under which the countries agreed to appreciate their currencies against the US$. This had been a problem in international exchange rates since 1944 when dollar was based on the gold standard.

Nevertheless, the gold value of the dollar was realigned again in 1973. This system changed again in March 1973 when the major currencies began to float against each other. (Britannica, 2012)

Table 42. European countries with high debt as % of GDP

No.	EU17 (2012)	% of GDP
01	GREECE	166.1
02	ITALY	119.4
03	IRELAND	117.9
04	PORTUGAL	107.4
No	EU Candidate	
01	CROATIA	99.0
02	SERBIA	80.0
03	FYROM	60.0
04	TURKEY	36.0

Figure 26. Associated Press. From left, Federal Reserve Chairman Ben Bernanke, US Treasury Secretary Timothy Geithner and Deputy Assistant Secretary of Treasury Mark Sobel.

G7

G7 was established in 1975 by a financial group consisting of finance ministers of the big seven industrial economies: Canada, France, Germany, Italy, Japan, UK, US. Reports from Rome (Phillips, M.M., and Meichtry, S., (2009) refer to the summit of G7 where the US Treasury Secretary Timothy Geithner (figure 26) and the rest of the group (a) softened their criticism of China's currency policy and (b) signed on a statement by top economic officials of G7 welcoming "China's fiscal measures and continued commitment to move to a more flexible exchange rates".

Mr. Geithner further explained that his visit to Rome meant to urge the rest of the G7 "to take more aggressive action to address financial chaos and an economic contraction" "We are confronted with a broader, deeper slowdown in global growth than we've seen in decades, It is very important to the US and to the world economy as a whole that the major economies move to put in place as forceful a package of fiscal actions as they can to try to get broad recovery back," Finally, in order to enact a very powerful economic recovery problem he confirmed that "the new administration in the US move with unprecedented speed".

G7 Objectives are [1] To strengthen the leadership capability of CEO, [2] Provide accurate and relevant intelligence, [3] formulate strategies to seize the global initiative, develop an integrated planning process, design and implement an effective command and control system and execute corporate operation to secure global dominant position.

G8

Group of Eight (G8), formerly Group of 7 (G7), is an intergovernmental organization that originated in 1975 through informal summit meetings of the leaders of the world's leading industrialized countries .The president of the European Commission joined the discussions in 1977.

Beginning in 1994, Russia joined the discussions, and the group became known as the Group of 8 (G8), or the "Political Eight"; Russia officially became the eighth member in 1997. The Group of 7 (G7) still functions when the agenda is limited to economic and financial issues. French President Valéry Giscard d'Estaing issued the opening statement during the joint press conference of the economic summit, in Rambouillet, France on November 17, 1975, in the presence of Aldo Moro (Italy), Harold Wilson (UK), Ford (US), Helmut Schmidt (Germany), and Takeo Miti (Japan), (figure 27)

The purpose of this association is to create an informal forum to co-ordinate their macroeconomic policies during the recession and formulate a common strategy in dealing with the developing world.

On May 18-19, 2012 the leaders of G8 met with Barak Obama (figure 28) at Camp David to address major global economic, political and security challenges.

Topics included:

(a) Energy and Climate Change,
(b) Food Security,
(c) Support of political transitions in the Middle East and South Africa (US Department of State, 2012)

Highlights of Obama's closing address, as reported by Cabin, A., 2012 include the following: "For the past three years, our nations have worked together and with others first to rescue a global economy from free-fall, then to wrestle it back to a path of recovery and growth. Our progress has been tested at times by shocks like the disaster in Japan, for example. Today it's threatened once again by the serious situation in the eurozone. As all the leaders here today agreed, growth and jobs must be our top priority. A stable, growing European economy is in everybody's best interests -- including America's. Europe is our largest economic partner

Figure 27. Opening statement of G8 association.

Figure 28. President Barak Obama.

Even as we've confronted our own economic challenges over the past few years, we've collaborated closely with our European allies and partners. We still have a lot of work to do. to take a balanced approach to reducing our deficit and debt, while preserving our investments in the drivers of growth and job creation over the long term -- education, innovation, and infrastructure for the 21st century.

Europe's situation, of course, is more complicated. They've got a political and economic crisis facing Greece, slow growth and very high unemployment in several countries. What's more, when they want to decide on a way to move forward, there are 17 countries in the eurozone that need to come to an agreement. Today we agreed that we must take steps to boost confidence and to promote growth and demand while getting our fiscal houses in order. We agreed upon the importance of a strong and cohesive eurozone, and affirmed our interest in Greece staying in the eurozone while respecting its commitments. Of course, we also recognized the painful sacrifices that the Greek people are making at this difficult time".

G20

G20 The group consists of Finance Ministers and Governors of Central Banks of nineteenth countries and the EU. Of these members twelve members are also members of G8 and/or BRICS associations. The other members of the association are: Argentina, Australia, Indonesia, Mexico, Saudi Arabia, South Africa South Korea and Turkey. (Figure 29)

Figure 29. G20 summit at Los Cabos, Mexico.

The members pledged to step up efforts to revive growth and job creation. (The Telegraph, 2012) During the summit at Los Cabos, Brazil on June 20, 2012, the participating leaders in their declaration of 85 paragraphs committed to strengthen their efforts for recovery, address financial market tensions and express their concern to implement mechanisms related to the following sectors:

- Support of economic recovery.
- Protect employment.

- Foster financial inclusion.
- Enhance food security.
- Meet the Challenges of development.
- Intensify the fight against corruption.

Figure 30. President Vladimir Putin.

In his presentation in this summit, President Vladimir Putin (figure 30) focused on some achievements of Russia in national and international level and proceeded to recommendations.

- Thanks to the G20 measures were taken to enable banks and IMF to support the countries worst affected by the crisis.
- Household debt in Russia came to 10.6% of GDP as of April, 2012 against 60% in Germany and France, 87% in Spain and 92% in US.
- The G20 must fulfill its commitment to reform the international financial institutions, including the IMF and the World Bank.
- It is time to come to an honest agreement on the acceptable level of protectionist measures to protect jobs in times of global crisis.
- The sense and purpose of our work together is to lay down fair rules for sustainable development in the entire global economy.

On the other hand, President Barak Obama in his closing address he emphasized that:

> "Over the last three years, these G20 summits have allowed our nations to pull the global economy back on the path of recovery and growth."

Today we agreed that we must take steps to boost confidence and to promote growth and demand while getting our fiscal houses in order. We agreed upon the importance of a strong and cohesive eurozone, and affirmed our interest in Greece staying in the eurozone while respecting its commitments.

Of course, we also recognized the painful sacrifices that the Greek people are making at this difficult time, and I know that my European colleagues will carry forward these discussions as they prepare for meetings next week".

BRICS

This acronym denotes the original participatory countries of this association: Brazil, Russia, India, China and later South Africa which joined the group in 2010. The association was initiated by Russia in 2006. Despite the differences between them, they are distinguished as fast-growing economies.

As of 2012, they represent 3 billion people with a combined GDP of US$ 13.7 trillion according to IMF.

The purpose of the association is to encourage commercial, political and cultural co-operation between the members. Summits are held annually hosted by the member countries in rotation.

According to Lukov, V. (2012) "The growing potential of BRICS as a factor of global significance is accepted now in the West, although there has recently been an inclination "not to notice" the establishment of this group.

Main activities refer to co-ordination of international political stage, security stage, economic and financial stage .in an interview after the BRICS summit in Sanyic, China (2011) Dimitri Medvedev (figure 31) said:

> The BRICS is a crucial and new mechanism for international cooperation, especially after the Republic of South Africa joined in. All the members are powerful countries with great economic strength, including the People's Republic of China.

At first they were gathering to discuss some minor issues. When this mechanism came into effect, those problems disappeared, because the BRIC started to take joint actions to plan their economic policies.

Secondly, we will discuss issues like how to improve international financial systems, and the standard of communicating with international financial institutions.

Furthermore, during the 2012 summit in India delivering his acceptance speech for the Doctor of Philosophy award, he said: "BRICS is relatively a young entity. With the inclusion of South Africa, it has now become a global entity.

At international forums, we can consolidate our efforts and coordinate our activities.

The key indicators of the associations G10, G8, G20 and BRICS (Table 43) have been taken from the various authentic sources IMF, Central Bank, National Statistical Offices, Economy Watch (2012).

Figure 31. Dimitri Medvedev.

Indicators for GDP, GDP Per Capita, Imports and Exports

Table 43. Indicators for GDP, Imports and Exports of associations G20, G10, G8, BRICS

No	G20 MEMBERS	G10	G8	BRICS	GDP US billion	GDP US$ PER CAPITAL	IMPORT % +	EXPORT % +
01	Argentina				0.53	12.749	15.38	4.59
02	Australia				1.47	64.517	8.54	5.42
03	Brazil			X	2.58	13.109	5.45	5.37
04	Canada	X	X		1.81	51.572	5.26	5.58
05	China			X	7.21	13.872	14.99	2.80
06	Germany	X	X		3.60	44.293	3.86	4.50
08	India			X	1.86	1.448	9.11	14.71
09	Indonesia				0.91	3.776	8.39	7.41
10	Italy	X	X		2.25	36.890	4.09	6.33
11	Japan	X	X		5.92	46.507	9.84	8.19
12	Mexico				1.23	11.115	3.50	6.89
13	Russia		X	X	2.20	15.765	10.72	3.89
14	Saudi Arabia				0.61	22.225	11.23	1.45
15	South Africa			X	0.40	7.874	5.54	5.20
16	South Korea				1.20	24.414	11.78	11.06
17	Turkey				0.88	12.012	7.78	6.13
18	UK	X	X		2.60	41.267	3.22	5.88
19	US	X	X		15.88	50.273	4.74	5.86
20	EU27		X					
G10	Sweden	X			0.52	8.244	5.74	5.29
G10	Netherlands	X			0.85	51.024	4.58	5.06
G10	Belgium	X			0.59	63.045	4.60	4.73
G10	France	X			0.50	44.000	3.39	3.64

Indicators for Demogaphic and Economic Sectors

Table 44. Indicators for Population, Unemployment Rates, Governmental Revenues and Expenditure

	Sector	Argentina	Australia	Brazil	Canada	China	Germany
01	Population million	41.28	22.79	196.53	35.08	1.354.80	81.28
02	Unemployment % of labor force	8.53	4.80	6.70	7.27	4.00	6.49
03	Gover. Revenue % of GDP	38.59	34.85	36.22	8.62	21.56	3.31
04	Gover. Expenditure % of GDP	41.64	35.47	38.79	41.41	22.90	44.88
05	G-20	X	X	X	X	X	X
06	G- 10				X		X
07	G - 8				X		X
08	BRICS			X		X	

No	Sector	France	India	Indonesia	Italy	Japan	Mexico
01	Population - million	63.54	1.249.30	240.51	62.88	127.30	110.81
02	Unemployment % of labor force	9.09		6.50	8.30	4.65	3.90
03	Gover. Revenue % of GDP	50.17	18.78	17.24	45.43	32.15	23.24
04	Gover. Expenditure % of GDP	56.15	26.06	18.63	48.92	40.55	25.63
05	G-20	X	X	X	X	X	X
06	G- 10	X			X	X	
07	G - 8	X			X	X	
08	BRICS		X				

No	Sector	Russia	Saudi Arabia	South Africa	South Korea	Turkey	UK
01	Population million	139.41	27.27	51.12	49.21	72.97	63.06
02	Unemployment % of labor force	7.10		23.70	23.76		
03	Gover. Revenue % of GDP	35.85	30.26	27.95	11.00		
04	Gover. Expenditure % of GDP	37.56	37.16	32.82	7.70		
05	G-20	X					X
06	G- 10						X
07	G - 8	X					X
08	BRICS	X		X	X		

No	Sector	US	EU	Belgium	Netherlands	Sweden	
01	Population million	63.06		10.88	16.69	9.45	
02	Unemployment % of labor force	7.75		8.23	4.37	6.57	
03	Gover. Revenue % of GDP	32.31		50.66	45.91	52.48	
04	Gover. Expenditure % of GDP	39.84		53.06	20.90	11.86	
05	G-20	X					
06	G- 10	X		X	X	X	
07	G - 8	X					
08	BRICS	X					

Table 45. Increase in important sectors

No	Sectors	% increase
01	Construction sector	29.6
02	Communication field	7.0
03	Products and services	4.6
04	Institutions	27.7
05	Electricity and Water	6.7

Figure 32. Venezuela's President Hugo Chavez.

4. South America and Caribbean

Venezuela

The World Economic Forum admits that initiatives constitute a form of recovery of the country from the recession status though indicators as of early 2012 (2012 Q1) are not promising.

Nevertheless, the finance minister Jorge Giordani and the President of the Venezuelan Central Bank (2012) confirmed that the country's economy grew 5.6% in the first quarter of 2012.

Besides, other indicators reveal interesting improvement in a number of economic sectors (Table 45)

The administration of Hugo Chavez since 1999 has shifted from the neoliberal economy projects to new ones which supported a socialist income redistribution and a social welfare.

On the other hand, comparison between demographic and economic indicators shows an improvement of the country in the most important sectors starting from the year 2000 (one year after the election of the President Hugo Chavez (figure 32) through 2005 to 2010 and finally Q1 (quarter 1) to 2012. The fluctuation of the indicators show a slow but steady improvement of the economic status.

Venezuelan analysts explain that their country's most important immediate challenge refers to the oil prices which have fallen almost 70% from their peak in July 2011.

Yet, Venezuela's reserve is estimated at $82 billion which is adequate to finance a moderate current account deficit.

To get an authentic image of the countries development we thought it advisable to compare the main demographic and economic indicators from the year 2000 through 2005 to 2010 and finally to 2012. (Table 46)

During the period of almost twelve years the increase of the GDP indicator from 117.15 US$ (2000) to US$ 300.82 (2012) and the GDP per capita from US$ 4845(2000) to US$ 12.046 (2012) constitute an overwhelming improvement due to the peak increase of oil prices. In the expenditure sector, the 28.29% of the GDP (2000) corresponds in real value to US$ 24.15 million as of US$ 12.14 million, being the 36.51% of the GDP US$ 300.82 (2012).

Table 46. Comparison between indicators in the years 2000, 2005, 2010 and 2012

No	Indicators	2000 (a)	2005 (b)	2010 (c)	2012 (d)	Difference (a)/(d)	Difference (c)/(d)
01	GDP in US$ millions	117.15	124.13	290.68	300.82	156.58%	3.48%
02	GDP per capita in US$	4.845	5.453	9.960	12.046	248.63%	20.90%
03	Investments as % of GDP	24.17	23.00	20.65	21.20	-12.30%	2.50%
04	Government revenue in % of GDP	32.74	37.62	31.16	33.52	2.40%	7.60%
05	Government expenditure in % of GDP	28.29	33.53	37.16	36.51	29.05%	-1.70%
06	Population (million)	24.18	26.43	29.18	30.36	9.30%	4.04%
07	Unemployment as % of labor force	14.00	12.24	8.60	8.60	-61.43%	0.00%

The President of the National Institute of Statistics of Venezuela, Elias Eljuri (2012) reported that about four million jobs have been created in the last 11 years, which shows the development of the country's economy. In the sector of tourism, an increase of 49.5% has been recorded in early 2012, since 312.725 international tourists visited the country between January and April 2012.

The economic strategy of Chavez can be seen through the initiatives expressed in key sectors as (a) economic reforms, (b) employment policies, (c) education and Health strategies (Figure 33).

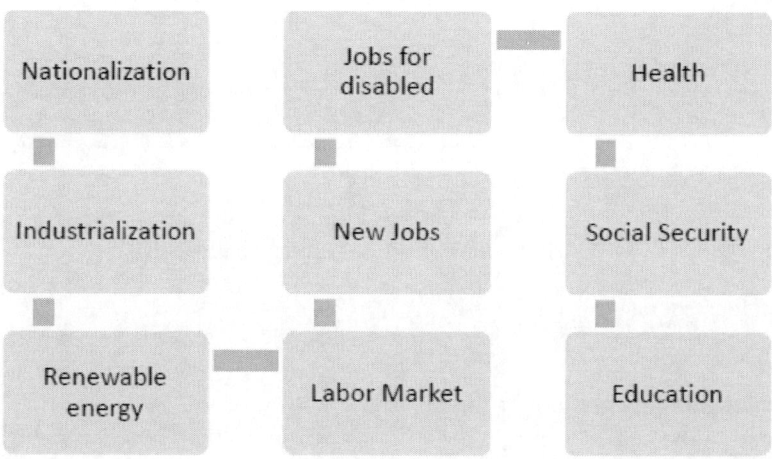

Figure 33. Sequence of reforms.

1. Nationalization

The implementation of the nationalization concept started with Steel Company (2012) in order the government to have control of natural resources so that people may benefit. The nationalization of Steel Company is anticipated to complement other public owned firms (Pearson, T., (2012))

2. Industrialization

Early in May, 2012, the government revealed its attempt to develop the country's light industrial sector. They plan (a) to export mobile phones, (b) in co-operation with the Vietnamese government to produce 80.000 light bulbs per year, (c) to construct another mobile phone factory. (Boothroyd, R., (2012))

3. Renewable Energy Development

A series of inauguration complexes took place within a short period of time: A new petrochemical complex and a thermoelectric power plant on July 22, 2012.

On the same date, a new thermoelectric power plant with the capacity to produce 100 megawatts was inaugurated in Zulia. According to minister Hector Navarro, a wind power project with 400 MW capacity will be incorporated into Zulia's power grid. (Robertson, E., (2012))

4. Labor Market

Indicators show a substantially better labor market than a decade ago. The unemployment rate has dropped from 14% of the labor force (2000) to 8.60% (2012), (figure 34).

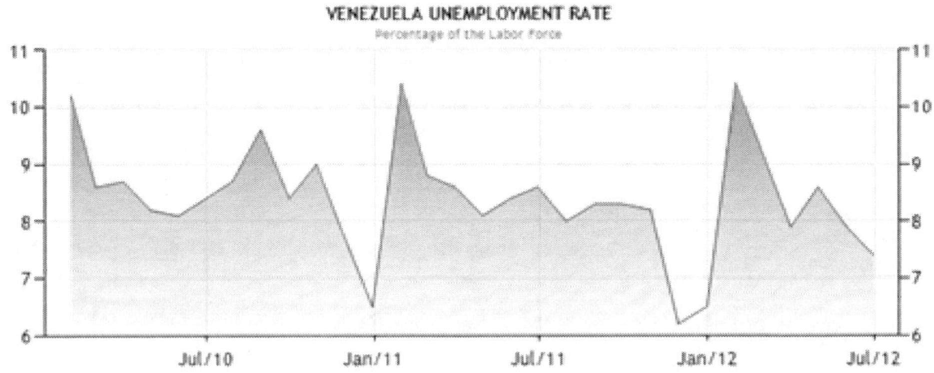

Source: www.TRADING ECONOMICS.COM/ Instituto National de Estadista.

Figure 34. Unemployment rate.

5. New Jobs

Governmental officials claim that thanks to the policy of the President four million jobs have been created in the last 11 years.

6. Jobs for Disabled Persons

In 2006, the relative law for people with disabilities provides that they should (a) have access to free medical assistance, (b) better access to public transport, (c) adequate representation in the work place Besides, as President Chavez declared (2009) "In Venezuela not one single disabled person can be left without the attention that corresponds to her/him."

7. Health

Following the government's policy special care is given to the infant mortality which from 26.17 per 1.000 live births in 2.000 has dropped to 20.18 in 2012 and children under five mortality from 23.2 (2000) to 13.1 (2012) per 1000 live births.(table 47)

In parallel, child mortality has fallen from 23.3% in 1999 to 17.0% in 2006. Furthermore, an improvement in health outcomes has been the access to drinking water and sanitation. (Press release, 2012).

Table 47. Mortality rates per 1.000 live births

No	Category	2000	2003	2006	2009	2012
01	Infant Mortality	26.17	23.79	21.54	21.54	20.15
02	Children under 5 mortality	23.20	21.10	19.20	17.50	13.10

8. Social Security

The program of offering benefits to old age and disabled persons has been risen from 1.7% to 4.4% of population.

9. Education

Following President Chavez's election, a new constitution was drawn up. Article 102 of the Constitution considers education to be a "human right and a fundamental social duty," and is complimented by Article 103, which establishes that "every person has the right to a full, high-quality, ongoing education under conditions and circumstances of equality… education offered at State institutions is free of charge up to the undergraduate university level.

The reference to the education status in Venezuela has to do with politics rather than the specific cognitive area which is also the case with all other sectors, as well .Officials blame the neoliberal policies which excluded the 70% of population from higher education. It is true, though that expenditure for education was just 3% of GDP(1998) or US$ 0.78 billion against US$ 9.08 billion in 2007.

However three facts cannot be disputed: (a) It is an improved status of the various economic and social sectors between the pre-1999 and after-1999 periods, (b) There is a clear difference of policy in all aspects of everyday life be neoliberal and socialistic concepts, (c) the data provided by the Chavez government match squarely with the respective data of the World Economic Forum.

A promising event is the steady increase of expenditure on education which has risen from US$ 0.15 billion (1994) to US$ 9.04 billion in 2007 (Figure 35)

5. Africa

Burkina Faco

Politically the country has experienced a major challenge to the status quo since 1983 when Cap. Thomas Sankara got power and adopted radical left-wing policies and changed the name of the country from Upper Volla to the present name which means "Land of Honest Men: "After four years he was overthrown and killed by his colleague Blaise Compaore (figure 36) who introduced a multi-party system.

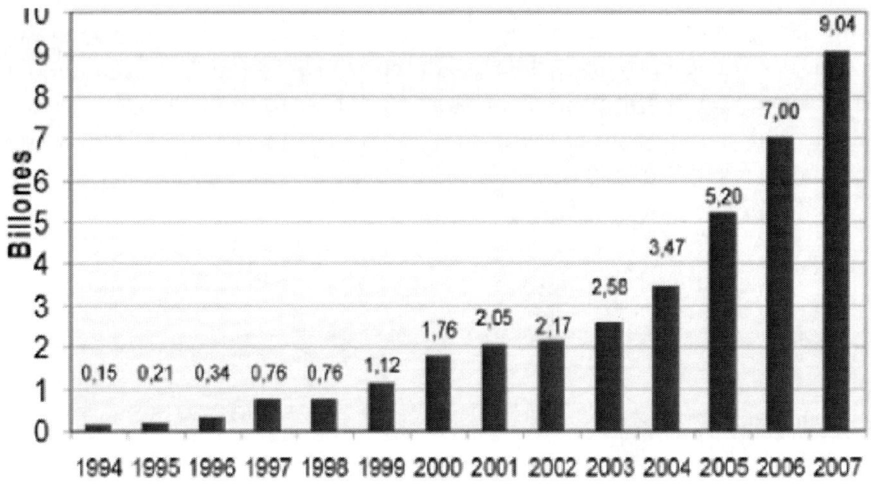

Figure 35. Expenditure data for education.

Figure 36. Blaise Compaore.

However, he has not demonstrated any attempt to strengthen democracy.

Key indicators:

- Population: 16.5 million
- Unemployment: 1.8% of population or 297.000
- GDP in US$: 10.30 billion
- GDP capita US$ 1.298
- Poverty rate: 81.3% of population.
- Investment 21.13% of GDP
- Imports value 5.08% of GDP
- Exports value 10.60% of GDP
- General revenue 21.81% of GDP
- General expenditure 24.56% of GDP

Regarding economic policy the government has made progress in the implementation of a poverty reduction, now 81.2% of population and the growth of the country through the support of World Bank and the International Monetary Funds.

Yet, the country's position in the state of development is in the number 1 (Figure 37) where position 1 means factor driven, position 1-2 transition, position 2 efficiency driven, position 2-3 transition and position 3 innovation driven.

Figure 37. states of development.

Figure 38. Agricultural activities.

Burkina Faso is one of the poorest nations in the world. It suffers from limited natural resources and large population. It really is an underdeveloped country in Africa.. Agriculture comprise approximately 35% of total GDP. More than 90% of the country's working population is engaged in agriculture. (Figure 38).

Principal crops include rice, millet, maize and sorghum. Principal export crop is cotton. Burkina Faso also produces and exports small quantities of sesame, groundnuts, garden vegetables, and shea nuts. Livestock industry has declined in recent years.

Industry

The mineral sector of Burkina is underdeveloped. However, the presence of new found gold in Poura gold reserves is already contributing to more than 10% of annual export earnings. Silver and Zinc are also mined in the vicinity. World Bank loans are used to modernize the Burkina Faso mining industry.

RECOMMENDATIONS

A government budget has two scales, the revenues made up by direct and indirect tax collections and profits from public properties and expenditure which includes purchase of goods and services, investments in infrastructure, social services, education, unemployment benefits and other.

Considering the latest figures of the difference between revenue and expenditures of the budgets of 82 countries of all continents related to the governmental revenues and expenditure provided by the Trading Economics Organization (2012) we observed that:

- Sixteen countries present surplus ranging from 0.30 (Sweden) to 13 (Norway)
- Sixty nine present deficit ranging from 0.50 (Mexico) to 13.10 (Ireland).
- Deficit figures of selected countries from Europe, Asia, North America are shown in Figure 39.

It is clear that all these countries have been spending more than their collection and the easy way to cover this deficit will be:

- Increase of direct and/or indirect taxes.
- Reduction of salaries and pensions
- Borrowing from free markets and/or other countries.

This practice takes us back to about 2400 years when Socrates through Plato in his work "The Republic" refers to the recourse of borrowing when "money-makers inject their poisoned loans wherever they find a victim and demand high rates of interest".

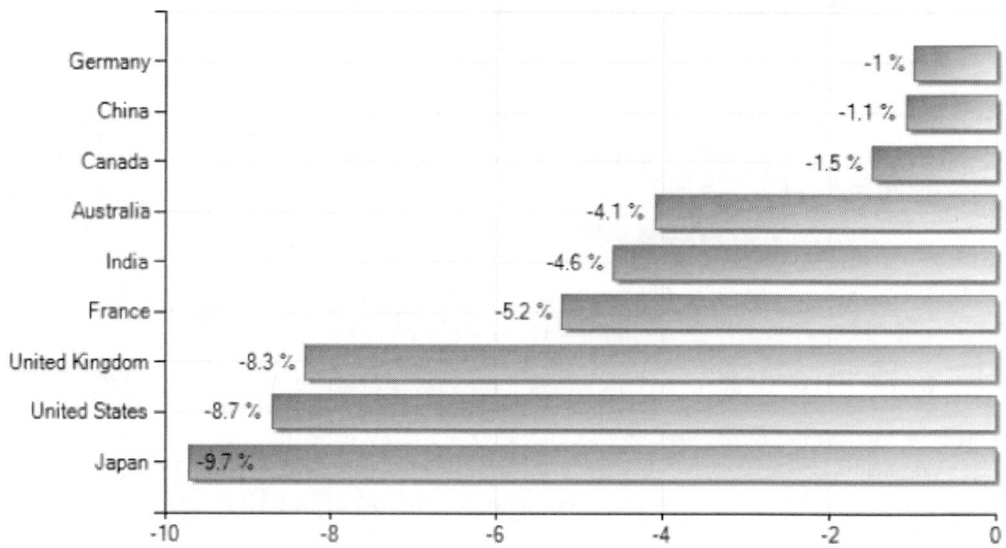

Source: Tradingeconomics.com.

Figure 39. Selected countries with deficits of their budgets.

To recommend mechanisms to combat the economic recession problems it is necessary to trace the causal reasons as referred to previous pages of this chapter. Besides, five factors are consider responsible for the economic recession (Table 48).

(a) Economists and policy makers when formulating their economic strategies should consider that not all countries could implement them for social, political, psychological and geographical reasons. Palm trees cannot grow in Northern Siberia.

(b) Politicians in a number of countries are only vote-hunters and their policies are geared towards unfair enrichment more often than not. The solution is exclusively based the citizens-voters provided they meant to be responsible citizens as a result of appropriately taught culture.

(c) Excessive borrowing creates serious problems since it is the output of improper appraisal which leads to high debt. The deficit created by uncontrolled spending cannot be covered by reducing salaries and pensions and increasing taxes because consumption is reduced followed by decreased revenue. The correct policy is to lay emphasis on investments and reduce administrative and not inelastic expenses like salaries.

(d) Illiteracy: It has been discussed in the previous pages that illiteracy rates play a very crucial role in the development of economic recession. Illiterate people quite often are reluctant to work hard in order to attain higher productivity rates through the use of advanced technologies. They are, also, indifferent or unable to comprehend and/or perform innovative practices at work. To increase the literacy rates we should, through quality, motivate learners and reduce drop-out rates. To this end, our proposal related to curriculum design is still valid. The responsible persons for designing teaching strategies and provide effective and reliable curricula should be a group of sociologists, researchers, policy-makers, economists and educationalists before trusting the delivery procedure to educators (teachers or faculty members at all levels.

(e) A recent tax debate between Republicans and Democrats in US is a proof of the political responsibility as concerns economic recession. The Republicans project provides taxation of 114 million middle-class families with payment of US$ 2.000 while at the same time they refuse to ask the wealthy to pay their fair share. On the other hand the Democrats propose a plan for income tax rates for high income households which will return the country in the 90s when 23 million jobs were created and a budget surplus was attained.

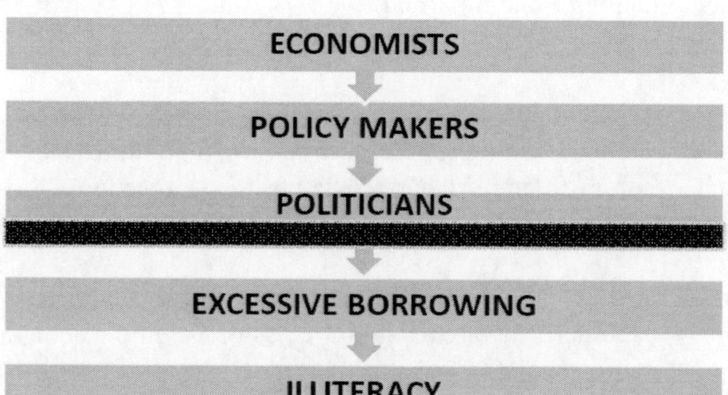

Table 48. Main factors for economic recession.

Conclusion

In the abstract of this chapter we had concluded that what matters is not the concept but the effectiveness of strategies influenced by the prevailing environments which would translate concepts into practical business activities.

For this reason:

1. In outlining the contribution of a number of scientists to the revolution of scientific principles we meant to emphasize that revolution is not the result of any single change but requires structured initiatives even at the expense of their lives. For example, the concept that the earth is the center of the Universe was replaced by the Heliocentrism, even though it is considered an atheist doctrine. Yet, Galileo got into trouble on the part of the clergy for supporting that the earth moves.

2. Recognizing the scientific qualifications of economists, we intended to overturn the opinion which some people have about economists. It is true that economists support different opinions related to the combat of recession , but it is equally correct that not all citizens in any country globally are receptive to new ideas. On the other hand, political controversies conclude to diversified strategies which prevent the proper and effective approach to the solution of an economic problem.

3. In the sector of Practices a reliable source of evaluation is the World Economic Forum (WEF) which provides detailed measurable data for the 192 countries – members which pertain to 12 pillars among which the following sectors are key factors: Innovation, Health and Primary Education, Technological development, Infrastructure, Education and Market structure.

4. Strategies: To distinguish the short-to-medium-term-view countries from the long-term-view the WEF will integrate the concept of sustainability into the Global Competitive Index (GCI) .Thus, the GCI has been organized in five sub-indexes, i.e.

(a) Human Capital
(b) Market Conditions
(c) Technology and Innovation
(d) Environment Enabling Conditions and
(e) Physical Environment

Authoritative documentation about most important sectors in view of establishing a new business units or expanding existing ones, such as corruption, inadequately educated workforce, bureaucracy, infrastructure, labor regulations and procedures linked to the time required for completion constitute a most important contributors to the success of investments.

Various countries from EU27, North America, Asia, Latin America and Africa have formulated strategies for the purpose of combating recession, attaining economic growth and expanding globally in the framework of globalization.

5. Challenges: To attain the effective implementation of strategies, almost all countries worldwide proceed to planning methodologies which will contribute to the attainment of targets. The challenges refer to:

(a) free trade philosophy
(b) information structure,
(c) labor markets and
(d) education

The establishment of these associations constitute an example of effective teamwork and a quicker and more effective mechanism for combating recession and developing economy.

On the basis of the differentiated GDP data of the members of the association (Table 49) the doubt which arises is how far these coalitions will last. The motivation of the economically strong countries to associate with the poorer countries is to expand their markets globally and control them financially. The "altera pars" benefits from the technological advances of the wealthier countries and the easier support of their investment projects. To succeed in this targets and meet the requirements of global competitiveness and understanding, countries characterized by the leading economic and industrial strength have established global economic associations known as G10, G7, G8, G20 and BRICS under a common strategy and well-structured organizational plan to combat recession and achieve economic growth.

Table 49. Lower and Higher GDP in Associations members

No	Associations	Lower GDP (US billion)	Higher GDP (US$ billion)
01	G8	Canada 1.81	US 15.88
02	G10	Sweden 0.51	US 15.88
03	G20	Sweden 0.51	US 15.88
04	BRICS	South Africa 0.41	China 7.11

Policy makers, economists and researchers, supported by international managers should take into consideration that the route from the identification of needs to the economic growth should follow eight steps (Figure 40).

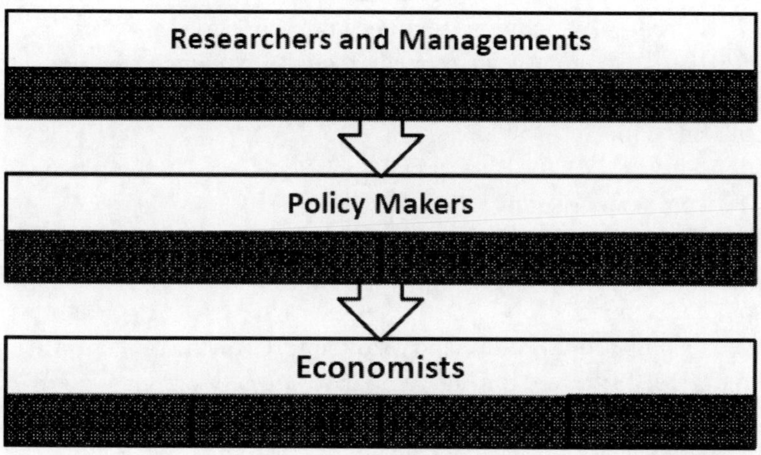

Figure 40. From needs identification to the economic growth.

REFERENCES

Almeida, R. and Cameiro, P., (2009): "Mandated Benefits, employment and inequality in a dual economy", *Social Ptotection Discussion Papers* 45051, The World Bank, retrieved on July 1, 2012 from ideas.repec.org/p/ wbk/hdnspu/45051.html

All About Creation. org (2012): "Origin of the Universe, the Atheist Mechanism, retrieved on June 2, 2012 from www.allaboutcreation.org/ origin-of-the universe.htm

All About Science.org (2012): Bing Bang Theory: The Premise, retrieved on June 21, 2012,from www.bing-bang-theory.com

American Physical Society (2010): "Physics and Economic Development World Conference in Physics and Sustainable Development, retrieved on June 7, 2012 from www.wcpsd.org/economics/

"Asia," *Chambers World Gazetter (5th ed.), 1988*

"Asia," *The Concise Columbia Encyclopedia, 1989*

Becker, G. S., (1993): "Human Capital: A Theoretical and Empyrical Analysis with Special Reference to Education", Chicago, University of Chicago Press.

Blind, K., (2011): "The Internet as Enabler for New Forms of Innovation: New Challenges for Research", *paper prepared for the 1st Berlin Symposium on Internet and Society, retrieved on July 16, 2012* from berlinsymposium.org/sites/berlinsymposium.org/files/innoindex.pdf

Bloom, D., et al. (2005): "Higher Education and Economic Development in Africa, retrieved on July 4, 2012 from siteresources.worldbank.org/ EDUCATION/Resources

Bransomb, L. M., (1996): "Social Capital: The Key Element in Science-based Economic Development, Case Studies around the World, Susan Raymond, Ed. New York Academy of Sciences.

Britannica (2012): "International payments and exchange, retrieved on August 5, 2012 from http://www.britannica.com/eb/article-61797

Brudloss, I. P. (2003): Contributions of Physics to the Information Age, Department of Physics, *UCLA, article retrieved on June*, 8, 2012 from www.physics.ucla.edu?-icub/history/

Burritt, J. (2011): The Universe, evolution and religion, *Religious Tolerance ORC,* retrieved on June, 3, 2012 from www.religious tolerance.org/ burritton.htm

Cabbanes, A., (2012): "The European monetary crisis explained" La Passerelle des Art, retrieved on July 10, 2012 from www.lapasserelle.com/billets/ greek_crisis.html.

Concise Columbia Encyclopedia

Current Population Survey, (2009): Bureau of Labor Statistics, US Government, retrieved on June 19, 2012 from www.bls.gov.

Davis, M. (2008): "The Impact of Recession on Business, INVESTPEDIA, retrieved on June, 14, 2002 from www.investopedia.com/articles/ ecobnomics/08/recession-affecting-business.asp#ax221.xm71JMMN

Dube, C. (2008): " Competition Policy and Economic Growth. Is there a Causal Factor?" *CUTS International*, retrieved on July 6, 2012 from www.cuts-international.org/pdf/CCIER-2-2008.pdf

Eljuri, Elias (2012), President of Venezuela's National Institute of Statistics.

Elliott, L., (2012) "Who to blame for the Great Recession? So many big names are in the frame", *The guardian*, 3 February, 2012

European Commission (2011): *Commission launches highlevel expert group on literacy*, retrieved on June 18, 2012 from ec.europa.eu/education/news/ news 2773-eu.htm.

European Foundation for the Improvement of Living and Working Condition (2006) "How to succeed as an SME in the internal market: Innovation stratyegies to cross-border business, *EMCC retrieved on July 26*, 2012 from www.eurofound.europa.eu/emcc/ content/source/eu 06025 a. htm?pl-reportsandp2=null

ESI International Inc.(2011): "*Importance of Training Staff*", retrieved on July 17, 2012 from http://www.corporateinnovationonline.com

Fiske, F. B., (1988), "National Policy Urged to Combat Adult Illiteracy", *The New York Times*, retrieved on June 19, 2012 from www.my times.com/1988/09/09/us/national-policy-urged-to-combat-adult-illiteracy.html

Gahiji, I. (2012) "*Rwanda: Government continuous to combat illiteracy, News of Rwanda, All Africa Global Media*" retrieved on June 19, 2012 from http://newsof rwanda.com/

Hammond, H. (2005): Illiteracy in the Arab World, *Background paper prepared for the Education for All Global Monitoring Report* 2006, "Literacy for Life", UNESCO

Hasan, A., (2012): "Way to combat illiteracy", *The Truth international weekly magazine*, retrieved on June 19, 2012 from www.thetruth mag.com

Heakel, R., (2010) "What is the Quantity Theory ofc Money" (QTM), *INVESTOPEDIA*, retrieved on June, 10, 2012 from www.investopedia.com/articles/os/090705.asp#ax221 xOEB-zgi

Herman von Romuy (2012): *"Europe needs urgent anti-recession strategy"*, article on Public Service Europe.

Hume, D.(1748): "An equiry concerning Human Understanding, section X "Of Miracloes"

Inter-Parliamentary Union (1997): "Overall Review of Parliamentary Action to Implement Agenda", report adopted by consensus by the Inter-Parliament council at its 160th session (Seoul, 12 April, 1997) retrieved on June, 2012 from www.ipu.org/cnl-e/160-no/htm

Johnson, P., (2012) "The Top Five Reasons Why Strategic Plans Fail", *Attard Communications,* Inc. retrieved on July 2, 2012 from www.businessknowhow.com/ manage/splanfail.htm

Kalra, N., Lauwers, T., Dewey, T., Stepleton, T., Dias, M., B.,(2012): " A Braille Writing Tutor to Combat Illiteracy in Developing Communities, Carnegy Mellon University, Robotics Institute.

Kebede, K. Mulderik (2008): *"NeedsAssessment and Technology Assessment: Crucial Steps in technology Transfer to cdeveloping countries"* retrieved on July 27, 2012 from upcommons.upc.edu/revistes/bit stream/2099/ 7078/1/kebede.pdf

Keynes, J. M., (1936): "The General Theory of Employment, Interest and Money" Macmillan Cambridge University Press, for TRoyal Economic Society

Keynes, John Maynard (1946): "Opening Remarks: "The Galton Lecture", Eugenics Review, 38(1)

Kingsley, Kob (2009): "Africa's Population now 1 billion, AFRICA News, retrieved n June 26, 2012 from www.africanews.com/site/Africas_population_now_1_billion/list_ messages/26588

Kreis, S. (2008): "The Age of Ideologies-reflexions on Kral Marx, The History Guide", *Lectures on Modern European Intellectual History,* retrieved on June 9, 2012 from www.historyguide.org/intellect/Lecture24a.html#marx

Krugman, P., (2007) "Who was Milton Frieman?" *The New York Review of Books,* vol. 59, number 11

Leadbester, C. and Meadway, J., (2003): "Attacking the Recession-How Innovation Can Fight the Downturn", discussion paper, National Endowment for Science. *Technology and thew Arts* (NESTA)

Lukov, V., (2012): "The Role of the BRICS and the Prospects for the Future," *A Global Forum for the New Generation*, retrieved on August 2, 2012 from www.brics.utoronto.ca/analysis/Lukov_Global_Forum.html

Magubane, K., (2008): "Major plans to combat illiteracy", *Media Club South Africa.com* retrieved on June 19, 2012 from www.mediaclubsouth africa.com

Malaysian Institute of Economic Research, (2012): *"Malaysian Economic Outlook",* retrieved on July 21, 2012 from www.mier.org.my/outlook/

Mathru School for the Blind, Bangalore, India: "All About Science.org (2010): *Bing Bang Theory: "The Premise",* retrieved on June 21, 2012 from www.bing-bang-theory.com

Merchant, N. (2008) "*Five Reasons Why Strategy Fails*", retrieved on July 5, 2012 from www.entrepreneur.com/article/195354

Mintz, S., (2012): "Digital History, retrieved o July 20, 2012 from www.Digitalhistory.uk.edu

Muranzi, R., (2005): "The Importance of Physics for the Economic Development, article appeared in the Nov. 2005 issue of Interactions, published by the Institute of Physics, retrieved on June, 7, 2012 from www.scidev.net/eu/opinions/physics-hyas-a-key-role-in-dedvelopment.html

National Geographic (2012): *Origin of the Universe*", issue 2012, retrieved on June 12, 2012 from science.nationalgeographic.com

Oshikoya, T. W. and Hussain, M. N. (2012): "Information Technology and the Strategy of Economic Development in Africa (1998), African development Review. Bloom, D., et al. 2001. "Globalization, Liberalization and Sustainable Human Development: Progress and Challenges in Jamaica." *Jamaica Country Assessment Occasional Paper*, UNCTAD/EDM/Misc.176. New York: United Nations Development Program

Pope, C., (2012): America's Dirty War Against Manufacturing", Bloomberg McFeatters,Dalr(2010): "Saluting 154 million in workforce on labor Day" *Napa Valley TRegister.*

Premi, M. (2002): "India"s Literacy Panorama Proceedings of Seminar on Progress of Literacy in India: What the Census 2001 Reveals, retrieved on June 18u, 2012 from www.education forallindia.com/page172.html

Robertson, J., Sorbello, T., Unsworth, K., (2008): "Innovation Implementation: The Role of Technology Diffusion Agencies", *Journal of technology, management and innovation*, Santiago.

Ross, R., (2006): *"The Role of Economists"* retrieved on June 25, 2012 from cafehay ek.com/2006/09 the_role_of_eco_html

Sayre, April Pulley (1997): *"Africa: Twenty First Century"* rretrieved on June 26, 2012 from http://books.google.gr/books/about/Africa.html?/id= v9ziwQ.P26 uw CRC.redir_ese=y

Scheib, A., (2012): *"Franco Modigliani, "The American Israeli: Cooperative Enterprise*, retrieved on June 10, 2012 from www.jewishvirtuallibrary. org/source/biography/ Fmontiliani:html

Scheth, A (2012) *"Franco Modigliani: Biography, Jewish Virtual Library,* retrieved on June 12, 2012 from www.jewishvirtuallibrary.org/jsource/ biography/Fmodigliani/htm/

Schumpeter, J. M. (1942) *"Capitalsim, Socialism and Democracy",* Harper and Row publishers, N. Y., retrieved on July 1, 2012 from www.sxcribd.com/ doc/8676091/j-a-schum peter-capitalism-socialism-and-Democracy.

Schwab, K. (editor, 2012): "The Global Competitiveness Report 2011-2012", Geneva World Economic Forum.

Skidelski, R., (2003): "John Maynard Keynes, 1883-1940, *Economist*, Philosopher, Stateman, Pan Macmillan Ltd.,

The New York Times and Bartholomew, Edinburgh (1992): *The New York Times Atlas of the World"* New York: Times Books

The Wall Street Journal/Econbomy (2009): "G-7 Softens Criticism of China's Currency Policy", reportage by Phillips, M. M. and Meichtry, S.

Theodoracopoulos, I. N., (1996): *"The CAPITAL of Karl Marx*, published by ESTIA Publishers, Athens.

Thorpe, E., (2011): Shawick Thorpe. *The Pearson General knowledge Manual,* India: Dorling Kindersley.

Timimi, K., (2010): *"Famous Economists Biographies",* retrieved on June,7, 2012 from www.economywatch.com/economist/

UNECA (2007): *"Technology and econbomic best practices to help drive Africa's growth"* retrieved on June 26, 2007 from http://www.uneca.org/

UNESCO (1965): *"Literacy as a factor in development",* World Congress of Ministers of Education on the Evaluation of Illiteracy, Teheran, 8-19 Sept. 1965, retrieved on June, 18, 2012 from unes.doc.unesco/org/images/0014// 001481/148164 eb.pdf

United Nations Development Program UNDP, 2011): "Volunteer teachers combat illiteracy in Burkina Faso, retrieved on June 19, 2012 from www.un.dp.org/content/undp/eu/home/ outwork/povertyreduction/successstories/volunteer-teachercombat/illiteracyin burkinafaso/

US Department of State, Diplomacy in Action (2012): *"Summit G8"* retrieved on August 6, 2012 fromwww.state-gov/e/eb/ecosum/2012g8/index.htm

US Department of State (q2009): *"Kazakhstan"* by Bureau of South and Central Asian Africa, retrieved on June 29, 2012 from www.state.gov/r/pa/ei/bgn/5487.htm

US Departgment of State (2011) "Economy and Business-Conditions and Resources", retrieved on June 22, 2012 from krakow.usconsulate.gov/ conditions_resources.html

Virgo, F. M. (2012) *"US Manufacturing Remains World's Largest"* Shopfloor, retrieved on June 28 from www.shopfloor.org/2011/03/u.s-manufacturing-remains-worlds-largest /18756

Wolfe, M., (2012) *"Economist Responsibilities"* retrieved on June 2, 2012 from www.ehow.com/list_6734737_economist_responsibilities.html

Xavier Sala-I- Maretin et al. (2012): "The Global competitiveness Index 2011-2012 setting the foundation for strong productivity, *World Econbomic Forum* retrieved on July 1, 2012 from www3,wo forum.org/docs/WEF_ acR_Report_201102012 pdf

Yang, Y., (2010): The Influence of Karl Marx on Economics, *EcoGuru*, retrieved on June, 9, 2012 from www.econguru.com/the-influence-of-Karl-Marx-on-economics/

Yavlinsky, G., (2012) *"The Hidden Cause of the Great Recession, extract from Realeconomik"* Yale Books, Yale University Press, London.

Yusof, Z. A. and Bhattaseli, D (2011): "Fostering Sustainable Development , The sector Investment Program" World Bank Discussions, paper No 363

Youth, E. (2008): "Survey: 9 to 10 Americans Believe in Miracles", *The Christian Post, Politics"*, retrieved on June 20, 2012 from www.christianpost.com/news/survey#9-10-americans-believe-in-miracles-35857/

Zalinski, E., (2010): *"Causes of Economic recession, Business News*, newspaper, retrieved on July, 12, 2012 from http://www.love-a-recession.com/Causes-of-Recession.html

In: Economic Developments and Emerging Markets ...
Editors: Guy Boscarino and Derek Notte
ISBN: 978-1-62417-986-0

Chapter 2

PARAMETRIC CONTROL OF ECONOMIC GROWTH PROCESSES AND CHARACTERISTICS OF SHORT-TERM CYCLES

Abdykappar A. Ashimov, Rakhman A. Alshanov, Yuriy V. Borovskiy* and Bahyt T. Sultanov
Kazakh National Technical University, named after K. Satpayev, Kazakhstan

"... the economic theory is used for predicting effects of alternative political rules and choosing one of them with good operating characteristics."

F. Kydland, E. Prescott [1]

ABSTRACT

In this work we briefly consider the components of the parametric control theory and list the main results of its component-wise development. The results of macroeconomic analysis and application of the parametric control theory to solving the problems of economic growth and attenuation of effects of various shocks on the base of a number of mathematical models are presented.

The general representation of computable models of general equilibrium (CGE models) as well as description of the CGE model of the branches of economy is considered. The method of parametric identification of the CGE model of the branches of economy is presented on the base of statistical data on evolution of the national economy of the Kazakhstan Republic.

Within the framework of application of the parametric control theory, the results of estimations of the stability indexes for the estimated CGE model of the branches of economy and examples of macroeconomic analysis along various cuts on the base of this model are given.

* Corresponding author's e-mail: yuborovskiy@gmail.com.

We choose the instruments of economic policy in the sphere of economic growth on the base of analysis of the sources of economic growth within the framework of the CGE model of the branches of economy and give a solution to one problem of economic growth with application of the parametric control approach based on the considered model.

Also, in this work we consider the problem of choice of the effective coordinated public policy in the sphere of economic growth by means of multiobjective optimization on the base of two CGE models with common economic instruments of optimization.

The results of macroeconomic analysis and parametric control of the processes of economic growth on the base of the Smets-Wouters dynamic stochastic general equilibrium model (DSGE model) estimated by the statistical data of the Kazakhstan Republic are described. In particular, we present the estimation of effect of internal shocks on the economic indexes of the Kazakhstan Republic and give a solution to one problem of economic growth on the base of the model under studying.

In the rest of the paper we present the results of parametric control of the characteristics of the short-term cycles on the base of the CGE model of the branches of economy and the model of Kondratiev wave, as well as the results of parametric control of effect of shocks on the evolution of national economy on the base of the Smets-Wouters DSGE model estimated by the statistical data of economy of the Kazakhstan Republic.

INTRODUCTION

It is known that the need for participation of State in regulation of national economy is not discussed now. Vice versa, possible internal instability of an economic system and duration of its natural adaptation to shocks of various kinds accompanied by undesirable social after-effects condition the active public economic policy based on the appropriate macroeconomic instruments [2, 3].

An important direction of macroeconomic theory development, which contains analysis tools and evaluations of state policy efficiency, lies in application of adequate mathematical models of national economy.

[4, 5] contains an original parametrical regulation theory for dynamical systems with a property of structural stability. The application of the proposed theory to the national economy regulation, taking into account various opportunistic situations, demands the followings:

- formation of a library of national economy mathematical models;
- provision of this library with tools for assessing sustainability indicators and weak structural stability of the models;
- implementation the means of parametrical regulation of the corresponding mathematical models;
- development of methods for making recommendations on effective state economic policy based on the analysis of the results of parametrical regulation on the basis of appropriate mathematical models.

The present work presents a brief description of the components of parametrical regulation theory and lists some results on component wise development of elements of the parametrical regulation theory for national economy based on the set of mathematical models.

Content of this work to some extent illustrates the possibility to evaluate the versions of alternative policy rules in the field of stabilization policies and policies in the sphere of economic growth.

The authors are grateful to N. Yu. Borovskiy and M. A. Onalbekov for their help in carrying out computer simulation experiments.

1. Elements of Theory of Parametric Control of National Economy

The theory of parametric control [4, 5] was developed on the base of the following classes of mathematical models, which (after some transformations) the descriptions of economic system of a country can be expressed by:

- nonlinear systems of ordinary differential equations (continuous-time dynamic systems);
- nonlinear systems of difference equations (discrete-time dynamic systems);
- nonlinear systems of different equations with additive noises (discrete-time dynamic stochastic systems).

It is known that

- a solution of continuous- or discrete-time dynamic systems [which can include both vectors of controlled parameters, instruments of the public policy (u) and vectors of uncontrolled parameters (a)] depends on vectors of initial conditions and parameters (coefficients) of this systems;
- to make a judgement on an object described by a dynamic system based on the research results, this system must possess the property of structural stability (or robustness);
- the conditions of stability of macroeconomic model (represented by one of the aforesaid dynamic systems) must be satisfied under small disturbance of initial statistical data for parametric identification of the model (input parameters).

On the base of the aforesaid, in [4, 5] the theory of parametric control of the market economy development was proposed. It consists of the following components:

1. The methods for forming the set (library) of the macroeconomic mathematical models of national economy. These mathematical models are oriented towards the description of various specific socio-economic situations.
2. The methods for estimating stability indexes and conditions of robustness (structural stability) of the models of national economic systems from the library without parametric control.

3. The methods of control or attenuation of non-robustness (structural instability) of the mathematical models of economic systems. The choice (synthesis) of the algorithms of control or attenuation of the structural instability of the respective mathematical model of a national economic system.
4. The methods of choice and synthesis of the laws of parametric control of the market economy development based on the mathematical models of a national economic system.
5. The methods of estimating robustness (structural stability) of the mathematical models of national economic systems from the library with parametric control.
6. The methods of the adjustment of the constraints on the parametric control of the market economic development in the case of structural instability of the mathematical models of a national economic system with parametric control. Adjustment of the constraints on the parametric control of market economic development.
7. Methods of study of influence of variation of uncontrolled parameters (uncontrolled factors) on results of solution of variational calculus problems of synthesis and choice (in the environment of given finite set of algorithms) of parametric control laws. Analysis of bifurcation of extremals of the variational calculus problems of choice of optimal parametric control laws.
8. Approach to choice of recommendations on estimation of political rules within the framework of application of the respective economic instruments of regulation of a national economy on the base of analysis of dependencies of optimal values of criteria of the respective problems of parametric control on the values of uncontrolled factors.

Let us cite the main results [5] of component-wise development of the parametric control theory.

- In the context of methods for forming the set (library) of the macroeconomic mathematical models, the algorithm of parametric identification of large-size macroeconomic models comprising joint application of two criteria of identification is proposed.
- In the framework of development of methods for studying stability of the mathematical models, the stability indexes and numerical algorithms of their estimation are proposed.
- In the context of development of methods for studying weak structural stability, the proposed numerical algorithm on the base of Robinson theorem on sufficient conditions of weak structural stability of the mathematical models is considered.
- Within the framework of methods for choice and synthesis of parametric control of a national economy, on the base of continuous- and discrete-time dynamic systems, as well as discrete-time dynamic systems with additive noise, we formulate and prove the respective theorems on existence conditions for solutions of the variational calculus problems of synthesis and choice (in the environment of given finite set of algorithms) of optimal parametric control laws.

- In the context of methods for studying the influence of variation of uncontrolled factors on the results of solution of variational calculus problems of synthesis and choice of parametric control laws, we formulate and prove the theorems on conditions of continuous dependence of optimal values of criteria of the considered variational calculus problems on uncontrolled parameters (values of uncontrolled functions).
- In the limits of studying bifurcations of extremals of the variational calculus problems of choice of optimal parametric control laws, we formulate and prove the theorems on sufficient conditions of existence of the respectively defined bifurcation points of the extremals of the considered variational calculus problems.
- The approach to choice of recommendations on estimation of political rules within the framework of application of the respective economic instruments of regulation of a national economy on the base of analysis of dependencies of optimal values of criteria of the respective problems of parametric control on the values of uncontrolled factors is proposed.

2. Macroeconomic Analysis, Parametric Control of Economic Growth Processes Based on Computable Models of General Equilibrium

As is well known [6], in the context of implementing economic policy estimating the values of the economic instruments ensuring the uniform growth (dynamic equilibrium) is required, which provides such economic development that the supply and demand in the macroeconomic markets increasing from one period to another are always equal when the use of labour and capital is full.

Consideration of the known mathematical models of economic growth starting from the Solow model up to various models of endogenous growth [7], model of cross generations [8], intersectoral balance model [9] and others [10] shows that the computable models of general equilibrium (CGE models) [11] and dynamic stochastic general equilibrium models (DSGE models) [12] satisfy the requirements listed above to the greater extent.

In Sections 2 and 3 we present the results of macroeconomic analysis and applications of the elements of the parametric control theory by the examples of the CGE model of economy branches and Smets-Wouters DSGE model. Also, in Section 2 we consider the problem of consistent optimization based on the set of macroeconomic models by the example of two models (CGE model of economy branches and CGE model with sector of knowledge).

2.1. Representation of CGE Model

A non-autonomous CGE model in general form is described by the following system of relations.

1) A subsystem of difference equations connecting the values of endogenous variables for two consecutive years

$$x_1(t+1) = f_1\ x_1(t), x_2(t), x_3(t), u(t), a(t)\ . \quad (1)$$

Here $t = 0, 1, \dots, n-1$ is a number of year in discrete time; $x(t) = \ x_1(t), x_2(t), x_3(t)\ \in R^m$ is the vector of endogenous variables of the system

$$x_i(t) \in X_i(t) \subset R^{m_i}, i = 1, 2, 3. \quad (2)$$

Here the variables

- $x_1(t)$ include the values of the key assets, account balances of the agents, etc.;
- $x_2(t)$ includes the values of the demand and supply of the agents in various markets etc.;
- $x_3(t)$includes various kinds of the market prices and shares of budget in the markets with governmental prices for various economic agents; $m_1 + m_2 + m_3 = m$;
- $u(t) \in U(t) \subset R^q$ is the vector function of controlled (regulated) parameters. The values of the coordinates of this vector correspond to the various instruments of the public economic policy, for example such that the shares of the state budget and budgets of the economic agents, various taxation rates, rates of the state bonds, etc.;
- $a(t) \in A \subset R^s$ is the vector function of uncontrolled parameters (factors). The values of coordinates of this vector characterize various time-dependent endogenous and exogenous socio-economic factors such as the prices of exported and imported goods, population of a country, parameters of the production functions, etc.;
- $X_1(t)$, $X_2(t)$, $X_3(t)$, $U(t)$ are the compact sets with nonempty interiors; $X_i = {}_{t=1}^{n} X_i(t), i = 1, 2, 3$; $X = {}_{i=1}^{3} X_i$; $U = {}_{t=0}^{n-1} U(t)$; A is an open connected set;
- $f_1: X \times U \times A \to R^{m_1}$ is a continuous function.

2) A subsystem of algebraic equations describing behaviour and interaction of the agents in various markets during a selected year. These equations admit expressing the variables $x_2(t)$ via the exogenous parameters and other endogenous variables

$$x_2(t) = f_2\ x_1(t), x_3(t), u(t), a(t)\ . \quad (3)$$

Here $f_2: X_1 \times X_3 \times U \times A \to R^{m_2}$ is a continuous function.

3) A subsystem of recurrent relations for iterative computations of the equilibrium values of the market prices in various markets and the shares of budget in the markets with governmental prices for various economic agents

$$x_3\ t\ [Q+1] = f_3\ x_2\ t\ [Q], x_3\ t\ [Q], L, u(t), a(t)\ . \quad (4)$$

Here $Q = 0,1,2, \dots$ is the number of iteration. L is a collection of positive numbers (adjustable iteration constants). As their values are decreased, an economic system goes to the equilibrium state faster, although this increases the danger of the prices' going to the negative region. $f_3: X_2 \times X_3 \times (0, +\infty)^{m_3} \times U \times A \to R^{m_2}$ is a continuous mapping (being contractive with fixed t; $x_1\ t\ \in X_1\ t\ ; u\ t\ \in U\ t\ ; a(t) \in A$ and some fixed L. In this case the mapping f_3 has a unique stationary point, which iterative process (3), (4) converges to.

With some fixed values of functions $u(t)$ and $a(t)$ for each moment of time t, CGE model of general equilibrium (1), (3), (4) defines the values of the endogenous variables $x(t)$ corresponding to the equilibrium of the demand and supply prices in the markets of goods and services of the agents by means of the following algorithm.

1) At the first step, we assume $t = 0$ and give initial values $x_1\ 0$ to the variables.
2) At the second step, for current t we give initial values $x_3\ 0\ [0]$ to the variables in various markets and for different agents; compute the values $x_2\ t\ [0] = f_2\ x_1(t), x_3(t)[0], u(t), a(t)$, (initial values of demand and supply of the agents in the markets of goods and services) by (3).
3) At the third step, for current time t we start iterative process (4). At that, for any Q the current values of the demands and supplies are found from (3): $x_2\ t\ [Q] = f_2\ x_1(t), x_3(t)[Q], u(t), a(t)$, via adjustment of the market prices and shares of the budgets in the markets with the governmental prices of the economic agents.
   The iterative process is stopped if the values of demands and supplies in various markets are equal accurate to 0.01%. As a result, we determine the equilibrium values of the market prices in each market and the shares of budget in the markets with governmental prices for various economic agents. The index Q for such equilibrium values of the endogenous variables is omitted.
4) At the subsequent step, based on the derived equilibrium solution for the moment t, we compute the values of the variables $x_1(t+1)$ for the next moment of time by means of difference equations (1). The value of t is incremented by one. Go to step 2.

A number of reiterations of steps 2, 3, 4 is defined in accordance with the problems of calibration, prediction and control in the time intervals chosen beforehand.

The considered CGE-model can be represented in form of a continuous mapping $f: X \times U \times A \to R^m$ defining a transformation of values of the system endogenous variables for the zero year to the respective values of the next year according to the algorithm described above. Here the compacts $X\ t\ = X_1\ t\ \times X_3\ t\ \times X_3\ t$ defining the compact set X in the state space of the endogenous variables is defined by the set of possible values of the variables x_1 and respective equilibrium values of the variables x_2 and x_3 computed by relations (3) and (4).

Let us suppose that for a chosen point $x_1(0) \in \mathrm{Int}(X_1)$ and the respective point $x(0) = (x_1(0), x_2(0), x_3(0))$ computed by (3), (4), the inclusion $x(t) \in \mathrm{Int}(X(t))$ is correct with fixed $u(t) \in \mathrm{Int}(U(t))$, $a(t) \in A$ for $t = 0, \ldots, n$ (n is some fixed natural number).

On the base of this description, the specific CGE model of economy branches is considered below.

2.2. Description of CGE Model of Economy Branches

Model Agents

The considered model [11], [5] describes the behaviour and interaction in 52 good markets and 16 labour markets of the following 19 economic agents (sectors).

Economic agent No. 1 represents the agriculture, hunt and forestry.

Economic agent No. 2 represents the fishery and fish-breeding.

Economic agent No. 3 is the mineral resource industry.

Economic agent No. 4 is the manufacturing and process industry.

Economic agent No. 5 is the production and distribution of electric power, gas and water.

Economic agent No. 6 is the construction.

Economic agent No. 7 represents the trade, repair services for cars and goods for household use.

Economic agent No. 8 represents the hotels and restaurants.

Economic agent No. 9 represents transport and communications.

Economic agent No. 10 corresponds to the financial activity.

Economic agent No. 11 represents the real estate activities, rent, and services for business.

Economic agent No.12 is the public administration.

Economic agent No. 13 represents the education.

Economic agent No. 14 is the public health and social services.

Economic agent No. 15 represents other public utilities, social and personal services.

Economic agent No. 16 represents the housekeeping services.

A part of the produced products of the economic agents producing goods and services (Economic agents No. 1–16) is used in production, other part is spent to the investment, and the remainder is sold to the households. The producing agents deal in the intermediate and investment products one with the other.

Economic agent No. 17 is the aggregate consumer joining the households.

The aggregate consumer purchases the consumer goods produced by the producing agents. Moreover, it purchases the imported goods offered by the outer world.

Economic agent No. 18 is the government represented by the aggregate of the central, regional, and local governments, as well as the off-budget funds. The government establishes the taxation rates and defines the sum of the subsidies to the producing agents, as well as the volumes of social transfers to the households. Besides, this sector includes the non-profit organizations servicing the households (the political parties, trade unions, public associations, etc.).

Economic agent No. 19 is the banking sector including the Central bank and commercial banks.

In addition we consider "Integral indexes" and "General indexes" in the model.

The considered model includes 698 endogenous variables and 2045 exogenous parameters to be estimated.

The following system of notations is used in this work for the exogenous and endogenous variables of the CGE models:

<Type><Parameter>_<Price and its code>_<Number of economic agent and market code >[<instant time or number of iteration (*Q*)>].

Here <Type> can take on two values, namely *C* is the exogenous parameter, and *V* is the endogenous variable.

<Parameter> corresponds to the action realized by the agent. The examples of such the actions can be given by *S* (product supply), *D* (product demand), *O* (determining the share of budget by the agent), and others.

For example, the notation *CO_p*3_1*z*[0] corresponds to the exogenous parameter, which is the share of budget of the first sector (agriculture, hunt and forestry) for purchasing the intermediate product produced by the third branch (mineral resource industry) at the price of *P*__3*z* for the intermediate products of the third branch in the zero year (2000).

Exogenous Variables of the Model

The exogenous parameters include the following:

- The coefficients of the production functions of the sectors;
- The various shares of the budgets of the sectors;
- The shares of the products for selling in the various markets;
- The depreciation rates of the capital assets and shares of the retired capital assets;
- The deposit interest rates;
- The various taxation rates;
- The coefficients reflecting the level of non-payments to the producing agents;
- The depreciation rates of the capital assets;
- The shares of the retired capital assets;
- The coefficient reflecting the level of wages liabilities for the employees in all branches;
- The export prices and governmental prices of the goods, services, and labour force, etc.

The list of the exogenous model parameters is given below in Table 1.

Table 1. Exogenous variables of the computable model of the economic branches

	Producing agents No. 1–16
CO_pi_il	The share of budget of the *i*-th branch spent for paying the labour force at the price of *P__il*.
CO_pj_iz	The share of budget of the *i*-th branch spent for purchasing the intermediate products produced by the branches *j* = 1,…,16 at the price of *P__jz*.
CO_pIm_iz	The share of budget of the *i*-th branch spent for purchasing the imported intermediate products.
CO_p_in	The share of budget of the *i*-th branch spent for purchasing the investment products at the price of *P__n*.
CO_pIm_in	The share of budget of the *i*-th branch spent for purchasing the imported investment products.
CE_pi_iz	The share of product produced by the *i*-th branch for selling in the markets of the intermediate products at the price of *P__iz*.
CE_p_ic	The share of product produced by the *i*-th branch for selling in the markets of the final products at the price of *P__ic*.

Table 1. (Continued)

Producing agents No. 1–16	
CE_p_in	The share of product produced by the *i*-th branch for selling in the markets of the investment products at the price of *P__in*.
CE_pexi_ic	The share of product produced by the *i*-th branch for selling in the markets of the exported products at the price of *P__exi*.
CA_r_i	The empirically determined coefficient of dimension of the production function.
CA_z_ji	The coefficients of the production function with the intermediate products of *j*-th branch consumed by the *i*-th branch ($j = 1,\ldots,16$).
CA_k_i	The coefficient of the capital in the production function.
CA_l_i	The coefficient of the labour in the production function.
CA_zIm_i	The parameter of the production function for import.
CO_y_i	The coefficient reflecting the level of non-payments to the producing agents.
CA_n	The depreciation rate of the capital assets.
CO_w_i	The coefficient reflecting the level of arrears of wages to the employees in all branches.
CR__i	The share of the retired capital assets.
Agent 17 Aggregate consumer	
*CO_p*_17*c*	The share of budget of the households spent for purchasing the final products at the price of *P__c*.
*CO_pIm*_17*c*	The share of budget of the households spent for purchasing the imported final products.
*CO_b*_17	The share of budget deposited at the banks.
*CS_pi*_17*l*	The number of employees employed in Sectors 1-16.
Agent 18 Government	
CT_vad	The VAT rate.
CT_pr	The organization profit tax rate.
CT_pod	The rate of physical body income tax.
CT_esn	The rate of single social tax.
*CO_s_i*_18	The shares of the consolidated budget for backing the producing agents.
*CO_tr*_18	The share of the consolidated budget for payment of social transfers to the inhabitants.
*CO_f*_18	The share of off-budget funds for payment of pensions, welfare payments, etc.
*CB_other*_18	The sum of the tax proceeds (not included into the considered ones), non-tax income, and other incomes of the consolidated budget.
Agent 19 Banking sector	
CP__bpercent	The deposit interest rate for enterprises.
CP_h_bpercent	The deposit interest rate for physical bodies.
General part of the model	
CP__exi	The price of the exported product produced by the *i*-th branch.
CP__Imn	The price of the imported investment products.
CP__Imc	The price of the imported final products.
CP__Imiz	The price of the imported intermediate products purchased by the *i*-th branch.
Technical parameters	
*Ceta*__1	The iteration constant applied in the case of the exogenous price.

Endogenous Variables of the Model

The endogenous variables include the following:

- the budgets of the sectors and their various shares;
- the remainders of the agents' budgets;
- the produced values added of the producing sectors;
- demands and supplies of the various products and services;
- the gains of the sectors;
- the capital assets of the producing sectors;
- the number of employees employed in Sectors 1–16;
- the wages of the employees;
- the various kinds of expenditures of the consolidated budget;
- the various kinds of prices of the products, services, and the labour force;
- the subsidies to the producing sectors;
- the social transfers to the citizens;
- the gross production of goods and services;
- the volume of production of the intermediate products;
- the volume of production of the final products;
- the GDP of the country.

The list of the endogenous variables of the model is given in Table 2 below.

Table 2. Endogenous variables of the computable model of economic branches

Producing agents No. 1–16	
VO_tc_i	The share of the budget of the producing agent spent for discharging the taxes to the consolidated budget.
VO_tf_i	The share of the budget of the producing agent spent for discharging the taxes to the off-budget funds.
VO_s_i	The remainder of the agent budget.
VD_pi_il	The demand of the labour power in the i-th branch at the price of *P__il.*
VD_pj_iz	The demand of the intermediate products produced by the branches $j = 1, \ldots, 16$ in the i-th branch at the price of *P__jz.*
VD_pIm_iz	The demand of the imported products in the i-th branch.
VD_p_in	The demand of the investment products in the i-th branch at the price of *P__in.*
VD_pIm_in	The demand of the imported invetsment products in the i-th branch.
VY__i	Production of products and services in the prices of the base period.
VY_g_i	The value added produced by the i-th branch.
VK__i	The capital assets of the producing agent.
VS_pi_iz	The supply of the intermediate products.
VS_p_ic	The supply of the final products.
VS_p_in	The supply of the investment products.
VS_pex_ic	The supply of the exported products.
VY_p_i	The gain of the producing agent.
VY_r_i	The profit of the producing agent.
VB__i	The budget of the producing agent.

Table 2. Continued

Producing agents No. 1–16	
VB_b_i	The balance of the banking accounts of the producing agent.
17 Population	
*VO_tc_*17	The share of the budget of the aggregate consumer for discharging the taxes to the consolidated budget.
*VO_s_*17	The remainder of the budget of Sector 17.
*VD_p_*17*c*	The households' demand of the final products.
*VD_pIm_*17*c*	The households' demand of the imported final products.
VW__i	The wages of the employees in Sectors 1–16.
*VB__*17	The budget of the households.
*VB_b_*17	The money of the households in banking accounts.
18 Government	
*VO_s_*18	The share of the retained consolidated budget.
*VO_s_*18*f*	The share of the retained off-budget funds.
*VG_s_i_*18	The subsidies to the producing sectors.
*VG_tr_*18	The social transfers to the inhabitants.
*VG_f_*18	The off-budget funds allocated for the inhabitants.
*VB__*18	The consolidated budget.
*VB_b_*18	The surplus (deficit) of the consolidated budget.
*VF__*18	The monetary assets of the off-budget funds.
*VF_b_*18	The remainder of the monetary assets of the off-budget funds.
Integral indexes of the model	
VY	The gross production of goods and services (in prices of the base period).
VS__z	The volume of production of the intermediate products (in prices of the base period).
VS__c	The volume of production of the final products (in prices of the base period).
VY_g	GDP
VP	The Consumer Price Index.
General part of model	
VP__il	The price of the labour force in the *i*-th branch.
VP__iz	The price of the intermediate product produced by the *i*-th branch.
VP__n	The price of the investment products.
VP__c	The price of the consumer products.
VP	The Consumer Price Index.
VD_ps_il	The total demand of the labour force at the price of *P__il.*
VD_ps_iz	The total demand of the intermediate products at the price of *P__iz.*
VD_ps_n	The total demand of the investment products at the price of *P__n.*
VD_ps_c	The total demand of the consumer products at the price of *P__c.*
VS_ps_il	The total supply of the labour force at the price of *P__il.*
VS_ps_iz	The total supply of the intermediate products at the price of *P__iz.*
VS_ps_n	The total supply of the investment products at the price of *P__n.*
VS_ps_c	The total demand of the consumer products at the price of *P__c.*

The Model Markets

The equilibrium prices are formed in 68 markets of the model as a result of levelling the supplies and demands of the various kinds of products, services, and labour force.

The described model has:

- 16 markets of the intermediate products and services produced by the producing agents;
- 16 foreign markets of the intermediate products and services imported by the producing agents;
- 1 market of the investment products;
- 1 market of the imported investment products;
- 1 market of the final products;
- 1 market of the imported final products;
- 16 foreign markets of the exported products produced by the producing agents; and
- 16 markets of the labour force.

In the internal markets, the market mechanism of price forming is used. The prices of the foreign markets enter the model exogenously. Let us now consider the formulas reflecting the process of changing the prices in the domestic markets (below, i means the agent number).

The price of the labour force in the i-th branch is given by

$$VP\_il[t, Q+1] = VP\_il[t, Q] \times VD\_ps\_il[t, Q] / VS\_ps\_il[t, Q]. \quad (5)$$

The price of the intermediate product produced by the i-th branch is as follows

$$VP\_iz[t, Q+1] = VP\_iz[t, Q] \times VD\_ps\_iz[t, Q] / VS\_ps\_iz[t, Q]. \quad (6)$$

The price of the investment products is determined by

$$VP\_n[t, Q+1] = VP\_n[t, Q] \times VD\_ps\_n[t, Q] / VS\_ps\_n[t, Q]. \quad (7)$$

The price of the consumer products is as follows

$$VP\_c[t, Q+1] = VP\_c[t, Q] \times VD\_ps\_c[t, Q] / VS\_ps\_c[t, Q]. \quad (8)$$

That is, we have 16 + 16 + 1 = 33 prices of the products sold in the domestic markets in the model.

The notations of the prices in the foreign markets are given below.

CP_exi	- The price of the exported product produced by the i-th branch.	
CP_Imn	- The price of the imported investment products.	(9)
CP_Imc	- The price of the imported final products.	
CP_Imiz	- The price of the imported intermediate product purchased by the i-th branch.	

Let us now proceed to the formulas describing the mechanism of forming the demand and supply of the products produced by the agents No. 1–16 at the governmental and market prices.

The final formulas determining the demand and supply of each economic agent in the product markets included to the model are presented below.

The total demand of the labour force at the price of $VP\_\_il[t]$ is given by

$$VD\_ps\_il[t] = VD\_pi\_il[t]. \quad (10)$$

For simplicity, we do not consider the demand of the labour force in the i-th branch from the other branches. In this connection, the total demand of the labour force at the price of $VP\_\_il[t]$ is defined by the demand in the single i-th branch.

The total supply of the labour force at the price of $VP\_\_il[t]$ is as follows:

$$VS\_ps\_il[t] = CS\_pi\_17l[t]. \quad (11)$$

The total demand of the intermediate product at the price of $VP\_\_jz[t]$ produced by the j-th branch is determined as

$$VD\_ps\_jz[t] = \mathrm{SUM}(VD\_pj\_iz[t])). \quad (12)$$

Here and below, SUM($X\_\_i[t]$) corresponding to $\sum_{i=1}^{16} X\_i[t]$, $i = 1, \ldots, 16$ is the economic agent number.

As can be seen, the total demand of the intermediate product at the price of $VP\_\_jz[t]$ consists of the demands of the intermediate products in the j-th branch ($j = 1,\ldots,16$) from the direction of all 16 branches.

The total supply of the intermediate product at the price of $VP\_\_iz[t]$ is given by

$$VS\_ps\_iz[t] = VS\_pi\_iz[t]. \quad (13)$$

The total demand of the investment products at the price of $VP\_\_n[t]$:

$$VD\_ps\_n[t] = \mathrm{SUM}(VD\_p\_in[t]). \quad (14)$$

The total supply of the investment products at the price of $VP\_\_n[t]$:

$$VS\_ps\_n[t] = \mathrm{SUM}(VS\_p\_in[t]). \quad (15)$$

The total demand of the consumer products at the price of $VP\_\_c[t]$:

$$VD\_ps\_c[t] = VD\_p\_17c[t]. \quad (16)$$

The total supply of the consumer products at the price of $VP\_\_c[t]$:

$$VS\_ps\_c[t] = \mathrm{SUM}(VS\_p\_ic[t]). \quad (17)$$

Let us present the notations defining the total supply and demand of the exported products.

The total demand of the exported products at the price of $CP\_pex\_ic[t]$ (given) is

$VD\_pex\_c[t] = \mathrm{SUM}(VD\_pex\_ic[t])$ (18)

The total supply of the exported products at the price of $CP\_pex\_ic[t]$ is

$VS\_pex\_c[t] = \mathrm{SUM}(VS\_pex\_ic[t])$ (19)

Let us proceed to describing the activity of the economic agents participating in the model.

Economic Agents No. 1 – 16 Producing Products and Services

The equation of the production function of an economic producing agent is given by

$$VY\_\_i[t+1] = CA\_r\_i[t] \times \mathrm{Exp}(VD\_p1\_iz[t] \times CA\_z\_1i) \times \mathrm{Exp}(VD\_p2\_iz[t] \times CA\_z\_2i) \times \mathrm{Exp}(VD\_p3\_iz[t] \times CA\_z\_3i) \times \mathrm{Exp}(VD\_p4\_iz[t] \times CA\_z\_4i) \times \mathrm{Exp}(VD\_p5\_iz[t] \times CA\_z\_5i) \times \mathrm{Exp}(VD\_p6\_iz[t] \times CA\_z\_6i) \times \mathrm{Exp}(VD\_p7\_iz[t] \times CA\_z\_7i) \times \mathrm{Exp}(VD\_p8\_iz[t] \times CA\_z\_8i) \times \mathrm{Exp}(VD\_p9\_iz[\mathrm{t}] \times CA\_z\_9i) \times \mathrm{Exp}(VD\_p10\_iz[t] \times CA\_z\_10i) \times \mathrm{Exp}(VD\_p11\_iz[t] \times CA\_z\_11i) \times \mathrm{Exp}(VD\_p12\_iz[t] \times CA\_z\_12i) \times \mathrm{Exp}(VD\_p13\_iz[t] \times CA\_z\_13i) \times \mathrm{Exp}(VD\_p14\_iz[t] \times CA\_z\_14i) \times \mathrm{Exp}(VD\_p15\_iz[t] \times CA\_z\_15i) \times \mathrm{Exp}(VD\_p16\_iz[t] \times CA\_z\_16i) \times \mathrm{Power}(((VK\_\_i[t] + VK\_\_i[t+1])/2), CA\_k\_i) \times \mathrm{Power}(VD\_pi\_il[t], CA\_l\_i) \times \mathrm{Exp}(VD\_pIm\_iz[t] \times CA\_zIm\_i). \quad (20)$$

Here CA_r_i, ,CA_z_ji (j = 1, ..., 16), CA_k_i, CA_l_i are the parameters of the production function, Power(X, Y) corresponds to X^Y, Exp(X) corresponds to e^X.

The following formulas determine the demand of the production factors by the i-th agent.

The demand of the labour force:

$VD\_pi\_il[t] = CO\_pi\_il[t] \times VB\_\_i[t]/VP\_\_il[t].$ (21)

The demand of the intermediate products produced by all producing agents:

$VD\_p1\_iz[t] = CO\_p1\_iz[t] \times VB\_\_i[t]/VP\_\_1z[t];$ (22)

$VD\_p2\_iz[t] = CO\_p2\_iz[t] \times VB\_\_i[t]/VP\_\_2z[t];$ (23)

$VD\_p3\_iz[t] = CO\_p3\_iz[t] \times VB\_\_i[t]/VP\_\_3z[t];$ (24)

$VD\_p4\_iz[t] = CO\_p4\_iz[t] \times VB\_\_i[t]/VP\_\_4z[t];$ (25)

$VD\_p5\_iz[t] = CO\_p5\_iz[t] \times VB\_\_i[t]/VP\_\_5z[t];$ (26)

$VD\_p6\_iz[t] = CO\_p6\_iz[t] \times VB\_\_i[t]/VP\_\_6z[t];$ (27)

$VD\_p7\_iz[t] = CO\_p7\_iz[t] \times VB\_\_i[t]/VP\_\_7z[t];$ (28)

$VD\_p8\_iz[t] = CO\_p8\_iz[t] \times VB\_\_i[t]/VP\_\_8z[t];$ (29)

$$VD\_p9\_iz[t] = CO\_p9\_iz[t] \times VB\_\_i[t]/VP\_\_9z[t]; \tag{30}$$

$$VD\_p10\_iz[t] = CO\_p10\_iz[t] \times VB\_\_i[t]/VP\_\_10z[t]; \tag{31}$$

$$VD\_p11\_iz[t] = CO\_p11\_iz[t] \times VB\_\_i[t]/VP\_\_11z[t]; \tag{32}$$

$$VD\_p12\_iz[t] = CO\_p12\_iz[t] \times VB\_\_i[t]/VP\_\_12z[t]; \tag{33}$$

$$VD\_p13\_iz[t] = CO\_p13\_iz[t] \times VB\_\_i[t]/VP\_\_13z[t]; \tag{34}$$

$$VD\_p14\_iz[t] = CO\_p14\_iz[t] \times VB\_\_i[t]/VP\_\_14z[t]; \tag{35}$$

$$VD\_p15\_iz[t] = CO\_p15\_iz[t] \times VB\_\_i[t]/VP\_\_15z[t]; \tag{36}$$

$$VD\_p16\_iz[t] = CO\_p16\_iz[t] \times VB\_\_i[t]/VP\_\_16z[t]. \tag{37}$$

The demand of the imported intermediate products:

$$VD\_pIm\_iz[t] = CO\_pIm\_iz \times VB\_\_i[t]/CP\_\_Imiz[t]. \tag{38}$$

The demand of the investment products:

$$VD\_p\_in[t] = CO\_p\_in[t] \times VB\_\_i[t]/VP\_\_n[t]. \tag{39}$$

The demand of the imported investment products:

$$VD\_pIm\_in[t] = CO\_pIm\_in[t] \times VB\_\_i[t]/CP\_\_Imn[t]. \tag{40}$$

The following formulas determine the supply of the products and services produced by the producing agent.

The supply of the intermediate products:

$$VS\_pi\_iz[t] = CE\_pi\_iz[t] \times VY\_\_i[t]. \tag{41}$$

The supply of the final products:

$$VS\_p\_ic[t] = CE\_p\_ic[t] \times VY\_\_i[t]. \tag{42}$$

The supply of the investment products:

$$VS\_p\_in[t] = CE\_p\_in[t] \times VY\_\_i[t]. \tag{43}$$

The supply of the exported products:

$$VS\_pex\_ic[t] = CE\_pexi\_ic[t] \times VY\_\_i[t]. \tag{44}$$

The following formula calculates the gain of the producing agent:

$$VY\_p\_i[t] = VS\_pi\_iz[t] \times VP\_\_iz[t] + VS\_p\_ic[t] \times VP\_\_c[t] + VS\_p\_in[t] \times VP\_\_n[t] + VS\_pex\_ic[t] \times CP\_\_exi[t]. \quad (45)$$

The profit of the producing agent:

$$VY\_r\_i[t] = CO\_y\_i[t] \times VY\_p\_i[t] - (VD\_p1\_iz[t] + VD\_p2\_iz[t] + VD\_p3\_iz[t] + VD\_p4\_iz[t] + VD\_p5\_iz[t] + VD\_p6\_iz[t] + VD\_p7\_iz[t] + VD\_p8\_iz[t] + VD\_p9\_iz[t] + VD\_p10\_iz[t] + VD\_p11\_iz[t] + VD\_p12\_iz[t] + VD\_p13\_iz[t] + VD\_p14\_iz[t] + VD\_p15\_iz[t] + VD\_p16\_iz[t] + (VW\_\_i[t] \times CO\_w\_i[t]) + CA\_n[t] \times (VK\_\_i[t] \times VP\_\_n[t])). \quad (46)$$

Here CO_y_i is the coefficient reflecting the level of non-payments; CA_n is the depreciation rate of the capital assets. Here we calculate the profit of the sector consisting of the gain corrected by the level of non-payments. The assets spent for the intermediate product, wages (without taking into account the debt, the coefficient CO_w_i) and amortization of the capital assets are subtracted.

The value added produced by the i-th sector is given by

$$VY\_g\_i[t] = VY\_r\_i[t] + VW\_\_i[t]. \quad (47)$$

The value added consists of the profit received in the current period and wages really paid to the sector employees.

The budget of the producing agent is as follows:

$$VB\_\_i[t] = VB\_b\_i[t-1] \times (1 + CP\_\_bpercent[t-1]) + CO\_y\_i[t] \times VY\_p\_i[t] + VG\_s\_i18[t-1]. \quad (48)$$

The agent budget consists of the following:

1) the funds on the banking accounts (taking into consideration interests on deposits);
2) the gain received in the current period;
3) the subsidies received from the consolidated budget VG_s_i18.

The dynamics of the banking account balance of the producing agent is as follows:

$$VB\_b\_i[t] = VO\_s\_i[t] \times VB\_\_i[t]. \quad (49)$$

The capital assets are determined by

$$VK\_\_i[t+1] = (1 - CR\_\_i[t]) \times VK\_\_i[t] + VD\_p\_in[t] + VD\_pIm\_in[t]. \quad (50)$$

This formula calculates the volume of the capital assets taking into account their retirement. The asset put into operation enters the formula with the plus sign.

The share of the budget of the producing agent for discharging the taxes to the consolidated budget is given by

$$VO\_tc\_i[t] = VY\_g\_i[t] \times CT\_vad[t]/VB\_\_i[t] + VY\_r\_i[t] \times CT\_pr[t]/VB\_\_i[t]. \quad (51)$$

This formula takes into consideration the value added tax (VAT) and profit tax.

The share of the budget for discharging the single social tax to the off-budget funds is described as

$$VO\_tf\_i[t] = VW\_\_i[t] \times CT\_esn[t]/VB\_\_i[t]. \quad (52)$$

The remainder of the budget of the producing agent is given by

$$VO\_s\_i[t] = 1 - (CO\_pi\_il[t]+CO\_p\_in[t] +VO\_tc\_i[t] + VO\_tf\_i[t] + VO\_p1\_iz[t] + CO\_p2\_iz[t] + CO\_p3\_iz[t] + CO\_p4\_iz[t] + CO\_p5\_iz[t] + CO\_p6\_iz[t] + CO\_p7\_iz[t] + CO\_p8\_iz[t] + CO\_p9\_iz[t] +CO\_p10\_iz[t] + CO\_p11\_iz[t] + CO\_p12\_iz[t] + CO\_p13\_iz[t] + CO\_p14\_iz[t] + CO\_p15\_iz[t] + CO\_p16\_iz[t] + CO\_pIm\_iz[t] + CO\_pIm\_in[t]). \quad (53)$$

Economic Agent No. 17. Aggregate Consumer

Let us proceed to the formulas determining the behaviour of the aggregate consumer.

The households' demand of the final products is given by

$$VD\_p\_17c[t] = CO\_p\_17c[t] \times VB\_\_17[t]/VP\_\_c[t]. \quad (54)$$

The households' demand of the imported final products is

$$VD\_pIm\_17c[t] = CO\_pIm\_17c[t] \times VB\_\_17[t]/CP\_\_Imc[t]. \quad (55)$$

The wages of the employees of Sectors No. 1–16:

$$VW\_\_i[t] = VD\_pi\_il[t] \times VP\_\_il[t]. \quad (56)$$

The budget of the households is determined as follows:

$$VB\_\_17[t] = VB\_b\_17[t - 1] \times (1 + CP\_h\_bpercent[t - 1]) + VB\_\_17[t - 1] \times VO\_s\_17[t - 1] + VG\_tr\_18[t - 1] + VG\_f\_18[t - 1] + \mathrm{SUM}(VW\_\_i[t]). \quad (57)$$

The agent's budget is formed from the following:

1) the funds on the banking accounts (subject to interests on deposits);
2) the retained money in cash kept from the preceding period;
3) the pensions, welfare payments, and subsidies received from the off-budget funds;
4) the wages received from the producing agents No. 1–16.

The dynamics of the banking account balance of the households is as follows:

$$VB\_b\_17[t] = CO\_b\_17[t] \times VB\_\_17[t]. \quad (58)$$

The share of the budget for discharging the income tax is given by

$$VO\_tc\_17[t] = \mathrm{SUM}(VW\_\_i[t]) \times CT\_pod[t]/VB\_\_17[t]. \quad (59)$$

The remainder of the money in cash is as follows:

$$VO\_s\_17[t] = 1 - CO\_p\_17c[t] - VO\_tc\_17[t] - CO\_b\_17[t] - CO\_pIm\_17c[t]. \quad (60)$$

Economic Agent No. 18. Government

As it is presented above, this economic agent is represented by the aggregate of the federal, regional, and local governments, as well as the off-budget funds. Besides that, it includes the non-profit organizations servicing the households.

Let us proceed to the formulas determining the behaviour of economic agent No. 18.

The consolidated budget is given by

$$VB\_\_18[t] = \mathrm{SUM}(VO\_tc\_i[t] \times VB\_\_i[t]) + VO\_tc\_17[t] \times VB\_\_17[t] + CB\_other\_18[t] + VB\_b\_18[t] \times (1 + CP\_\_bpercent[t - 1]). \quad (61)$$

This formula sums up the money collected as the taxes from the producing agents, as well as from the inhabitants. The value *CB_other_18* entering the model exogenously is the sum of the other taxes (not included in the list of taxes considered in the model), non-taxable income, and other income of the consolidated budget. The obtained sum is incremented by the funds on the banking accounts (subject to the deposit interests).

The dynamics of the banking account balance of the consolidated budget is determined by

$$VB\_b\_18[t + 1] = VO\_s\_18[t] \times VB\_\_18[t]. \quad (62)$$

The cash assets of off-budget funds are as follows:

$$VF\_\_18[t] = \mathrm{SUM}(VO\_tf\_i[t] \times VB\_\_i[t]) + VF\_b\_18[t] \times (1 + CP\_\_bpercent[t-1]). \quad (63)$$

This formula calculates the sum collected from the producing agents in the form of the single social tax entering the accounts of the following off-budget funds: the pension fund, the social insurance fund, the Federal and territorial funds of obligatory medical insurance. The derived sum is added by the funds on the banking accounts (subject to the deposit interests).

The dynamics of the banking account balance of the off-budget funds is determined by

$$VF\_b\_18[t + 1] = VO\_s\_18f[t] \times VF\_\_18[t]. \quad (64)$$

The subsidies to the producing sectors are as follows:

$$VG\_s\_i18[t] = CO\_s\_i18[t] \times VB\_\_18[t]. \quad (65)$$

The social transfers to the inhabitants:

$$VG\_tr\_18[t] = CO\_tr\_18[t] \times VB\_\_18[t]. \quad (66)$$

The assets of the off-budget funds made available for the inhabitants:

$$VG\_f\_18[t] = CO\_f\_18[t] \times VF\_\_18[t]. \quad (67)$$

This includes the assets of the pension fund and social insurance fund for paying out the pensions and welfare payments.

Integral Indexes of the Model

Let us present the formulas for calculating some integral indexes of the economy of Kazakhstan.

The gross production of goods and services (in prices of the base period):

$$VY[t] = \mathrm{SUM}(VY\_\_i[t]). \quad (68)$$

The total supply of the intermediate products (in prices of the base period):

$$VS\_\_z[t] = \mathrm{SUM}(VS\_pi\_iz[t]). \quad (69)$$

The total supply of the final products (in prices of the base period):

$$VS\_\_c[t] = \mathrm{SUM}(VS\_p\_ic[t]). \quad (70)$$

GDP of Kazakhstan:

$$VY\_g[t] = \mathrm{SUM}(VY\_g\_i[t]). \quad (71)$$

The Consumer Price Index:

$$VP[t] = 100 \times (VP\_\_c[t]/VP\_\_c[t - 1]). \quad (72)$$

The considered model is presented in the context of the common relations:

- relations (1) are presented by $m_1 = 67$ expressions;
- relations (2) are presented by $m_2 = 597$ expressions;
- relations (3) are presented by $m_3 = 34$ expressions.

2.3. Parametric Identification of CGE Model of Economy Branches

The problem of parametric identification of the considered macro-economic mathematical model consists in finding estimations for unknown values of its parameters which minimize the value of the objective function that characterizes the deviations of the values of the output model variables from the respective measured values (known statistical data for the period of time $t = t_1, t_1 + 1, \ldots, t_2$). This problem reduces to finding the minimal value of the function in several variables (parameters) in some close region D of the Euclidean space under constraints like (2) imposed onto the values of the model endogenous variables. In the case of large dimensions of the region of the possible values of the parameters to be found, the standard methods for finding extremum of a function are often inefficient since the objective function can have several local minimum points. Below we propose an algorithm taking into account peculiarities of the problem of parametric identification of macro-economic models that allows avoiding the mentioned problem of "local extremums".

The region $D = \prod_{i=1}^{q+s\ t_2-t_1\ +m_1}[a^i, b^i]$ where $[a^i, b^i]$ is the interval of possible values of the parameter p^i; $i = 1, \ldots, (q + s)(t_2 - t_1 + 1) + m_1$ was considered as the region $D \subset \prod_{t=t_1}^{t_2}[U(t) \times A(t)] \times X_1(t_1)$ for estimation of possible values of the exogenous parameters (values of the exogenous functions $u(t)$, $a(t)$ and initial conditions of the dynamic equations (1)). At that, the estimations of the parameters with the respective measured values available were searched within the intervals $[a^i, b^i]$ with the centres in the respective measured values (in case of one such value) or within some intervals covering the measured values (in case of several values). Other intervals $[a^i, b^i]$ for searching the parameters were chosen by means of indirect estimations of their possible values. For searching the minimal values of the continuous function in severable variables K: $D \rightarrow R$ with the additional constraints on the endogenous variables in computational experiments we applied the Nelder-Mead directed search algorithm [13]. Using this algorithm for an initial point $p_1 \in D$ can be interpreted as a sequence $\{p_1, p_2, p_3, \ldots\}$ (converging to the point of local minimum $p_0 = \operatorname{argmin}_D K$ of the function K, where $K(p_{j+1}) \leq K(p_j)$, $p_j \in D$, $j = 1, 2, \ldots$ While describing the next algorithm we will consider that the point p_0 can be found with sufficient precision.

To solve the problem of parametric identification of the considered computable model based on the obvious assumption about disagreement (in general case) of the minimum points of two different functions, the following two criteria are proposed:

$$K_A\ p = \sqrt{\frac{1}{n_\alpha(t_2-t_1+1)} \sum_{t=t_1}^{t_2} \sum_{i=1}^{n_A} \alpha_i \left(\frac{y^i\ t - y^{i*}\ t}{y^{i*}\ t}\right)^2},$$

$$K_B\ p = \sqrt{\frac{1}{n_\beta(t_2-t_1+1)} \sum_{t=t_1}^{t_2} \sum_{i=1}^{n_B} \beta_i \left(\frac{y^i\ t - y^{i*}\ t}{y^{i*}\ t}\right)^2} \qquad (73)$$

Here $\{t_1, \ldots, t_2\}$ is the identification time interval; $y^i(t)$, $y^{i*}(t)$ are computed and measured values of the model output variables; $K_A\ p$ is the auxiliary criterion; $K_B\ p$ is the main criterion; $n_B > n_A$; $\alpha_i > 0$ and $\beta_i > 0$ are some weighting coefficients whose values are determined while solving the problem of parametric identification of the dynamic system; ${}_{i=1}^{n_A}\alpha_i = n_\alpha$, ${}_{i=1}^{n_B}\beta_i = n_\beta$.

The problems of minimization of the respective criteria will be referred to as the problem *A* and problem *B*. The algorithm of solving the parametric identification problem was chosen in form of the following stages.

1. For some vector of the initial values of the parameters $p_1 \in D$, in parallel solve the problems *A* and *B*. As a result, find the minimum points p_{A0} and p_{B0} of the criteria K_A and K_B, respectively.
2. If $K_B \; p_{B0} \; < \varepsilon$ holds for some sufficiently small number ε, then the model parametric identification problem is solved.
3. Otherwise, choose the point p_{B0} as the initial point p_1, solve problem *A* and, choosing the point p_{A0} as the initial point p_1, solve problem *B*. Go to stage 2.

Sufficiently large number of reiterations of stages 1, 2, 3 gives the searched values of the parameters an opportunity to leave neighbourhoods of non-global minimum points of one criterion with the help of other one and thereby solve the parametric identification problem.

As a result of joint solution of problems *A* and *B* according to the described algorithm with application of the Nelder-Mead algorithm we obtain the values $K_A = 0.015$ and $K_B = 0.0063$. At that, the relative value of deviations of the computed values of the variables used in the main criterion from the respective measured values constitutes less than 0.63%.

The results of computation and retrospective prediction of the model for 2008 year partially presented in Table 3 demonstrate the computed (VY, VY_g, VP), measured values, and deviations of the computed values of the main output variables of the model from the respective measured values. Here the time interval 2000-2007 years corresponds to the period of the model parametric identification; 2008 year is the retrospective prediction period; VY is the gross output ($\times\, 10^{12}$ tenge in prices of 2000 year, tenge is the monetary unit of Kazakhstan); VY_g is GDP ($\times\, 10^{12}$ in prices of 2000 year); $VP$is the Consumer Price Index in percents with respect to a preceding year; sign "*" corresponds to the measured values; sign "Δ" corresponds to the deviations (in percents) of the computed values from the respective measured values.

Table 3. Measured, computed values of the model output variables and respective deviations

Year	2000	2001	2002	2003	2004	2005	2006	2007	2008
VY^*	5.44	6.32	6.47	6.86	7.72	8.52	9.25	9.69	9.84
VY	5.38	6.32	6.47	6.86	7.72	8.52	9.27	9.64	9.82
ΔVY	-1.22	-0.02	0.00	0.00	0.05	0.08	0.21	-0.51	-0.26
VY_g^*	2.45	2.78	3.05	3.36	3.72	4.09	4.55	5.01	5.18
VY_g	2.47	2.78	3.05	3.35	3.72	4.09	4.55	5.01	5.20
ΔVY_g	0.88	0.07	-0.04	-0.02	-0.02	-0.02	-0.04	-0.15	0.38
VP^*		106.4	106.6	106.8	106.7	107.5	108.4	118.8	109.5
VP		107.6	106.8	106.9	106.7	107.3	108.2	118.6	109.4
ΔVP		1.13	0.18	0.08	-0.05	-0.23	-0.22	-0.24	-0.05

The mean square error of the retrospective prediction of all 491 endogenous variables of the model for 2008-2009 years equals to 5.86%.

The results of verification show acceptable adequacy of the CGE model of the economy branches.

2.5. Estimation of Stability Indexes of CGE Models of Economy Branches

By definition of Orlov [14], the mathematical model of an economic system in general form is some function $y = f(p)$

$$f: A \to B, \tag{74}$$

transferring values of initial (exogenous) data $p \in A$ to solutions (values of endogenous variables) $y \in B$.

After constructing the mathematical model of some real-life phenomenon or process f and defining some actual value of the point p by known measured data or solving the parametric identification problem, the question about the adequacy of the analysed model. The condition of the model stability against admissible perturbations of the initial data [31] is a one of the conditions of the model adequacy. In the case of such stability, small perturbations of the model initial data result in small changes of its solution. In monograph [14], the definitions of the basic stability indicators are introduced (these definitions are presented below). But monograph [14] does not propose any algorithm for computing the considered indicators of the mathematical model stability.

Below we present the developed algorithms for evaluating the mathematical model stability indicators which characterize the stability of the solutions of the mathematical model against the initial data perturbations. At that, all of the model parameters and variables must be made dimensionless beforehand.

Let $X = \ X^1, X^2, \dots, X^k$ be some vector of values of the model exogenous parameters for a time interval $t \in \{0, \dots, T\}$. Let $X_0 = \ X_0^1, X_0^2, \dots, X_0^k$. denote the respective vector of the base values for the same time interval. The vector that incorporates the values of the parameters and the initial values of the variables of differential (or difference) equations is considered as the vector X. The vector of the measured statistical data used for finding the model equation coefficients is considered as the vector X for the econometric models.

Let $p = \ p^1, p^2, \dots, p^k$ be a vector of the normalized input data of the mathematical model where $p^i = \frac{X^i}{X_0^i}$, $i = 1, \dots, k$. The vector $p_0 = (1, 1, \dots, 1)$.

Let A be a space of the normalized input data vectors which includes all admissible sets p, $A \subset R^k$ is a metric space with the Euclidean metric defined by the space R^k, $p_0 \in A$.

Let $Y = Y(p) = (Y^1, Y^2, \dots, Y^n)$ be a selected vector of the values of the endogenous variables for some chosen interval (or moment) of time obtained for the selected values of p. The vector that incorporates the values of some selected set of the model endogenous variables for the aforesaid interval (or moment) of time is considered as the vector Y for the dynamic models. The vector of the coefficients of the model equations or vector of the values of some selected set of the model endogenous variables for the aforesaid interval (or moment) of time is considered as the vector Y for the econometric models.

In particular, with $p = p_0$, introduce the notation $Y_0 = Y(p_0) = (Y_0^1, Y_0^2, \dots, Y_0^n)$. The normalized vector of the values of the endogenous variables for the moment of time T_1 is denoted by $y = y(p) = (\frac{Y^1}{Y_0^1}, \frac{Y^2}{Y_0^2}, \dots, \frac{Y^n}{Y_0^n})$; $y_0 = y(p_0) = (1, 1, \dots, 1)$.

Let $B \subset R^n$ be a region which contains all possible output values y for $p \in A$ with the Euclidean metric of the space R^n, $y_0 \in B$. The considered model defines the mapping of the set A into the set B.

For the selected point $p \in A$ and number $\alpha > 0$, let $U_\alpha(p)$ denote the intersection of a neighbourhood of the point p with radius α with the set A:

$$U_\alpha \; p \; = \; x_1 \in A: \rho \; p_1, p \; \leq \alpha \; .$$

Here and below ρ denotes the Euclidean distance between two points of the Euclidean space.

For some subset $B_1 \subset B$, let $d(B_1)$ denote the diameter of the set B_1, i.e.

$$d(B_1) = \sup(\rho \; y_1, y_2 \; : y_1, y_2 \in B_1).$$

Definition 1. The number

$$\beta \; p, \alpha \; = d(f(U_\alpha(p))). \tag{75}$$

is called *the stability indicator* of the econometric model in the point $X \in A$ for $\alpha > 0$.

Algorithm 1 for evaluating the model stability indicator $\beta \; p, \alpha$ by the Monte Carlo method is as follows.

1. Choose the sets of the input parameters (X) and output variables (Y). Compute their normalized values.
2. Define the vector of the normalized input data $p = (p^1, p^2, \dots, p^k)$, number $\alpha > 0$, and set $U_\alpha \; p$.
3. Generate the set of sufficiently large number M of pseudo-random points (p_1, p_2, …, p_M) uniformly distributed in $U_\alpha \; p$.

For this purpose consecutively generate the coordinates p_j^i $(i = 1, \dots, k; j = 1, \dots, M)$ of the point p_j in the numerical intervals $p^i - \alpha, p^i + \alpha$ covering $U_\alpha \; p$ using a sensor of pseudo-random numbers distributed uniformly. If the inequality ${}^{k}_{i=1}(p_j^i - p^i)^2 \leq \alpha^2$ holds (i.e. $x_j \in U_\alpha \; p$)), this point is added to the created set.

4. For each point p_j of the set, define the point $y_j = f(p_j)$, $j = 1, \dots, M$, by simulation.
5. Evaluate

$$\beta = \max(\rho \; y_i, y_j \; : i, j = 1, \dots, M).$$

6. Stop.

With $\alpha = 0.01$, the obtained number $\beta/2$ characterizes the (maximum) percent of changing the values of the model output variables under perturbing the input data by 1%.

In this work we estimate the stability index β of the CGE model of economy branches.

Table 4. Stability indexes of CGE model of economy branches

Time in computations	β/2
2000	0.215287
2001	1.532906
2002	0.803227
2003	0.014729
2004	0.053558
2005	0.018151
2006	0.032543
2007	0.060006
2008	0.011421
2009	0.609306
2010	0.042315
2011	0.153076
2012	0.444828
2013	0.960665
2014	2.442443
2015	16.26265

We consider the initial values (for 1999 year) of endogenous variables, products of all 16 producing sectors of the model, as the set of input parameters A. The computed values of all endogenous variables of the model for chosen year are considered to be the values of output variables of the model. The results of estimation of the stability indexes of the model are given in Table 4.

Analysis of Table 4 shows that the model has sufficiently high stability for computations up to 2014 year.

2.5. Macroeconomic Analysis Based on CGE Model of Economy Branches

Macroeconomic analysis based on the CGE model of the economy brunches was implemented by means of the following tools:

1. Retrospective analysis at the level of the economy branches.
2. Retrospective analysis of the integral indicators of the national economy, indicators of government, households.
3. Retrospective comparative analysis of indicators of the economy branches.
4. Perspective analysis at the level of the economy branches.
5. Perspective analysis of the integral indicators of the national economy, indicators of government, households.
6. Perspective comparative analysis of indicators of the economy branches.

7. Analysis of the elasticity coefficients of the endogenous variables by exogenous factors.
8. Analysis of sources of economic growth.

2.5.1. Retrospective and Prospective Analysis of Indicators of Economic Agents and National Economy

Below we present several results of the retrospective analysis at the level of the branches and national economy.

The Results of the Retrospective Analysis of the Exogenous Variables at the Level of the Economy Branches

The retrospective analysis of the exogenous indicators by the economy branches for 2000-2009 years show that within the said period

- The numbers of employed in all branches excepting branch 16 (housekeeping services) increased on average. The maximum average growth (computed by linear trend) is observed for branch 6 (construction) and equal to 4.2×10^4 people per year. The negative grow (decrease on 2.3×10^3 people per year on average) was observed in branch 16 (housekeeping services).
- The trend of the indicator (Share of intermediate products in branch's output) had positive slope for branches 8, 11, 10, 2, 4, 1, and negative slope for branches 16, 9, 13, 14, 5, 12, 6, 7, 15, 3. The maximum growth of trend by 4.4% per year is observed in branch 8 (hotels and restaurants). The maximum decrease by 3.4% per year is observed in branch 3 (mineral resource industry).
- The trend of the indicator (Share of final products in branch's output) had positive slope for branches 5, 12, 16, 1, 4, 14, 11, 13 and negative slope for branches 3, 6, 15, 10, 9, 7, 2, 8, 6. The maximum growth of trend by 4.9% per year is observed in branch 5 (production and distribution of electric power, gas, and water). The maximum decrease by 3.8% per year is observed in branch 8 (hotels and restaurants).
- The trend of the indicator (Share of export in branch's output) had positive slope for branches 7, 3, 4, 9, 15, 10, 2, 5 and negative slope for branches 6, 13, 8, 11, 12, 1. In branches 14 and 16 the export component was close to zero. The maximum growth of trend by 3.6% per year is observed in branch 7 (trade, repair services for cars and goods for household use). The maximum decrease by 1.8% per year is observed in branch 1.
- The trend of the indicator (Price of export supplies of branch's products) had positive slope for all 16. The maximum growth of trend by 62% per year is observed in branch 5 (production and distribution of electric power, gas, and water). The minimum increase by 5.9% per year is observed in branch 13 (Education).
- The trend of the indicator (Share of budget of branch spent for paying labour force) had positive slope for branches 15, 13, 7, 4, 6, 5, 9, 18, 2 and negative slope for branches 14, 10, 11, 3, 16. The maximum growth of trend by 1.9% per year is observed in branch 15 (other public, social, and personal services). The maximum decrease by 1.4% per year is observed in branch 16 (Housekeeping services).

- The trend of the indicator (Share of budget of branch spent for purchasing investment products) had positive slope for branches 12, 6, 11, 5, 4, 1, 9, 14, 2, 7, 10 and negative slope for branches 13, 15, 8, 3. This share was zero for branch 16. The maximum growth of trend by 10% per year is observed in branch 12 (Public administration). The maximum decrease by 2% per year is observed in branch 3 (Mineral resource industry).
- The indicator "Share of investment product in branch's output" was nonzero in 8 branches of 16. At that, in branches 6, 7, 2, 1, 9 (hereinafter the branches are ordered by decrease of the trend angular coefficient) this indicator had positive slope of linear trend (the maximum value 2.9% per year was observed in branch 6 (construction)). This share was equal to 78% for branch 6 in 2008 year. Branches 15, 4, 11 demonstrated negative slopes of linear trends of the considered indicator. The maximum decrease of the trend is observed in branch 11 (real estate activities, rent, and services for business) and constitutes 0.94% per year.

Below (Figure 1) we give a graphic illustration of the dynamics (and its trend) of one of the economic agents' indicators for retrospective macroanalysis is presented by the example of the variable "Share of investment products in the output of economic agent 1 (Agriculture, hunt, and forestry)".

The analysis of Figure 1 shows positive slope of linear trend of the considered indicator thereby demonstrating growth (on average) of share of the investment products in the output of economic agent 1 (Agriculture, hunt, and forestry) within the considered period.

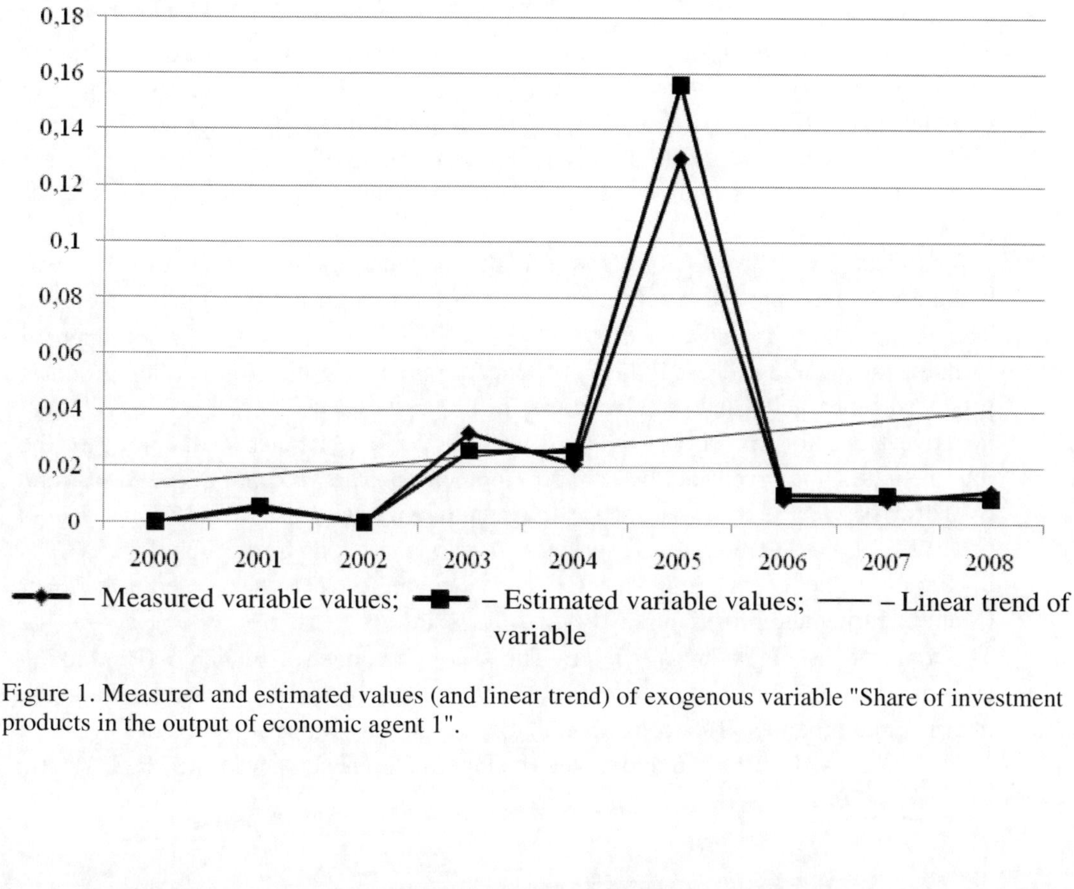

Figure 1. Measured and estimated values (and linear trend) of exogenous variable "Share of investment products in the output of economic agent 1".

In the context of the retrospective analysis of the indicators of economy branches on the base of the considered model, we also obtained the measured and computed values (and their linear trends) of such exogenous indicator for each of 16 economy branches as *CO_pj_iz* (The share of budget of the *i*-th branch spent for purchasing the intermediate products produced by the branch *j*; *i*, *j* = 1 ,..., 16). Particularly, the retrospective analysis of the said indicators of branches for 2000-2009 years show that

- The trend of the indicator *CO_pj_*1z (The share of budget of branch 1 (Agriculture, hunt, and forestry) spent for purchasing the intermediate products produced by branch *j* had positive slope for branches *j* = 1, 8, 2, 13 and negative slope for branches *j* = 11, 3, 15, 14, 10, 6, 12, 5, 7, 9. Branch 1 did not buy the intermediate goods produced by economy branch 16. The maximum growth trend of 1.7% per year is observed for *j* = 1. The maximum decrease of 0.63% per year is observed for *j* = 3 (Mineral resource industry). The maximum value of the indicator *CO_pj_*1*z* of 36% was noted for *j* = 1.
- The trend of the indicator *CO_pj_2z* (The share of budget of branch 2 (Fishery and fish-breeding) spent for purchasing the intermediate products produced by branch *j* had positive slope for branches *j* = 9, 11, 5, 10, 3, 6, 8 and negative slope for branches *j* = 14, 15, 1, 12, 4, 7, 2. Branch 2 did not buy the intermediate goods produced by economy branches 13 and 16. The maximum growth of trend by 1.6% per year is observed for *j* = 9. The maximum decrease by 3.2% per year is observed for *j* = 2 (Mineral resource industry). The maximum value of the indicator *CO_pj_2z* equal to 18% was noted for *j* = 4 (Manufacturing and process industry) in 2008 year.
- The trend of the indicator *CO_pj_3z* (The share of budget of branch 3 (Mineral resource industry) spent for purchasing the intermediate products produced by branch *j* had positive slope for branches *j* = 11, 10, 7, 12, 1, 2, 16 and negative slope for branches *j* = 8, 15, 13, 4, 14, 6, 5, 3, 9. Branch 3 did not buy the intermediate goods produced by economy branch 16. The maximum growth of trend by 1.3% per year is observed for *j* = 11 (Real estate activities, rent, and services for business). The maximum decrease by 1.1% per year is observed for *j* = 9 (Transport and communications). The maximum value of the indicator *CO_pj_3z* equal to 19% was noted for *j* = 11 (Real estate activities, rent, and services for business) in 2008 year.
- The trend of the indicator *CO_pj_4z* (The share of budget of branch 4 (Manufacturing and process industry) spent for purchasing the intermediate products produced by branch *j* had positive slope for branches *j* = 4, 3, 11, 1, 10, 6, 8, 2 and negative slope for branches *j* = 14 12, 13, 15, 5, 9, 7. Branch 4 did not buy the intermediate goods produced by economy branch 16. The maximum growth of trend by 0.46% per year is observed for *j* = 4. The maximum decrease by 0.46% per year is observed for *j* = 7 (Trade, repair services for cars and goods for household use). The maximum value of the indicator *CO_pj_4z* equal to 25% was noted for *j* = 4 (Manufacturing and process industry) in 2008 year.
- The trend of the indicator *CO_pj_5z* (The share of budget of branch 5 (Production and distribution of electric power, gas and water) spent for purchasing the intermediate products produced by branch *j* had positive slope for branches *j* = 4, 5, 11, 3, 10, 8, 14, 1, 12, 2, 16 and negative slope for branches *j* = 15, 13, 6, 9, 7. The

maximum growth of trend by 2.0% per year is observed for $j = 4$ (Manufacturing and process industry). The maximum decrease by 1.0% per year is observed for $j = 7$ (Trade, repair services for cars and goods for household use). The maximum value of the indicator *CO_pj_5z* equal to 19% was noted for $j = 4$ (Manufacturing and process industry) in 2008 year.

- The trend of the indicator *CO_pj_6z* (The share of budget of branch 6 (Construction) spent for purchasing the intermediate products produced by branch j had positive slope for branches j = 4, 6, 10, 8, 2, 14 and negative slope for branches j = 13, 12, 1, 15, 11, 5, 7, 9, 3. Branch 6 did not buy the intermediate goods produced by economy branch 16. The maximum growth of trend by 2.3% per year is observed for $j = 4$ (Manufacturing and process industry). The maximum decrease by 0.79% per year is observed for $j = 3$ (Mineral resource industry). The maximum value of the indicator *CO_pj_6z* equal to 37% was noted for $j = 4$ (Manufacturing and process industry) in 2008 year.
- The trend of the indicator *CO_pj_7z* (The share of budget of branch 7 (Trade, repair services for cars and goods for household use) spent for purchasing the intermediate products produced by branch j had positive slope for branches j = 11, 3, 10, 8, 13 and negative slope for branches j = 2, 14, 12, 15, 1, 5, 6, 9, 4, 7. Branch 7 did not buy the intermediate goods produced by economy branch 16. The maximum growth of trend by 2.3% per year is observed for $j = 11$ (Manufacturing and process industry). The maximum decrease by 0.79% per year is observed for $j = 3$ (Real estate activities, rent, and services for business). The maximum value of the indicator *CO_pj_7z* equal to 17% was noted for $j = 11$ (Real estate activities, rent, and services for business) in 2008 year.
- The trend of the indicator *CO_pj_8z* (The share of budget of branch 8 (Hotels and restaurants) spent for purchasing the intermediate products produced by branch j had positive slope for branches j = 1, 9, 10, 6 and negative slope for branches j = 3, 14, 13, 12, 4, 2, 15, 7, 5, 8, 11. Branch 8 did not buy the intermediate goods produced by economy branch 16. The maximum growth of trend by 0.92% per year is observed for $j = 1$ (Agriculture, hunt, and forestry). The maximum decrease by 0.62% per year is observed for $j = 11$. The maximum value of the indicator *CO_pj_8z* equal to 20% was noted for $j = 4$ (Manufacturing and process industry) in 2008 year.
- The trend of the indicator *CO_pj_9z* (The share of budget of branch 9 (Transport and communications) spent for purchasing the intermediate products produced by branch j had positive slope for branches j = 9, 11, 10, 8, 2 and negative slope for branches j = 14, 13, 15, 12, 1, 3, 6, 5, 4, 7. Branch 9 did not buy the intermediate goods produced by economy branch 16. The maximum growth of trend by 0.69% per year is observed for $j = 9$. The maximum decrease by 1.0% per year is observed for $j = 7$ (Trade, repair services for cars and goods for household use). The maximum value of the indicator *CO_pj_9z* equal to 15% was noted for $j = 4$ (Manufacturing and process industry) in 2008 year.
- The trend of the indicator *CO_pj_10z* (The share of budget of branch 10 (Financial activity) spent for purchasing the intermediate products produced by branch j had positive slope for branches j = 4, 9, 11, 8, 1, 6, 5, 15, 13, 2 and negative slope for branches j = 16, 14, 12, 3, 12, 3, 7, 10. The maximum growth of trend by 0.15% per

year is observed for j = 4 (Manufacturing and process industry). The maximum decrease by 0.98% per year is observed for j = 10. The maximum value of the indicator *CO_pj_*10z equal to 16% was noted for j = 4 (Manufacturing and process industry) in 2008 year.

- The trend of the indicator *CO_pj_11z* (The share of budget of branch 11 (Real estate activities, rent, and services for business) spent for purchasing the intermediate products produced by branch j had positive slope for branches j = 4, 7, 9, 5, 13, 14, 8, 2 and negative slope for branches j = 16, 12, 1, 15, 3, 6, 11. The maximum growth of trend by 1.6% per year is observed for j = 4 (Manufacturing and process industry). The maximum decrease by 0.70% per year is observed for j = 11. The maximum value of the indicator *CO_pj_11z* equal to 25% was noted for j = 11 in 2008 year.
- The trend of the indicator *CO_pj_12z* (The share of budget of branch 12 (Real estate activities, rent, and services for business) spent for purchasing the intermediate products produced by branch j had positive slope for branches j = 8, 10, 6, 11, 15, 1, 14, 13 and negative slope for branches j = 2, 9, 3, 5, 4, 7, 12. Branch 12 did not buy the intermediate goods produced by economy branch 16. The maximum growth of trend by 0.92% per year is observed for j = 8 (Hotels and restaurants). The maximum decrease by 0.76% per year is observed for j = 11. The maximum value of the indicator *CO_pj_12z* equal to 18% was noted for j = 4 (Manufacturing and process industry) in 2008 year.
- The trend of the indicator *CO_pj_13z* (The share of budget of branch 13 (Education) spent for purchasing the intermediate products produced by branch j had positive slope for branches j = 4, 6, 9, 15, 1, 8 and negative slope for branches j = 2, 10, 3, 12, 14, 7, 13, 5, 11. Branch 13 did not buy the intermediate goods produced by economy branch 16. The maximum growth of trend by 2.9% per year is observed for j = 4 (Manufacturing and process industry). The maximum decrease by 0.66% per year is observed for j = 11 (Real estate activities, rent, and services for business). The maximum value of the indicator *CO_pj_13z* equal to 24% was noted for j = 4 (Manufacturing and process industry) in 2008 year.
- The trend of the indicator *CO_pj_14z* (The share of budget of branch 14 (Public health and social services) spent for purchasing the intermediate products produced by branch j had positive slope for branches j = 4, 6, 1, 7, 8 and negative slope for branches j = 2, 3, 12, 11, 10, 15, 9, 13, 5, 14. Branch 14 did not buy the intermediate goods produced by economy branch 16. The maximum growth of trend by 3.2% per year is observed for j = 4 (Manufacturing and process industry). The maximum decrease by 1.1% per year is observed for j = 14. The maximum value of the indicator *CO_pj_14z* equal to 37% was noted for j = 4 (Manufacturing and process industry) in 2008 year.
- The trend of the indicator *CO_pj_15z* (The share of budget of branch 15 (Other public utilities, social and personal services) spent for purchasing the intermediate products produced by branch j had positive slope for branches j = 4, 8, 10, 5, 7, 2, 16 and negative slope for branches j = 6, 1, 3, 13, 12, 9, 14, 15, 11. Branch 15 did not buy the intermediate goods produced by economy branch 16. The maximum growth of trend by 0.22% per year is observed for j = 4 (Manufacturing and process industry). The maximum decrease by 0.83% per year is observed for j = 11 (Real

estate activities, rent, and services for business). The maximum value of the indicator *CO_pj_15z* equal to 21% was noted for $j = 4$ (Manufacturing and process industry) in 2008 year.

- The trend of the indicator *CO_pj_16z* (The share of budget of branch 15 (Housekeeping services) spent for purchasing the intermediate products produced by branch j had positive slope for branches $j = 11, 4$ and negative slope for branches $j = 1, 14, 16, 8, 15, 6, 9, 13$. Branch 16 did not buy the intermediate goods produced by economy branches 2, 3, 7, 10, 12. The maximum growth of trend by 1.2% per year is observed for $j = 11$ (Real estate activities, rent, and services for business). The maximum decrease by 0.41% per year is observed for $j = 13$ (Education). Branch 16 did not buy the intermediate goods in 2008 year.

Below (Figure 2) we give a graphic illustration of the initial data for retrospective macroanalysis at the level of the economy branches within the framework of the exogenous variable "The share of budget of branch 1 (Agriculture, hunt, and forestry) spent for purchasing the intermediate products produced by branch 4 (Manufacturing and process industry)". Negative slope of the linear trend at Figure 2 is evidence of decrease (on average) of the mentioned share of expenses within the considered time interval.

Also, we carried out the retrospective analysis of the endogenous variables at the level of the economy branches. Below (Figures 3, 4) we give a graphic illustration of the dynamics, its rates and trends of one of the economic agents' indicators for retrospective macroanalysis is presented by the example of the indicator "Gross values added in prices of 2000 year" of the branch "Manufacturing and process industry".

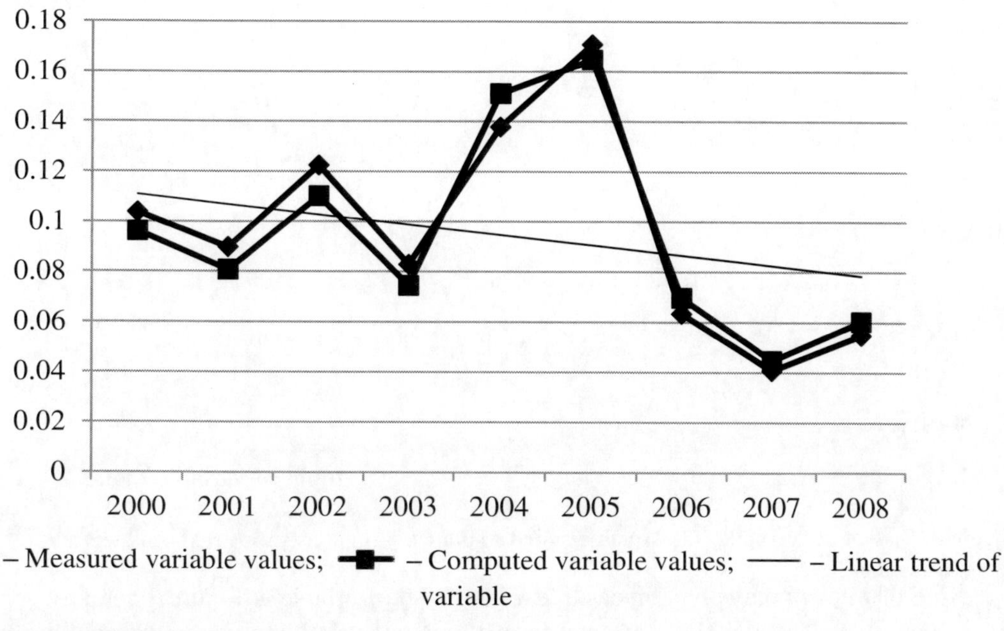

Figure 2. Exogenous variable "Share of budget of branch 1 (Agriculture, hunt, and forestry) spent for purchasing the intermediate products produced by branch 4 (Manufacturing and process industry)", its trend and measured values.

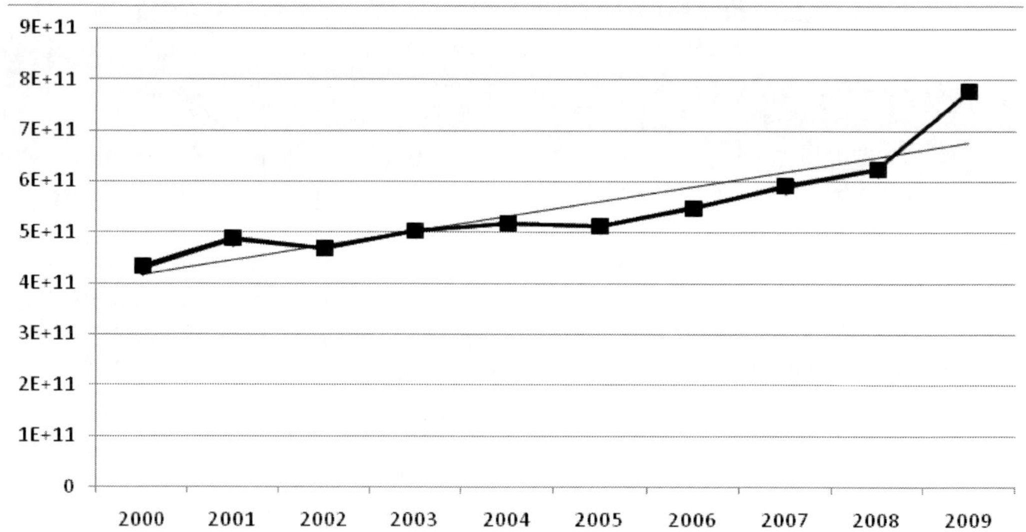

– Measured variable values; – Computed variable values; – Linear trend of variable

Figure 3. Gross value added of branch 4 (in prices of 2000 year), its trend and measured values.

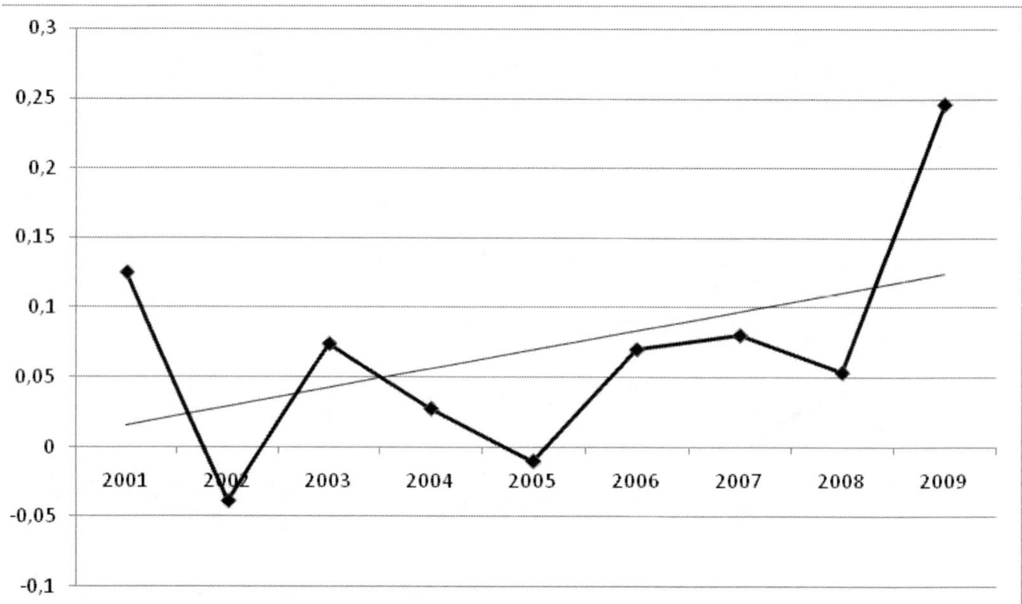

– Variable rate values; – Linear trend of variable rate

Figure 4. Rate of the variable "Gross value added of branch 4 (in constant prices)" and its trend.

Note that the computed and measured values of the variable at Figure 3 actually coincide with each other. It is explained by quality of the parametric identification of the considered model. Analysis of data presented in Figures 3 and 4 (the positive slope of linear trends of both variable "Gross value added of branch 4" and its rate) shows accelerated growth of the said indicator within the interval of 2000-2009 years.

Table 5. Gross value added of branch

No.	No. of branch	Value of indicator GVA in year 2008 (tenge, in prices of 2000)
1	3	9.92923×10^{11}
2	11	7.94163×10^{11}
3	7	6.49315×10^{11}
4	4	6.24578×10^{11}
5	9	5.83432×10^{11}
6	6	$4,29188\times10^{11}$
7	10	2.79863×10^{11}
8	1	2.78125×10^{11}
9	13	1.49005×10^{11}
10	12	9.00073×10^{10}
11	5	8.88712×10^{10}
12	15	8.88331×10^{10}
13	14	7.84887×10^{10}
14	8	4.36384×10^{10}
15	16	4.40276×10^{9}
16	2	4.14457×10^{9}

Table 6. Fixed assets of branch

No.	No. of branch	Value of indicator "Fixed assets of branch" in year 2008 (tenge, in prices of 2000)
1	3	1.55078×10^{12}
2	11	1.15043×10^{12}
3	4	6.43257×10^{11}
4	9	6.41518×10^{11}
5	12	3.21504×10^{11}
6	6	2.94698×10^{11}
7	5	2.75815×10^{11}
8	1	2.33045×10^{11}
9	7	2.21753×10^{11}
10	13	9.67223×10^{10}
11	14	8.86825×10^{10}
12	10	7.66333×10^{10}
13	15	5.98872×10^{10}
14	8	3.36888×10^{10}
15	2	1.44711×10^{9}
16	16	0

Retrospective Comparative Analysis

Within the framework of retrospective comparative analysis of the economy branches' indicators for each year from 2000 to 2008 we also carried out ranking of the branches with respect to the following indicators:

- gross value added (GVA) of a branch (in prices of year 2000);
- fixed assets of a branch (in prices of year 2000);
- capital productivity of a branch;
- demand of labour force in a branch;
- labour productivity in a branch.

Analysis of the results of ranking with respect to the said indicators allows to conclude, e.g., the following.

- The maximum GVA over all the economy branches is observed within the branch in 2000-2002 years, in branch 11 (Real estate activities, rent, and services for business) in 2003-2004 years, and in branch 3 (mineral resource industry) in 2005-2008 years.
- The maximum fixed assets was discovered in branch 3 (Mineral resource industry) for all time of observations.
- The maximum capital productivity was observed in branch 7 (Trade, repair services for cars and goods for household use) in 2000-2005 years, and in branch 10 (Financial activity) in 2006-2008 years.
- The maximum number of employees (about 30% of the whole number of employees) was noted in branch 1 (Agriculture, hunt and forestry) for all period of observations.
- At that, branch 1 demonstrated the minimum labour productivity in 2000-2008 years. The maximum labour productivity was demonstrated by branch 3 (Mineral resource industry) for practically all considered period with growing gap with respect to other branches.

The results of the Kazakhstan Republic economy branches ranking within the framework of the said indicators in year 2008 are presented in Tables 5-9.

The Results of the Retrospective Analysis of the Main Integral Indicators of the Country Development, Indicators of Public Administration and Households

Also, we present the results of retrospective analysis of the said indicators for 2000-2009 years.

Below (see Figures 5, 6) we present an example of initial data for the retrospective macroanalysis of the national economy indicators in the context of the endogenous variable "Volume of production of investment products".

Analysis of the trends presented in Figures 5 and 6 allows concluding that in 2000-2009 one can observe average growth of real output of the investment products in the country with some average deceleration of this growth.

The Results of Perspective Analysis of the Endogenous Variables at the Level of the Economy Branches

The perspective analysis of the endogenous variables of the Kazakhstan Republic was carried out on the base of computation of the economy branches' model up to 2015 year with use of extrapolation of the measured values of the exogenous variables of the model up to 2015 year.

Table 7. Capital productivity of branch

No.	No. of branch	Value of indicator "Capital productivity of branch" in year 2008
1	10	3.65197
2	7	2.928106
3	2	2.864025
4	13	1.540538
5	15	1.48334
6	6	1.456368
7	8	1.295338
8	1	1.193439
9	4	0.970963
10	9	0.909455
11	14	0.885053
12	11	0.690321
13	3	0.640272
14	5	0.322213
15	12	0.279957
16	16	0

Table 8. Branch demand of labour force

No.	No. of branch	Value of indicator "Branch demand of labour force" in year 2008
1	1	2349730
2	7	1150307
3	13	754299
4	9	588847
5	4	572886
6	6	548895
7	11	378129
8	12	352489
9	14	347329
10	15	205425
11	3	200283
12	5	164816
13	8	103048
14	10	96165
15	16	24360
16	2	19971

Table 9. Labour productivity in branch

No.	No. of branch	Value of indicator "Labour productivity in branch" in year 2008 (in prices of 2000)
1	3	4957609
2	10	2910234
3	11	2100238
4	4	1090231
5	9	990804
6	6	781913.2
7	7	564471
8	5	539213,9
9	15	432435
10	8	423477.3
11	12	255348
12	14	225978.1
13	2	207533.5
14	13	197540.5
15	16	180739
16	1	118364.8

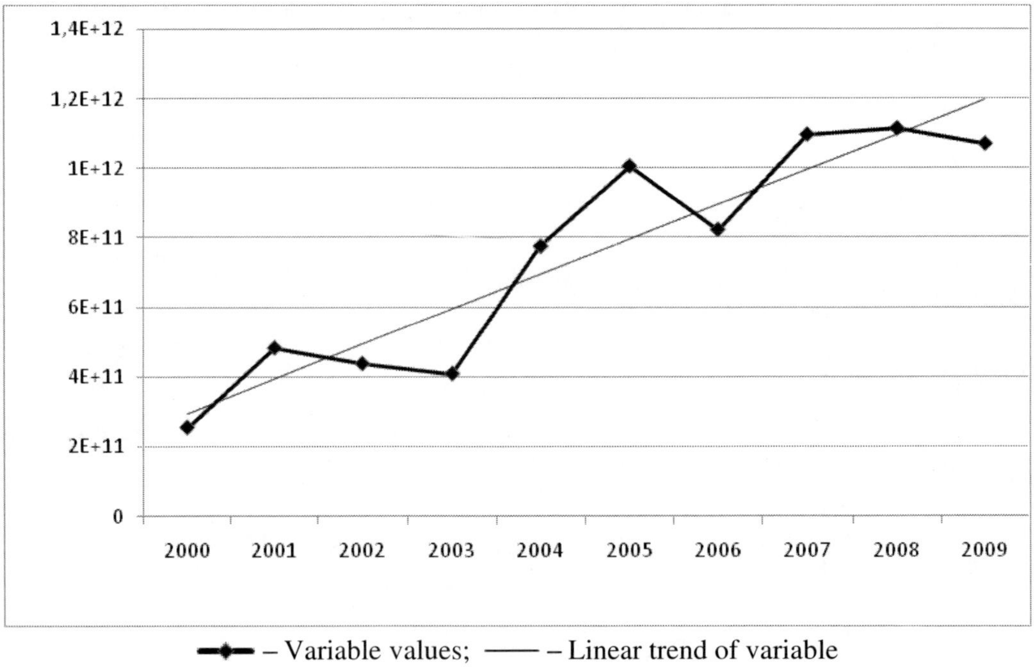

Figure 5. Endogenous variable "Volume of production of investment products" and its trend.

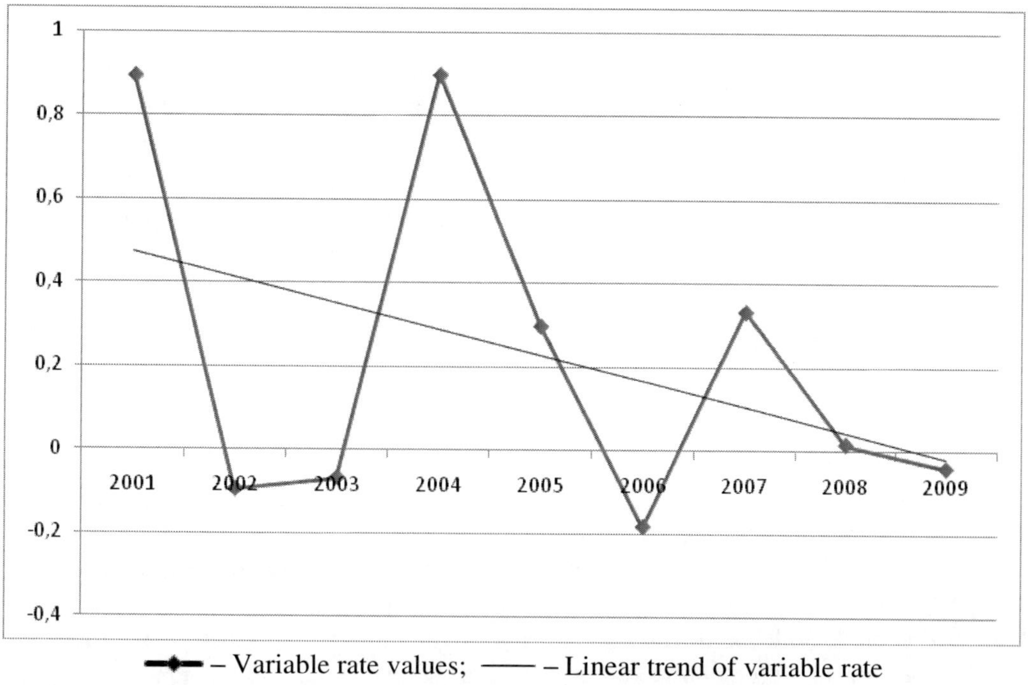

Figure 6. Rate of endogenous variable "Volume of production of investment products" and its trend.

The perspective analysis of the endogenous indicators for 2000-2015 years by the branches shows that:

- The trends of the indicator "Budget of branch's enterprises (in prices of 2000 year)" increase for all branches. At that the maximum growth by 7.37×10^{10} tenge per year is observed in branch 10, whereas the minimum growth by 3.3×10^{8} tenge per year is observed in branch 14. The trends of the rates of the considered indicator have negative slopes for almost all branches excepting 12 and 4. This shows that all branches excepting 12 and 4 are expected to demonstrate drop of the growth rate of the budgets (maximum drop by 3% per year by trend was observed in branch 14).
- The trends of the indicator "Gross value added by branch (in prices of 2000 year)" increase for almost all branches (excepting branches 12 and 14). The maximum growth by 7.4×10^{10} tenge per year is observed in branch 4, whereas the maximum drop by 2.2×10^{9} tenge per year is observed in branch 14. The trends of the rates of the considered indicator have negative slopes for almost all branches excepting 5, 15, 13, 2, 12, 14. This shows that all branches excepting 5, 15, 13, 2, 12, 14 are expected to demonstrate drop of the growth rate of GVA (maximum drop by 18% per year by trend was observed in branch 16; maximum growth by 1.6% was observed in branch 14).
- The trends of the indicator "Output of branch (in prices of 2000 year)" increase for almost all branches (excepting branches 16, 2, 3, 14, 11, 9). The maximum growth by 1.6×10^{11} tenge per year is observed in branch 6, whereas the maximum drop by 3.4×10^{10} tenge per year is observed in branch 9. The trends of the rates of the considered indicator have negative slopes for almost all branches excepting 4, 5, 12,

15. This shows that all branches excepting 4, 5, 12, 15 are expected to demonstrate drop of the growth rate of the Output of branch (maximum drop by 19% per year by trend was observed in branch 16; maximum growth by 1.1% was observed in branch 4).

- The trends of the indicator "Fixed assets of branch (in prices of 2000 year)" increase for almost all branches (excepting branches 2, 8, 14, 1, 5). The maximum growth by 5.5×10^{10} tenge per year is observed in branch 6, whereas the maximum drop by 6.2×10^{9} tenge per year is observed in branch 9. The trends of the rates of the considered indicator have negative slopes for almost all branches excepting 10, 2, 8, 14, 1, 5. This shows that all branches excepting 10, 2, 8, 14, 1, 5 are expected to demonstrate drop of the growth rate of the Fixed assets (maximum drop by 1% per year by trend was observed in branch 12; maximum growth by 0.58% was observed in branch 10).
- The trends of the indicator "Supply of investment products of branch (in prices of 2000 year)" increase for branches 6, 4, 7, 1, 9, 15, 2. The maximum growth by 1.1×10^{11} tenge per year is observed in branch 6, whereas the maximum drop by 8.8×10^{8} tenge per year is observed in branch 11. The trends of the rates of the considered indicator have negative slopes for almost all branches or almost zero slopes for all aforementioned branches. This shows that all these branches excepting 11 are expected to demonstrate drop of the growth rate of the investment products. The investment products are not produced in that branches not mentioned in this paragraph.
- The trends of the indicator "Supply of final products of branch (in prices of 2000 year)" increase for almost all branches (excepting branches 11, 2, 14, 9). The maximum growth by 3.1×10^{10} tenge per year is observed in branch 4, whereas the maximum drop by 6.9×10^{9} tenge per year is observed in branch 9. The trends of the rates of the considered indicator have negative slopes for almost all branches excepting 15, 7. This shows that all branches excepting 15, 7 are expected to demonstrate drop of the growth rate of the Supply of final products (maximum drop by 15% per year by trend was observed in branch 12; maximum growth by 0.28% was observed in branch 15).
- The trends of the indicator "Supply of intermediate products of branch (in prices of 2000 year)" increase for almost all branches (excepting branches 16, 2, 14, 3, 9). The maximum growth by 6.6×10^{10} tenge per year is observed in branch 4, whereas the maximum drop by 1.0×10^{10} tenge per year is observed in branch 9. The trends of the rates of the considered indicator have negative slopes for almost all branches excepting 5, 15. This shows that all branches excepting 5, 15 are expected to demonstrate drop of the growth rate of the Supply of intermediate products (maximum drop with considerable jumps was observed in branch 16; maximum growth by 2.9% was observed in branch 5).
- The trends of the indicator "Supply of export products of branch (in prices of 2000 year)" increase for almost all branches (excepting branches 2, 13, 11, 1, 9). The maximum growth by 3.0×10^{10} tenge per year is observed in branch 4, whereas the maximum drop by 1.6×10^{9} tenge per year is observed in branch 9. The trends of the rates of the considered indicator have negative slopes for almost all branches

excepting 12, 8, 13, and 1. This shows that all branches excepting 12, 8, 13, 1 are expected to demonstrate drop of the growth rate of the Supply of export products (maximum drop by 22% per year by trend was observed in branch 15; maximum growth by 3.3% was observed in branch 13). The export products are not produced by branches 14, 16.

- The trends of the indicator "Profit of branch (in prices of 2000 year)" increase for almost all branches (excepting branches 16, 14, 12, 13). The maximum growth by 6.6×10^{10} tenge per year is observed in branch 4, whereas the maximum drop by 9.7×10^{9} tenge per year is observed in branch 13. The trends of the rates of the considered indicator have negative slopes for almost all branches excepting 4, 5, 1, 15, 2, and 14. This shows that all branches excepting 4, 5, 1, 15, 2, 14 are expected to demonstrate drop of the growth rate of the Profit of branch (maximum drop by 15% per year by trend was observed in branch 12; maximum growth by 4.0% was observed in branch 15).

Below (see Figures 7, 8) we present an example of initial data for the retrospective macroanalysis at the level of the economy branches on the base of the endogenous variables and in the context of the endogenous variable "Gross value added (in prices of 2000 year) of economic agent 4 (Manufacturing and process industry)", its rate, and their trends.

Analysis of information presented in Figures 7, 8 and comparison of that figures with Figures 2, 3 show that in perspective up to 2015 one can expect growth (on average) GVA of branch 4 with growth (on average) of the rate of that growth. That is, the tendencies similar to that of Figure 2, 3 noted before remains the same (according to the base prediction) for 2010-2015 years. Nevertheless, the angular coefficient of the trend in Figure 7 is some higher than the respective angular coefficient in Figure 8. This is evidence to growth (on average) of the rate of that growth in the perspective period. This is also witnessed by greater (on average) value of the trend in Figure 8 in comparison with its values for the retrospective period (Figure 3).

Within the framework of the perspective analysis of the economy branches' indicators on the base of the considered model we also obtained the computed prediction values (and their linear trend) of the indicators "Demand of intermediate products of branch j in branch i (in constant prices of 2000 year)" ($i, j = 1, ..., 16$) for each of 16 economy branches.

The perspective analysis of the mentioned indicators of branches for 2000-2015 years shows that:

- The trend of the indicator "Demand of intermediate products of branch j in branch 1 (Agriculture, hunt, and forestry) (in constant prices of 2000 year)" had positive slope for branches j = 4, 5, 16, 10, 7, 12, 8, 16 and negative slope for branches j = 13, 2, 15, 14, 3, 11, 9. Branch 1 does not purchase intermediate products of economy branch 16. The maximum increase of the trend by 2.3×10^{9} tenge per year is observed for $j = 4$, whereas the maximum decrease by 2.3×10^{9} tenge per year is observed for $j = 9$. The trend of the rate of the considered indicator increases only for products of branch 5 and decreases for other branches.
- The trend of the indicator "Demand of intermediate products of branch j in branch 2 (Fishery and fish-breeding) (in constant prices of 2000 year)" had positive slope for

branches $j = 4, 5, 9, 10, 11, 12, 6, 3, 7, 8$ and negative slope for branches $j = 15, 14, 1, 2$. Branch 2 does not purchase intermediate products of economy branches 13 and 16. The maximum increase of the trend by 5.1×10^7 tenge per year is observed for $j = 4$, whereas the maximum decrease by 7.7×10^7 tenge per year is observed for $j = 2$. The trend of the rate of the considered indicator increases only for products of branches 12, 6, 8 and decreases for other branches.

- The trend of the indicator "Demand of intermediate products of branch j in branch 3 (Mineral resource industry) (in constant prices of 2000 year)" had positive slope for branches $j = 5, 4, 10, 7, 6, 8, 12, 15, 11$ and negative slope for branches $j = 2, 1, 16, 13, 14, 3, 9$. The maximum increase of the trend by 4.8×10^9 tenge per year is observed for $j = 5$, whereas the maximum decrease by 3.1×10^9 tenge per year is observed for $j = 9$. The trend of the rate of the considered indicator increases only for products of branches 3, 16 and decreases for other branches.
- The trend of the indicator "Demand of intermediate products of branch j in branch 4 (Manufacturing and process industry) (in constant prices of 2000 year)" had positive slope for branches $j = 4, 5, 10, 6, 11, 8, 12$ and negative slope for branches $j = 16, 15, 14, 13, 2, 1, 3, 7, 9$. The maximum increase of the trend by 1.5×10^{10} tenge per year is observed for $j = 4$, whereas the maximum decrease by 2.3×10^9 tenge per year is observed for $j = 9$. The trend of the rate of the considered indicator increases only for products of branches 4, 5, 6, 12, 16 and decreases for other branches.
- The trend of the indicator "Demand of intermediate products of branch j in branch 5 (Manufacturing and process industry) (in constant prices of 2000 year)" had positive slope for branches $j = 5, 4, 6, 10, 3, 7, 11, 12, 15, 8, 1, 2$ and negative slope for branches $j = 16, 9, 13, 14$. The maximum increase of the trend by 1.2×10^{10} tenge per year is observed for $j = 5$, whereas the maximum decrease by 1.3×10^7 tenge per year is observed for $j = 14$. The trend of the rate of the considered indicator increases only for products of branches 7, 2, 16 and decreases for other branches.
- The trend of the indicator "Demand of intermediate products of branch j in branch 6 (Construction) (in constant prices of 2000 year)" had positive slope for branches $j = 4, 6, 5, 7, 10, 11, 8, 12, 1, 15, 13$ and negative slope for branches $j = 2, 14, 3, 9$. Branch 6 does not purchase intermediate products of economy branch 6. The maximum increase of the trend by 1.0×10^{10} tenge per year is observed for $j = 4$, whereas the maximum decrease by 3.7×10^8 tenge per year is observed for $j = 9$. The trend of the rate of the considered indicator decreases for all 15 branches.
- The trend of the indicator "Demand of intermediate products of branch j in branch 7 (Trade, repair services for cars and goods for household use) (in constant prices of 2000 year)" had positive slope for branches $j = 10, 4, 5, 11, 8, 7, 12$ and negative slope for branches $j = 13, 2, 14, 15, 1, 3, 9$. Branch 7 does not purchase intermediate products of economy branch 16. The maximum increase of the trend by 3.8×10^9 tenge per year is observed for $j = 10$, whereas the maximum decrease by 1.9×10^9 tenge per year is observed for $j = 9$. The trend of the rate of the considered indicator increases only for products of branches 4, 5, 6, 15, 1 and decreases for other branches.
- The trend of the indicator "Demand of intermediate products of branch j in branch 8 (Hotels and restaurants) (in constant prices of 2000 year)" had positive slope for

branches j = 4, 5, 10, 1, 6, 7, 8, 9, 15, 11, 12, 3 and negative slope for branches j = 14, 13, 2. Branch 8 does not purchase intermediate products of economy branch 16. The maximum increase of the trend by 1.0×10^9 tenge per year is observed for j = 4, whereas the maximum decrease by 1.2×10^7 tenge per year is observed for j = 2. The trend of the rate of the considered indicator increases only for products of branches 4, 8, 15, 12 and decreases for other branches.

- The trend of the indicator "Demand of intermediate products of branch j in branch 9 (Transportation and communications) (in constant prices of 2000 year)" had positive slope for branches j = 4, 5, 10, 6, 8, 12, 11 and negative slope for branches j = 16, 2, 15, 14, 13, 1, 7, 3, 9. The maximum increase of the trend by 4.5×10^9 tenge per year is observed for j = 4, whereas the maximum decrease by 1.2×10^9 tenge per year is observed for j = 9. The trend of the rate of the considered indicator increases only for products of branches 5, 16, 15, 7 and decreases for other branches.
- The trend of the indicator "Demand of intermediate products of branch j in branch 10 (Financial activity) (in constant prices of 2000 year)" had positive slope for all branches excepting 2 and 4. The maximum increase of the trend by 1.1×10^{10} tenge per year is observed for j = 10, whereas the maximum decrease by 4.8×10^5 tenge per year is observed for j = 14. The trend of the rate of the considered indicator increases only for products of branches 16, 2 and decreases for other branches.
- The trend of the indicator "Demand of intermediate products of branch j in branch 11 (Real estate activities, rent, and services for business) (in constant prices of 2000 year)" had positive slope for branches j = 4, 10, 5, 6, 7, 8, 12, 9, 15, 13 and negative slope for branches j = 18, 2, 11, 1, 14, 3. The maximum increase of the trend by 5.3×10^9 tenge per year is observed for j = 4, whereas the maximum decrease by 3.0×10^8 tenge per year is observed for j = 3. The trend of the rate of the considered indicator increases only for products of branch 15 and decreases for other branches.
- The trend of the indicator "Demand of intermediate products of branch j in branch 12 (Public administration) (in constant prices of 2000 year)" had positive slope for branches j = 5, 4, 10, 6, 8, 12, 15, 1, 11, 13 and negative slope for branches j = 2, 14, 3, 7, 9. Branch 12 does not purchase intermediate products of economy branch 16. The maximum increase of the trend by 3.2×10^9 tenge per year is observed for j = 5, whereas the maximum decrease by 2.8×10^8 tenge per year is observed for j = 9. The trend of the rate of the considered indicator increases only for products of branches 6, 12, 7 and decreases for other branches.
- The trend of the indicator "Demand of intermediate products of branch j in branch 13 (Public administration) (in constant prices of 2000 year)" had positive slope for branches j = 5, 4, 6, 15, 10, 13, 12, 8, 1 and negative slope for branches j = 2, 3, 9, 7, 14, 11. Branch 13 does not purchase intermediate products of economy branch 16. The maximum increase of the trend by 4.2×10^8 tenge per year is observed for j = 5, whereas the maximum decrease by 2.8×10^8 tenge per year is observed for j = 11. The trend of the rate of the considered indicator increases only for products of branch 13 and decreases for other branches.
- The trend of the indicator "Demand of intermediate products of branch j in branch 14 (Public health and social services) (in constant prices of 2000 year)" had positive slope for branches j = 4, 5, 6, 7, 10, 15, 8, 12, 13 and negative slope for branches j =

2, 3, 11, 1, 14, 9. Branch 14 does not purchase intermediate products of economy branch 16. The maximum increase of the trend by 5.1×10^9 tenge per year is observed for $j = 4$, whereas the maximum decrease by 7.0×10^8 tenge per year is observed for $j = 5$. The trend of the rate of the considered indicator increases only for products of branch 13 and decreases for other branches.

- The trend of the indicator "Demand of intermediate products of branch j in branch 15 (Other public utilities, social and personal services) (in constant prices of 2000 year)" had positive slope for branches j = 4, 5, 10, 15, 7, 6, 8, 12, 11, 13, 1 and negative slope for branches j = 16, 2, 3, 9, 14. The maximum increase of the trend by 2.8×10^9 tenge per year is observed for $j = 4$, whereas the maximum decrease by 6.3×10^7 tenge per year is observed for $j = 14$. The trend of the rate of the considered indicator increases only for products of branches 4, 16 and decreases for other branches.
- The trend of the indicator "Demand of intermediate products of branch j in branch 16 (Housekeeping services) (in constant prices of 2000 year)" had positive slope for branches j = 6, 4, 8 and negative slope for branches j = 1, 16, 14, 15, 9, 11, 13. Branch 16 does not purchase intermediate products of economy branches 10, 5, 12, 7, 2. The maximum increase of the trend by 8.8×10^6 tenge per year is observed for $j = 6$, whereas the maximum decrease by 1.0×10^7 tenge per year is observed for $j = 13$. The trend of the rate of the considered indicator increases only for products of branches 4, 16, 15 and decreases for other branches.

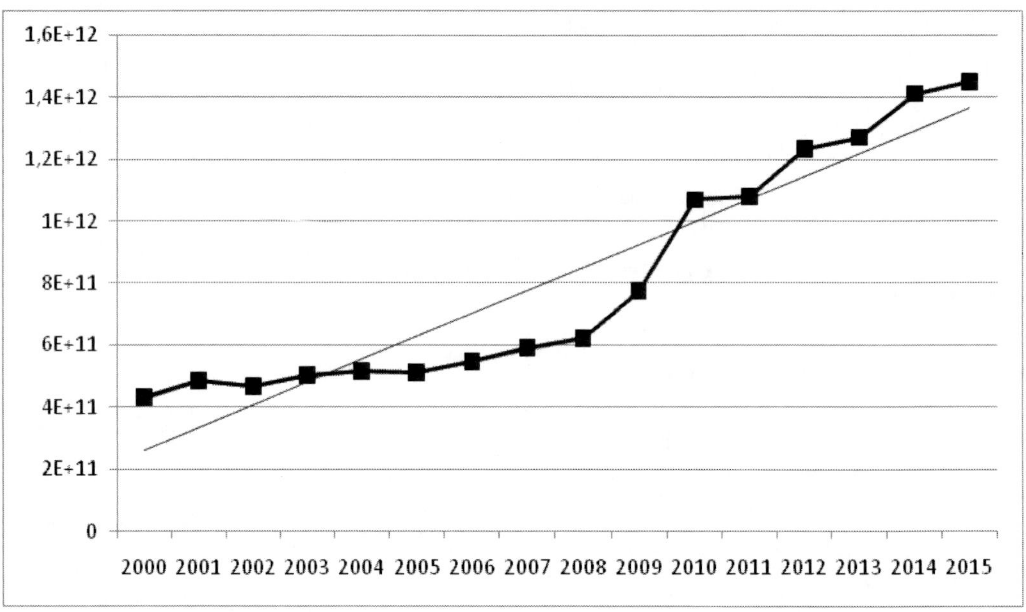

– Measured variable values; – Computed values of variable; – Linear trend of variable

Figure 7. Gross value added by branch 4 (in prices of 2000 year), its trend and measured values.

Below (see Figures 9, 10) we present an example of initial data for the perspective macroanalysis of the dynamics (and its trend) and rate (and its trend) of the economy branches in the context of the endogenous variable "Demand of intermediate products of branch 4 (Manufacturing and process industry) in branch 1 (Agriculture, hunt, and forestry) (in prices of 2000 year)".

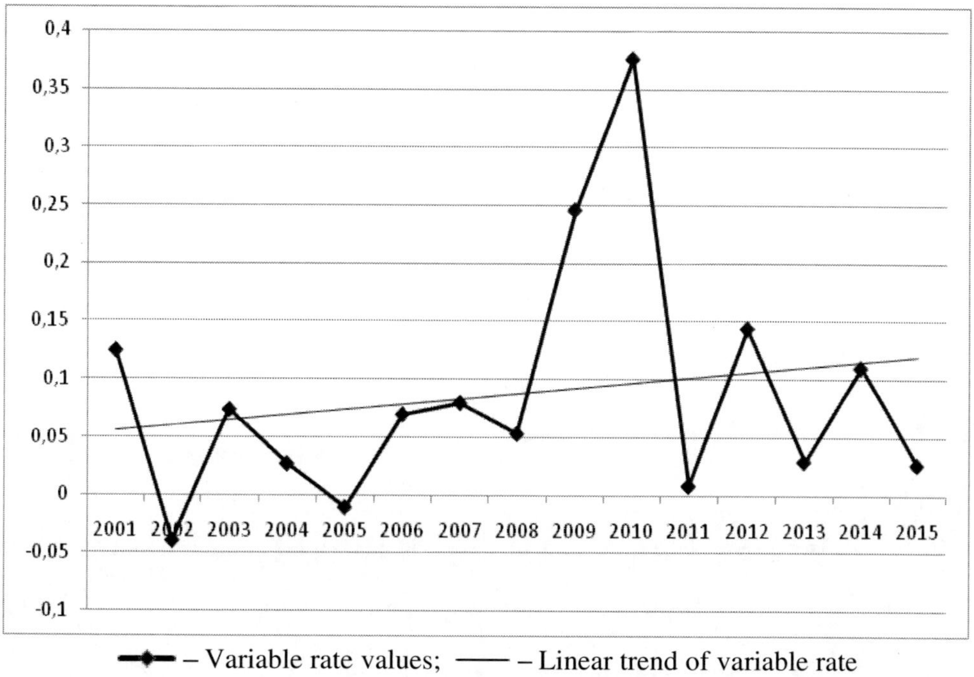

Figure 8. Rate of variable "Gross value added by branch 4 (in fixed prices)" and its trend.

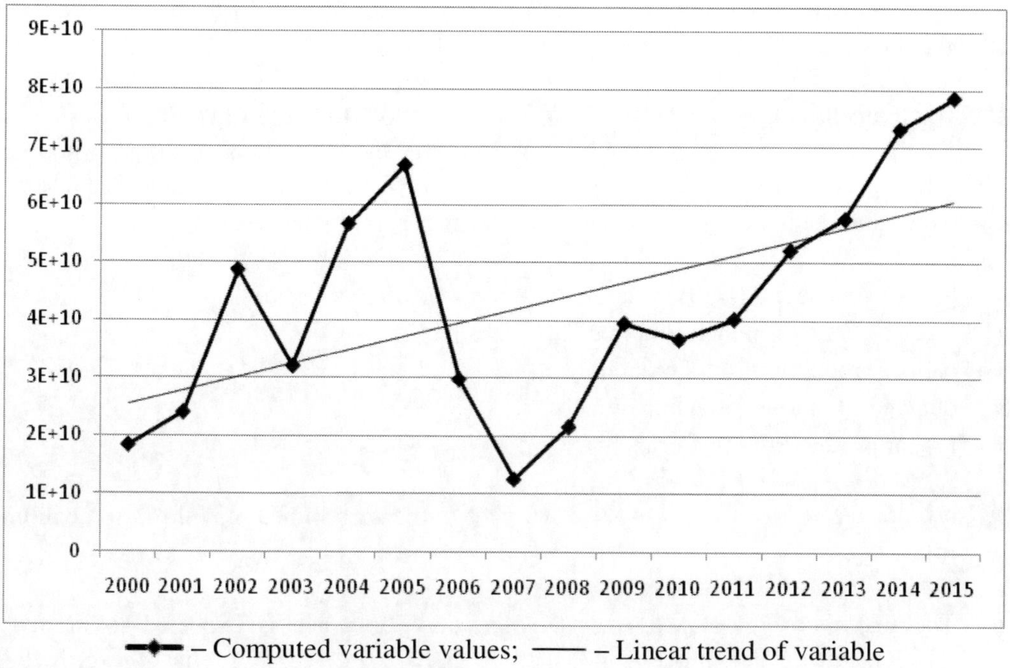

Figure 9. Demand of intermediate products of branch 4 in branch 1 (in prices of 2000 year) and its trend.

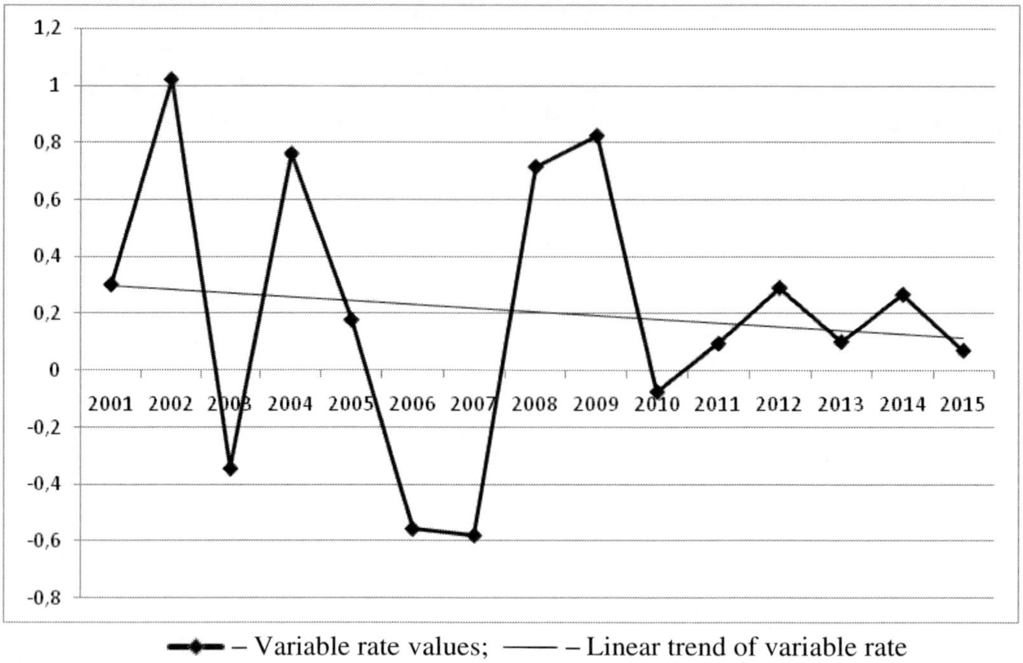

Figure 10. Rate of the variable "Demand of intermediate products of branch 4 in branch 1 (in prices of 2000 year)" and its trend.

Positive slope of the linear trend of the indicator (Figure 9) argues on increase (on average) of the considered branch demand within the considered time interval. At the same time, negative slope of the trend of the rate of this indicator argues on decrease (on average within the period of 2000-2015) of the indicator growth rate.

Perspective Comparative Analysis of the Indicators of the Economy Branches

Within the framework of perspective comparative analysis of the economy branches' indicators for each year from 2009 to 2015 we also carried out ranking of the branches with respect to the same indicators just as in the case of retrospective macroanalysis:

- gross value added (GVA) of a branch (in prices of year 2000);
- fixed assets of a branch (in prices of year 2000);
- capital productivity of a branch;
- demand of labour force in a branch;
- labour productivity in a branch.

Analysis of the results of ranking with respect to the said indicators allows to conclude, e.g., the following.

- The maximum real GVA over the economy branches is expected in branch 4 (Manufacturing and process industry) in 2010-2015 years. At that, the growth of GVA in this branch in 2015 year will be equal to about 86% in comparison with 2009 year.

- The maximum fixed assets over the economy branches is expected in branch 3 (Mineral resource industry) for the whole period of retrospective and perspective prediction. The growth of the fixed assets in this branch in 2015 year will be equal to about 9.2% in comparison with 2009 year.
- The maximum capital productivity over the economy branches is expected in branch 2 (Fishery and fish breeding) in 2010-2015 years. The growth of capital productivity in this branch in 2015 year will be more than 60 times greater than capital productivity of branch 12 (Public administration).
- The maximum number of employees is expected in branch 1 (Agriculture, hunt, and forestry) in 2000-2015 years. In 2015, the demand of labour force in this branch will constitute 2581607 people. The minimum number of people starting from 2012 year is expected in branch 16 (Housekeeping services); in 2015 it will constitute 24361.
- The maximum labour productivity is expected in branch 10 (Financial activity) in 2010-2015 years. In 2015, the branch labour of productivity will be 23 times greater than the same of branch 14 (Public health and social services).

The results of the Kazakhstan Republic economy branches ranking within the framework of the said indicators in year 2015 are presented in Tables 10-14.

Perspective Analysis of the Main Integral Indicators of the Country Development, Indicators of Public Administration and Households

The perspective analysis of the aforementioned indicators in 2000-2015 years shows the following.

- GVA of the country (in prices of 2000 year) in the considered period grows with average rate of the trend equal to 2.5×10^{11} tenge per year. At that, the rate of the GVA indicator decreases (by its trend) by 0.84% per year.
- Gross products output (in prices of 2000 year) in the considered period grows with average rate of the trend equal to 4.0×10^{11} tenge per year. At that, the rate of this indicator decreases (by its trend) by 0.45% per year.
- Investment products output (in prices of 2000 year) in the considered period grows with average rate of the trend equal to 1.2×10^{11} tenge per year. At that, the rate of this indicator decreases (by its trend) by 2.0% per year.
- Final products output (in prices of 2000 year) in the considered period grows with average rate of the trend equal to 9.9×10^{10} tenge per year. At that, the rate of this indicator decreases (by its trend) by 0.46% per year.
- Intermediate products output (in prices of 2000 year) in the considered period grows with average rate of the trend equal to 1.7×10^{11} tenge per year. At that, the rate of this indicator decreases (by its trend) by 1.2% per year.
- Export products output (in prices of 2000 year) in the considered period grows with average rate of the trend equal to 5.1×10^{10} tenge per year. At that, the rate of this indicator decreases (by its trend) by 1.4% per year.
- State budget (in prices of 2000 year) in the considered period grows with average rate of the trend equal to 1.0×10^{11} tenge per year. At that, the rate of this indicator decreases (by its trend) by 1.5% per year.

- Social transfers to households (in prices of 2000 year) in the considered period grows with average rate of the trend equal to 1.0×10^{11} tenge per year. At that, the rate of this indicator decreases (by its trend) by 1.5% per year.
- Fixed assets to households (in prices of 2000 year) in the considered period grows with average rate of the trend equal to 2.2×10^{10} tenge per year. At that, the rate of this indicator decreases (by its trend) by 0.79% per year.
- Capital productivity in the considered period grows with average rate of the trend equal to 1.4×10^{-2} $1/(\text{year})^2$. At that, the rate of this indicator decreases (by its trend) by 0.69% per year.
- Total number of employees employed within the said period grows with the average rate of trend equal to 1.9×10^5 people per year. At that, the rate of this indicator decreases (by its trend) by 0.14% per year.
- Labour productivity in the said period grows with the average rate of trend equal to 1.9×10^4 tenge/(people×year^2). At that, the rate of this indicator decreases (by its trend) by 0.68% per year.
- Consumer price index in the said period fluctuates within the limits of 6-8% per year along with insignificant drop of the trend of this indicator with the rate of 0.008% per year.
- Real prices of investment products in the considered period drop with the average rate of trend equal to 1.2% per year. At that, the rate of this indicator drops (by its trend) by 0.16% per year.
- Money of households on the banking accounts (in prices of 2000 year) in the said period increase with the average rate of trend equal to 4.3×10^9 tenge per year. At that, the rate of this indicator drops (by its trend) by 0.047% per year.

Table 10. Gross value added of branch

No.	No. of branch	Value of indicator GVA in year 2015 (tenge, in prices of 2000)
1	4	1.45122×10^{12}
2	6	6.65949×10^{11}
3	3	6.57340×10^{11}
4	7	5.87018×10^{11}
5	11	5.44444×10^{11}
6	10	4.38485×10^{11}
7	1	4.25092×10^{11}
8	9	4.12240×10^{11}
9	5	3.05231×10^{11}
10	15	2.09079×10^{11}
11	13	1.99452×10^{11}
12	8	9.18216×10^{10}
13	12	7.56724×10^{10}
14	14	5.51708×10^{10}
15	2	7.43652×10^{9}
16	16	5.15982×10^{9}

Table 11. Fixed assets of branch

No.	No. of branch	Value of indicator "Fixed assets of branch" in year 2015 (tenge, in prices of 2000)
1	3	1.72577×10^{12}
2	11	1.33253×10^{12}
3	9	7.32356×10^{11}
4	12	7.25682×10^{11}
5	4	6.67990×10^{11}
6	6	3.87501×10^{11}
7	5	2.88265×10^{11}
8	7	2.78329×10^{11}
9	1	2.08751×10^{11}
10	10	1.41096×10^{11}
11	13	1.16788×10^{11}
12	14	7.83038×10^{10}
13	15	7.73435×10^{10}
14	8	2.76785×10^{10}
15	2	1.16724×10^{10}
16	16	0

Table 12. Capital productivity of branch

No.	No. of branch	Value of indicator "Capital productivity of branch" in year 2015
1	2	6.371005
2	8	3.317435
3	10	3.107695
4	15	2.703250
5	4	2.172512
6	7	2.109079
7	1	2.036358
8	6	1.718572
9	13	1.707822
10	5	1.058857
11	14	0.704573
12	9	0.562895
13	11	0.408581
14	3	0.380897
15	12	0.104278
16	16	0

Below (see Figures 11, 12) we carry out the analysis of initial data for the perspective macroanalysis of the national economy indicators in the context of the endogenous variable "Investment products output".

Analysis of the trends presented in Figures 11 and 12 allows to conclude that in 2000-2015, just as in the retro-analysis period (Figures 5, 6) one can expect average growth of real output of the investment products in the country with some average deceleration of this growth. However decrease of negative slope of the rate trend (Figure 11) for the period of

2000-2015 years in comparison with the period of retro-analysis (Figure 5) argues on some increase of the rate of growth of investment products by the end of the considered period.

Table 13. Branch demand of labour force

No.	No. of branch	Value of indicator "Branch demand of labour force" in year 2015
1	1	2581608
2	7	1235941
3	13	963120
4	6	841841
5	9	604098
6	4	595420
7	11	516757
8	12	406500
9	14	401955
10	3	247131
11	15	225833
12	5	184281
13	8	150238
14	10	139326
15	2	27907
16	16	24361

Table 14. Labour productivity in branch

No.	No. of branch	Value of indicator "Labour productivity in branch" in year 2015 (in prices of 2000)
1	10	3147177
2	3	2659884
3	4	2437295
4	5	1656329
5	11	1053578
6	15	925811.3
7	6	791063.1
8	9	682406.4
9	8	611171.8
10	7	474956.1
11	2	266466.3
12	16	211806.2
13	13	207089.8
14	12	186155.8
15	1	164661.7
16	14	137256.4

2.5.2. Analysis of Elasticity Coefficients of Endogenous Variables of Computable Model of General Equilibrium of Economy Branches

Within the limits of the retrospective analysis of the economy branches we obtained the estimations of the elasticity coefficients of 18 endogenous variables for each of 16 economy branches, as well as 14 integral indicators of the country economy, public administration, and households in 2009 year over 341 exogenous indicators of the model for 2008-2014 years. These elasticity coefficients form the matrix of size 302×341 with the elements defined by

$$F_{pj}(t) = 100\frac{x_j^n(t)-x_j(t)}{x_j(t)}. \qquad (76)$$

Here p is the varied exogenous parameter; $x_j(t)$ is the value of j-th endogenous for t = 2009 obtained from base computation of the model; $x_j^n(t)$ is the value of the respective endogenous variable obtained from increase of the values of the varied exogenous parameter p in 2000-2008 years; at that, other values of the exogenous parameters remain unchanged in comparison with the base computation.

For example, the computed elasticity coefficient of the Households' budget by the Share of budget of branch 1 spent for paying the labour force appears to be equal to 0.0229. This means that if the share of budget of branch 1 spent for paying the labour force in each of 2000-2008 years increase by 1% in comparison with the base variant, than in 2009 the households' budget increase by 0.0229 in comparison with the base variant.

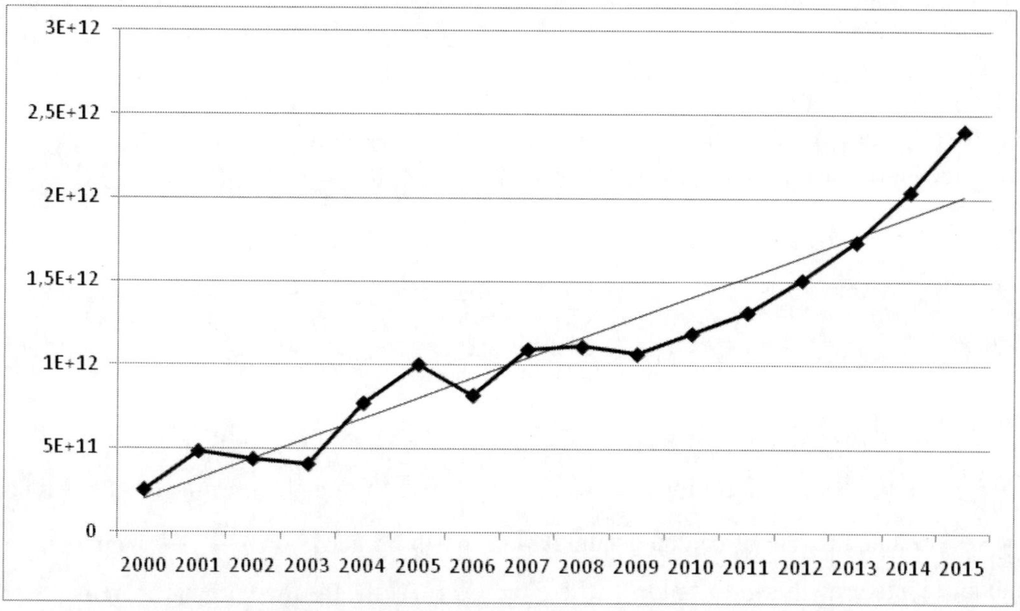

– Computed variable values; – Linear trend of variable

Figure 11. Investment products output (in prices of 2000 year) and its trend.

Within the limits of the perspective analysis of the economy branches we also obtained the similar elasticity coefficients of endogenous indicators in 2015 year by exogenous indicators in 2009 year.

Analysis of influence of the exogenous parameters and, in particular, the parameters *CO_pj_iz* "The share of budget of the i-th branch spent for purchasing the intermediate products produced by the branch j" on the country's GVA shows that

- the mean value of the absolute values of the GVA elasticity coefficients by the shares *CO_pj_iz* appears to be rather high and equal to 4.7450; and
- the mean value of the absolute values of the GVA elasticity coefficients by shares of budgets of other branches appears to be 2.379.

This remark allows recommending the parameters *CO_pj_iz* for use in parametric regulation of the national economy growth. We consider experiments with the parametric regulation below in Section 2.6.

2.5.3. Analysis of Sources of Economic Growth Based on Computable Model of General Equilibrium of Economy Branches

Let us now proceed to the analysis of economic growth sources by the economy branches on the base of the retrospective data for 2000-2015 years. For this purpose, using expressions (20) for the production functions, let us estimate the influence of variation of these functions' arguments on the rate of GVA growth of branches $VY\_\_i[t+1]$ under assumption about constancy of the coefficients of the intermediate products consumed by the branch $CA\_z\_ji[t]$, coefficients of the capital $CA\_k\_i[t]$ and coefficients of labour $CA\_l\_i[t]$. Such assumption is applied in extrapolation of the said functions within the prediction period 2010-2015 years of the model computation.

Taking the logarithm of both parts of (20), finding the total increment of the function $\ln(VY\_\_i)$ and discarding the terms of higher infinitesimal order, we obtain the following estimate of the growth rate y_i of the real GVA of i-th branch in dependence on the grow rates of the production function arguments: CA_r_i, VD_pj_iz, $Ki_m = (VK\_\_i[t] + VK\_\_i[t+1])/2$ and $VD\_pi\_il[t]$.

$$y_i = \frac{\Delta VY\_i}{VY\_i} = \frac{\Delta CA\_r\_i}{CA\_r\_i} + \sum_{j=1}^{16}(CA\_z\_ji \times VD\_pj\_iz)\frac{\Delta VD\_pj\_iz}{VD\_pj\_iz}$$
$$+CA\_k\_i\frac{\Delta Ki_m}{Ki_m} + CA\_l\_i\frac{\Delta VD\_pi\_il}{VD\_pi\_il}. \quad (77)$$

Let $a_i = \frac{\Delta CA\_r\_i}{CA\_r\_i}$ denote the rate of the technological progress in i-th branch; $z_{ij} = \frac{\Delta VD\_pj\_iz}{VD\_pj\_iz}$ denote the rate of the intermediate products of j-th branch consumed in i-th branch; $k_i = \frac{\Delta Ki_m}{Ki_m}$ denote the rate of capital accumulation in i-th branch; $l_i = \frac{\Delta VD\_pi\_il}{VD\_pi\_il}$ denote the rate of labour inputs growth in i-th branch, where the sign "Δ" means the variable increment. The time in (77) is omitted for brevity.

The coefficients in the right-hand side of formula (77) with the rates indicated above and characterize the degree of influence of the considered factors on the economic growth and allow comparing their influence with that of the technological progress with the coefficient equal to 1. Denoting these coefficients by $\alpha_{ij} = CA\_z\_ji \times VD\_pj\_iz$, $\beta_i = CA\_k\_i$, $\gamma_i = CA\_l\_i$, from (77) we obtain its abbreviated notation

$$y_i = a_i + \sum_{j=1}^{16}\alpha_{ij}z_{ij} + \beta_i k_i + \gamma_i l_i. \quad (78)$$

Table 15. Coefficients characterizing influence of factors of economic growth

Number of branch i	β_i	γ_i	α_{i1}	α_{i2}	α_{i3}	α_{i4}	α_{i5}	α_{i6}	α_{i7}
1	0.3089	0.9051	$1.345 \cdot 10^{-12}$	$9.480 \cdot 10^{-02}$	$2.897 \cdot 10^{-14}$	$2.171 \cdot 10^{-13}$	$1.602 \cdot 10^{-14}$	$2.028 \cdot 10^{-14}$	$1.345 \cdot 10^{-12}$
2	0.2426	2.4964	$1.590 \cdot 10^{-16}$	$7.308 \cdot 10^{-01}$	$7.087 \cdot 10^{-16}$	$9.884 \cdot 10^{-15}$	$1.390 \cdot 10^{-15}$	$1.120 \cdot 10^{-15}$	$1.590 \cdot 10^{-16}$
3	0.9650	0.6886	$2.886 \cdot 10^{-15}$	0.000	$2.970 \cdot 10^{-12}$	$8.269 \cdot 10^{-13}$	$9.894 \cdot 10^{-14}$	1.478	$2.886 \cdot 10^{-15}$
4	1.2900	0.0805	$2.227 \cdot 10^{-13}$	$1.343 \cdot 10^{-1}$	$8.634 \cdot 10^{-13}$	$9.989 \cdot 10^{-13}$	$9.029 \cdot 10^{-03}$	$2.641 \cdot 10^{-16}$	$2.227 \cdot 10^{-13}$
5	1.0×10^{-10}	2.4083	$3.199 \cdot 10^{-17}$	$5.537 \cdot 10^{-17}$	$8.913 \cdot 10^{-14}$	$6.720 \cdot 10^{-4}$	$5.406 \cdot 10^{-3}$	$5.300 \cdot 10^{-4}$	$3.199 \cdot 10^{-17}$
6	0.9343	0.7721	$2.078 \cdot 10^{-16}$	$9.186 \cdot 10^{-18}$	$3.313 \cdot 10^{-14}$	$3.940 \cdot 10^{-13}$	$2.467 \cdot 10^{-15}$	$1.324 \cdot 10^{-14}$	$2.078 \cdot 10^{-16}$
7	1.0×10^{-10}	1.8792	$8.133 \cdot 10^{-16}$	$1.061 \cdot 10^{-15}$	$3.115 \cdot 10^{-14}$	$2.660 \cdot 10^{-13}$	$4.124 \cdot 10^{-14}$	$2.508 \cdot 10^{-13}$	$8.133 \cdot 10^{-16}$
8	1.0691	0.4706	$2.003 \cdot 10^{-02}$	$1.498 \cdot 10^{-15}$	$4.343 \cdot 10^{-17}$	$6.171 \cdot 10^{-14}$	$3.375 \cdot 10^{-15}$	$7.288 \cdot 10^{-15}$	$2.003 \cdot 10^{-02}$
9	0.8660	0.2153	$4.289 \cdot 10^{-16}$	2.265	$1.666 \cdot 10^{-13}$	$8.753 \cdot 10^{-15}$	$3.057 \cdot 10^{-14}$	$4.005 \cdot 10^{-14}$	$4.289 \cdot 10^{-16}$
10	0.6702	0.5492	0.000·	0.000·	0.000	$4.397 \cdot 10^{-15}$	$5.602 \cdot 10^{-15}$	$8.811 \cdot 10^{-15}$	0.000
11	1.2022	0.1006	$2.929 \cdot 10^{-05}$	2.267	$5.064 \cdot 10^{-14}$	$1.030 \cdot 10^{-12}$	$1.664 \cdot 10^{-04}$	$5.308 \cdot 10^{-13}$	$2.929 \cdot 10^{-05}$
12	1.0×10^{-10}	2.5822	$5.415 \cdot 10^{-14}$	0.000	0.000	$4.255 \cdot 10^{-13}$	$2.573 \cdot 10^{-14}$	0.000	$5.415 \cdot 10^{-14}$
13	0.2635	1.7177	$2.370 \cdot 10^{-14}$	0.000	0.000	$8.847 \cdot 10^{-13}$	$1.142 \cdot 10^{-13}$	$5.739 \cdot 10^{-15}$	$2.370 \cdot 10^{-14}$
14	0.0227	1.7814	$6.736 \cdot 10^{-14}$	$2.018 \cdot 10^{-1}$	$6.280 \cdot 10^{-15}$	$1.288 \cdot 10^{-12}$	$1.919 \cdot 10^{-13}$	$4.809 \cdot 10^{-13}$	6.736·10-14
15	0.9304	0.2173	$1.041 \cdot 10^{-15}$	$1.408 \cdot 10^{-3}$	$7.681 \cdot 10^{-15}$	$3.712 \cdot 10^{-13}$	$3.926 \cdot 10^{-14}$	$1.579 \cdot 10^{-13}$	$1.041 \cdot 10^{-15}$
16	0	1.9372	0	0	0	0	0	0	0

Table 15. (Continued)

Number of branch i	α_{i8}	α_{i9}	α_{i10}	α_{i11}	α_{i12}	α_{i13}	α_{i14}	α_{i15}	α_{i16}
1	$4.325 \cdot 10^{-15}$	$4.595 \cdot 10^{-2}$	$1.807 \cdot 10^{-14}$	$1.177 \cdot 10^{-14}$	0	$1.681 \cdot 10^{-16}$	$1.178 \cdot 10^{-15}$	$9.304 \cdot 10^{-18}$	0
2	$8.054 \cdot 10^{-17}$	$1.087 \cdot 10^{-14}$	$2.616 \cdot 10^{-17}$	2.280	0	0	0	0	0
3	$4.633 \cdot 10^{-14}$	$2.221 \cdot 10^{-2}$	$1.436 \cdot 10^{-13}$	$4.626 \cdot 10^{-12}$	0	$1.581 \cdot 10^{-15}$	$4.991 \cdot 10^{-4}$	$1.571 \cdot 10^{-14}$	0
4	$8.038 \cdot 10^{-15}$	$1.444 \cdot 10^{-13}$	$4.353 \cdot 10^{-14}$	$6.338 \cdot 10^{-16}$	0	$1.247 \cdot 10^{-16}$	$3.379 \cdot 10^{-3}$	$5.445 \cdot 10^{-16}$	0
5	$1.642 \cdot 10^{-15}$	$2.760 \cdot 10^{-1}$	$6.132 \cdot 10^{-3}$	$2.512 \cdot 10^{-14}$	0	$3.273 \cdot 10^{-5}$	$2.026 \cdot 10^{-17}$	$5.494 \cdot 10^{-4}$	0
6	$7.977 \cdot 10^{-15}$	$1.117 \cdot 10^{-2}$	$6.525 \cdot 10^{-16}$	$2.014 \cdot 10^{-14}$	0	$8.436 \cdot 10^{-17}$	$1.554 \cdot 10^{-17}$	$1.089 \cdot 10^{-15}$	0
7	$6.067 \cdot 10^{-14}$	$4.548 \cdot 10^{-1}$	$3.407 \cdot 10^{-13}$	$1.505 \cdot 10^{-12}$	0	$5.394 \cdot 10^{-16}$	$8.576 \cdot 10^{-5}$	$4.404 \cdot 10^{-16}$	0
8	$1.528 \cdot 10^{-15}$	$8.267 \cdot 10^{-15}$	$3.053 \cdot 10^{-14}$	$1.507 \cdot 10^{-14}$	0	$2.339 \cdot 10^{-17}$	$1.338 \cdot 10^{-17}$	$2.244 \cdot 10^{-16}$	0
9	$3.346 \cdot 10^{-14}$	$1.002 \cdot 10^{-12}$	$3.203 \cdot 10^{-13}$	$4.339 \cdot 10^{-13}$	0	$3.657 \cdot 10^{-15}$	$4.803 \cdot 10^{-16}$	$1.091 \cdot 10^{-15}$	0
10	$1.991 \cdot 10^{-14}$	$3.270 \cdot 10^{-01}$	$2.862 \cdot 10^{-13}$	$5.408 \cdot 10^{-15}$	0	$1.347 \cdot 10^{-14}$	0	$8.805 \cdot 10^{-18}$	0
11	$6.664 \cdot 10^{-14}$	$6.596 \cdot 10^{-13}$	$2.965 \cdot 10^{-13}$	$2.646 \cdot 10^{-12}$	0	$9.260 \cdot 10^{-14}$	$1.661 \cdot 10^{-03}$	$9.599 \cdot 10^{-06}$	0
12	$4.185 \cdot 10^{-1}$	$1.548 \cdot 10^{-1}$	$9.506 \cdot 10^{-13}$	$7.964 \cdot 10^{-3}$	0	$2.126 \cdot 10^{-14}$	$1.408 \cdot 10^{-15}$	$2.886 \cdot 10^{-15}$	0
13	$4.820 \cdot 10^{-14}$	$5.807 \cdot 10^{-1}$	$1.794 \cdot 10^{-14}$	$1.331 \cdot 10^{-13}$	0	$3.939 \cdot 10^{-15}$	$7.738 \cdot 10^{-17}$	$1.835 \cdot 10^{-14}$	0
14	1.891	$2.118 \cdot 10^{-13}$	$4.749 \cdot 10^{-17}$	$2.743 \cdot 10^{-13}$	0	$2.219 \cdot 10^{-16}$	$1.782 \cdot 10^{-14}$	$6.646 \cdot 10^{-14}$	0
15	$6.285 \cdot 10^{-14}$	$2.139 \cdot 10^{-1}$	$9.739 \cdot 10^{-14}$	$8.436 \cdot 10^{-14}$	0	$9.538 \cdot 10^{-17}$	$1.416 \cdot 10^{-17}$	$1.222 \cdot 10^{-13}$	0
16	0	0	0	0	0	0	0	0	0

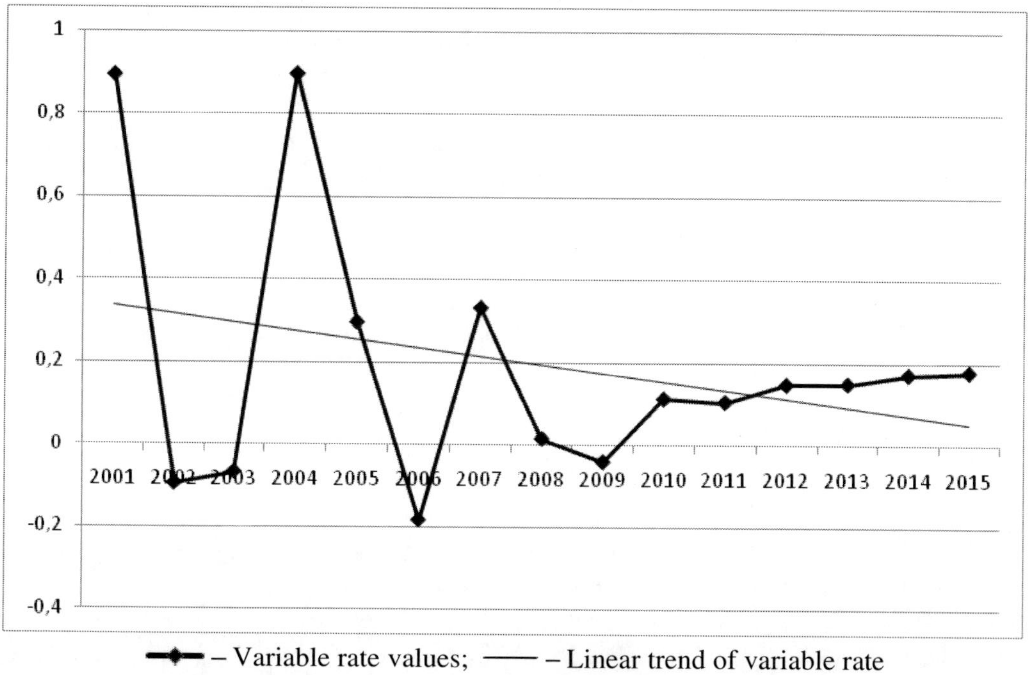

Figure 12. Rate of variable "Investment products output (in prices of 2000 year)" and its trend.

Let us present the values of the coefficients defining the contributions of the sources of economic growth of the branches on the base of the considered model in 2008 year. (See Table 15). The coefficients in the table show the percent of increase of the GVA growth rate as a result of increase of the growth factors (fixed assets, labour or demand of the intermediate products of the economic agents) by 1%.

Analysis of the table of coefficients β_i, γ_i, $\alpha_i^s = {}_{j=1}^{16}\alpha_{ij}$ in Table 14 shows that, excepting the rate of technological progress with equal influence on the rate of growth of all branch is the same in the considered model, the maximum influence on the growth of GVA of economy branches 1, 5, 7, 12, 13, 16 is exerted by the rate of growth of the labour inputs, GVA of branches 4, 6, 8, 10, 15 is exerted by the rate of the capital accumulation, GVA of the rest of the branches 2, 3, 9, 11 14 is exerted by the rate of the intermediate products of other branches consumed by the branch.

It should be also noted that for branches 5, 7, 12, 16 the rates of capital accumulation do not practically influence the respective rate of output growth, rates of labour inputs have nonzero effect on the rates of production growth of all branches. For branches 6, 16, the rates of the consumed intermediate products do not have effect on the respective rate of the output growth.

The results of analysis allow selecting the following share of budgets of 16 economy branches as the instruments for solving the problems of economic growth:

- O_{ij}, the share of budget of i-th branch spent for goods and services purchased in j-th branch;
- O_i^l, the share of budget of i-th branch spent for paying the labour force;
- O_i^n, the share of budget of i-th branch spent for purchasing the investment goods.

2.6. Finding Optimal Parametric Control Laws Based on CGE Model of Economy Branches

In computation experiments with the stochastic computable model of the economy branch, we use the following optimization criterion

$$Kr = \frac{1}{6} \sum_{t=2010}^{2015} VY[t]. \tag{79}$$

Here *Kr* is the mean value of the gross output of the country in the prices of 2000 year in 2010-2015 years.

In experiments with optimization criterion (79) we applied the constraints on the growth of the consumer prices of the following form:

$$VPr[t] \le 1.09VP[t], t = 2010, \dots, 2015.$$

Here $VP[t]$ is the computed level of the consumer prices in the model without parametric control, $VPr[t]$ is the consumer price level with parametric control.

In the computation experiments we realize regulation of 1536 exogenous parameters, shares of budgets of *j*-th producing agents spent for purchasing the goods and services produced by *i*-th producing agents in 2010-2015 years: $O_{ij}[t]$; t = 2010,…,2015; i, j = 1,…,16. Here $\sum_{i=1}^{16} O_{ij}[t] \le 1$ for the mentioned values of *t*. The base values of these share obtained as a result of solving the model parametric identification problem by the data of 2000-2008 years will be denoted by O_{ij}; i, j = 1,…,16.

We considered the following problem of finding optimal values of the adjusted parameters vectors. On the base of the stochastic computable model of the economy branches, find the mentioned values of the shares of the producing agents' budgets $O_{ij}(t)$ which provide the upper bound of the criterion *Kr* with additional constraints on these shares of the following form:

$$0.5 \le O_{ij}(t)/O_{ij} \le 2; i, j = 1, \dots, 16; t = 2010, \dots, 2015.$$

These optimization problems were solved by Nelder-Mead algorithm. After application of the parametric control of the shares of the stochastic model budgets, the value of the criterion appears to be equal to $Kr = 1.2453 \cdot 10^{13}$; its value increases by 25.89% in comparison with the base variant.

2.7. Choice of Economic Instruments by Means of Multiobjective Optimization

In this work we consider the problem of choice of an effective consistent public policy in the sphere of economic growth by means of multiobjective optimization based on two non-autonomous computable models of general equilibrium (CGE models) with common economic instruments of optimization. We consider the conditions of consistent optimization based on the couple of models and results of development of the elements of parametric control theory for the class of discrete-time non-autonomous controlled dynamic systems.

The efficiency of application of a method of joint parametric identification of two models under study, namely, the CGE model of the economy branches and CGE model with the sector of knowledge is shown. Parametric identification of the considered couple of the studied macroeconomic mathematical models consists in finding estimations of unknown parameters (values of the exogenous functions) in some given bounded regions, for which the minimal value of some objective function is attained.

We find an estimation of the Pareto set of the considered two-objective problem and its solution on the base of an optimization problem with respect to one utility function.

2.7.1. Choice and Joint Parametric Identification of Two CGE Models

The following requirements lie in the base of the choice of two or more mathematical models for solving the problems of consistent optimization:

1. Acceptable accuracy of the model description of one object region, in which the problem of consistent optimization is solved with the criteria $(K_1, K_2, ...)$ corresponding to some chosen goals of economic policy, as well as ability to obtain these criteria.
2. Fulfilment of conditions of statement and solution of the multiobjective optimization problem on the base of the considered models.
3. Efficient realization of solution of the consistent optimization problem.

In the context of the first requirement, let us consider ability of choice of two CGE models for the consistent optimization in the field of economic growth. The said ability is assessed by solving the joint parametric identification problem for the following CGE models [11].

Model 1: CGE model of the economy branch. The detailed description of this model is given above.

Model 2: CGE model with sector of knowledge is represented by the following economic agents (sectors): sector of science and education (knowledge); innovative sector represented by the set of innovationally active enterprises and organizations; sector of other economy branches; aggregated consumer joining the households; government; banking sector. Here first three economic sectors are the producing agents. The model with sector of knowledge consists of 110 equations, which are used for computations of the values of its 110 endogenous variables. This model also includes 86 estimated exogenous functions. Among them, 9 functions are common for both considered models.

The problem of parametric identification of the considered pair of the macroeconomic mathematical models under study consists in finding estimates of unknown values of their parameters in given region, which provide the minimal value of the objective function that chatacterizes:

- deviation of the model output variables from the respective observed values (known statistical data);
- divergence between the values of endogenous variables of two models with the same meaning;

and under the additional constraint of coincidence of the respective values s_0 of the estimated parameters common for both models.

This problem leads to finding minimum of a function of several variables (parameters) in some closed region D of the Euclidian space with the constraints (2) imposed to the values of endogenous parameters. In the case of high dimensionality of the region of possible values of the parameters to be found, standard methods for finding extremums of a function are often ineffective because of presence of several local minimums of the objective functions. Below we propose an algorithm that takes into account the particular features of the problem of joint parametric identification of two macroeconomic models and allows avoiding the problem of "local extremums."

The region $D = {}_{j=1}^{q+s_1+s_2-s_0+m_{1,1}+m_{2,1}} \; a_j, b_j$, where a_j, b_j is the interval of possible values of the parameter p_j, $j = 1, \dots, (q + s_1 + s_2 - s_0)(t_1 - t_2) + m_{1,1} + m_{2,1})$, is considered as the region in the set ${}_{t=t_1}^{t_2}(U(t) \times A_1(t) \times A_2(t)) \times X_{1,1}(t_1) \times X_{2,1}(t_1)$ (here the first subscript of the set X, subscripts of the sets A and variables s, m corresponds to a number of model, s_0 is a number of uncontrolled functions common for both models). At that, the estimations of the parameters with available observed values are searched within small intervals a_j, b_j with centers at the respective observed values (in the case of one such value) or within some interval that cover the observed values (in the case of several such values). Other intervals a_j, b_j for searching the parameters are chosen by means of indirect estimates of their possible values. To find the minimum values of the continuous function in several variables $D: \to R$ with additional constraints (2) on the endogenous variables we apply the Nelder-Mead algorithm of directional search [13] in the computational experiments.

To solve the problem of parametric identification of the considered models on the base of obvious assumption about noncoincidence (in general case) of the minimum points of two different functions, we propose two criteria of the following type:

$$K_A\ p = \overline{\frac{1}{n_\alpha(t_2-t_1+1)} \; {}_{t=t_1}^{t_2} \; {}_{i=1}^{2} \; {}_{j=1}^{n_{i,A}} \alpha_{i,j} \; \frac{y_i^j[t]-y_i^{j*}[t]}{y_i^{j*}[t]}^{2} + {}_{j=1}^{n_{u,A}} \alpha_{u,j} \; \frac{y_1^j[t]-y_2^j[t]}{y_2^j[t]}^{2}} \; ,$$

$$K_B\ p = \overline{\frac{1}{n_\beta(t_2-t_1+1)} \; {}_{t=t_1}^{t_2} \; {}_{i=1}^{2} \; {}_{j=1}^{n_{i,B}} \beta_{i,j} \; \frac{y_i^j[t]-y_i^{j*}[t]}{y_i^{j*}[t]}^{2} + {}_{j=1}^{n_{u,B}} \beta_{u,j} \; \frac{y_1^j[t]-y_2^j[t]}{y_2^j[t]}^{2}} \qquad (80)$$

Here $\{t_1, \dots, t_2\}$ is the identification time interval; $y_i^j[t]$, $y_i^{j*}[t]$ are computed and measured values of the model output variables; $K_A\ p$ is the auxiliary criterion; $K_B\ p$ is the main criterion; $n_{i,B}$ is the number of observed endogenous variables of i-th model, $n_{i,B} > n_{i,A}$; $n_{u,B}$ is the number of endogenous variables of two models that have similar meaning, $n_{u,B} > n_{u,A}$; $\alpha_{i,j} > 0$, $\alpha_{u,j} > 0$, $\beta_{i,j} > 0$and $\beta_{u,j} > 0$ are some weighting coefficients whose values are determined while solving the problem of parametric identification of the dynamic system; ${}_{i=1}^{2} \; {}_{j=1}^{n_{i,A}} \alpha_{i,j} + {}_{j=1}^{n_{u,A}} \alpha_{u,j} = n_\alpha$, ${}_{i=1}^{2} \; {}_{j=1}^{n_{i,B}} \beta_{i,j} + {}_{j=1}^{n_{u,B}} \beta_{u,j} = n_\beta$.

As a result of joint solution of problems A and B according to the described algorithm (Section 2.3) on the base of statistical data of the Kazakhstan Republic for 2000-2008 years with application of the Nelder-Mead algorithm [13] we obtain the values $K_A = 0.044$ and $K_B = 0.026$. At that, the relative value of deviations of the computed values of the variables used in the main criterion from the respective measured values, as well as deviations of the values of the variables (common for two models) of model 1 from the respective values of the variables of model 2 constitutes less than 0.26%.

The results of computation and retrospective prediction of the model for 2008 year partially presented in Table 16 demonstrate the computed (VY_1, VYg_i, VP_i), measured values, and deviations of the computed values of the main output variables of the model from the respective measured values. Here the time interval 2000-2007 years corresponds to the period of the model parametric identification; 2008 year is the retrospective prediction period; VY_1 is the gross output ($\times 10^{12}$ tenge in prices of 2000 myear, tenge is the monetary unit of the Kazakhstan Republic; model 2 does not have an analogue of the variable VY_1); VYg_i is GDP ($\times 10^{12}$ tenge in prices of 2000 year); VP_i is the Consumer Price Index in percents with respect to a preceding year; sign "*" corresponds to the measured values; sign "Δ" corresponds to the deviations (in percents) of the computed values from the respective measured values.

Table 16. Measured, computed values of the model output variables and respective deviations

Variable/Year	2000	2001	2002	2003	2004	2005	2006	2007	2008
VY_1^*	5.44	6.32	6.47	6.86	7.72	8.52	9.25	9.69	9.84
VY_1	5.36	6.28	6.41	6.80	7.68	8.49	9.23	9.55	9.66
ΔVY_1	-1.51	-0.55	-0.9	-0.87	-0.45	-0.33	-0.26	-1.4	-1.89
VYg_i^*	2.45	2.78	3.05	3.36	3.72	4.09	4.55	5.01	5.18
VYg_1	2.45	2.76	3.03	3.33	3.72	4.09	4.55	5.01	5.22
VYg_2	2.43	2.80	3.05	3.35	3.74	4.07	4.57	5.05	5.21
ΔVYg_1	-0.13	-0.69	-0.71	-0.72	-0.13	0.01	0.03	-0.07	0.86
ΔYVg_2	-0.70	0.75	0.08	-0.38	0.51	-0.39	0.46	0.85	0.53
VP_i^*		106.4	106.6	106.8	106.7	107.5	108.4	118.8	109.5
VP_1		107.8	108.8	108.4	107.5	107.2	108.1	117.4	109.3
VP_2		104.6	105.8	106.4	105.9	107.2	108.1	118.5	109.3
ΔVP_1		1.35	2.11	1.52	0.83	-0.24	-0.25	-0.30	-0.16
ΔP_2		-1.73	-0.77	-0.36	-0.72	-0.31	-0.32	-0.29	-0.15

2.7.2. Consistent Optimization Based on Two CGE Models

In the context of realization of the requirement 2 and 3 stated above (Section 2.7.1), in this work we consider multiobjective choice of economic instruments in the field of economic growth and innovation sphere with the two-dimensional criterion $K = (K_1, K_2)$. The components of the criterion K characterizes the mean values of the real GDP (K_1) of the first model and mean values of the real GVA (K_2) of the second model, respectively.

We consider the problem of two-criteria optimization (81)-(83):

$$\max_O K\ O\ . \tag{81}$$

Here $O = \ O_j^k[t]: j = 1, \ldots, 16;\ k = 1, 2, 3;\ t = 2011, \ldots, 2015$ is the set of common regulated parameters; $O_j^k[t]$ are the additional investments in year *t* from financing *j*-th sector of the first model within the limits of financing *k*-th sector of the second model (or, equivalently, additional investments in year *t* for financing *k*-th sector of the second models within the limits of financing *j*-th sector of the first model). The constraints of form (4) on the regulated parameters and conditions of their realization on the base of two models have the following form:

$$O_j^k[t] \geq 0;\quad {}_{t,j,k}\, O_j^k[t] \leq 11.5 \times 10^{12};\ O_j[t] = \ {}_k\, O_j^k[t];\ O^k[t] = \ {}_j\, O_j^k[t]. \quad (82)$$

Here 11.5×10^{12} is the total volume of investments in tenge for the period of 2011-2015 years; $O_j[t]$ are the additional investments in year *t* for financing *j*-th producing sector of the first model; $O^k[t]$ are the additional investments in year *t* for financing *k*-th producing sector of the second model.

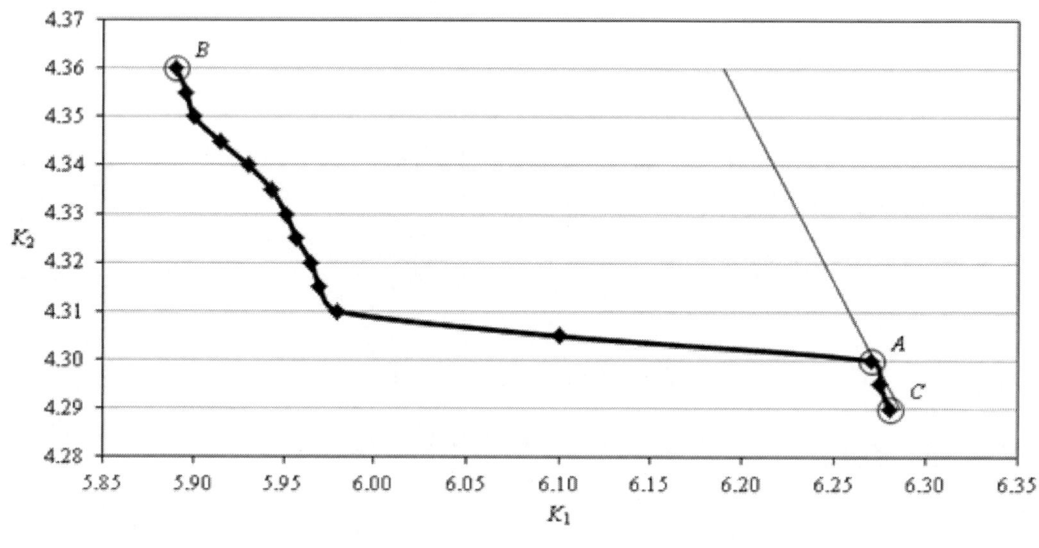

Figure 13. Estimate of the Pareto set of Problem 1.

These additional investments obviously satisfy the equality $\ {}_k\, O^k[t] = \ {}_j\, O_j[t] = \ {}_{j,k}\, O_j^k[t]$, which ensures equal values of the additional investments into the economy of the country for each year of regulation *t* in the context of each of two CGE models.

The constraints of form (2) on the vectors of endogenous variables $y_1[t]$ and $y_2[t]$ of the first and second models, respectively, as well as *l*-th coordinates of these vector that have similar meaning are given by

$$y_1[t] \in Y_1\ t\ , y_2[t] \in Y_2(t),\ \ y_1^l[t] - y_2^l[t]\ \leq \varepsilon^l. \quad (83)$$

Here $Y_1(t)$ и $Y_2(t)$ are the given sets; ε^l are some given small numbers, $l = 1, \ldots, 18$.

To formulate two-criteria problem with application of Nelder-Mead algorithm, we constructed the estimate of the Pareto set (see Figure 13) in form of the set of points, which are consecutively connected by the continuous line at the plane (K_1, K_2). The units of measurement of the criteria at the figures: K_1 - 10^{12} tenge, K_2 - 10^{10} tenge in prices of 2000 year.

Let the preferences of a person making decisions (PMD) in the context of the two-criteria (81)-(83) on the base of the considered models be defined via the linear utility function

$$U\ K_1, K_2 \ = \ {}_{i=1}^{2} W_i\, K_i/K_i^0 \rightarrow \max \qquad (84)$$

Here $W_i > 0$ are the weights, which are unknown apriori, but it is assumed that ${}_{i=1}^{2} W_i = 1$.

In the context of finding the values of the weights W_i, PMD communicates not the values of weights themselves, but the index of relative importance $a_{1,2}$ for the pair of criteria (K_1, K_2). In this case PMD determines the number $a_{1,2} = \frac{W_1}{W_2}$. Adding the condition $W_1 + W_2 = 1$, we obtain the system of two equations with two unknown variables that gives the values of weights:

$$W_1 = \frac{1}{1+a_{1,2}};\ W_2 = 1 - W_1. \qquad (85)$$

Let us formulate the following problem of optimization of the utility function on the base of two objective functionals of the two-criteria optimization problem of synthesis of optimal law of parametric control.

Problem 1

On the base of two estimated CGE models (model of the economy branches and model with the sector of knowledge) and for given indicator of relative importance $a_{1,2}$ for the pair of criteria (K_1, K_2), find such set of additional investments $O^k[t]$ and $O_j[t]$ to the national economy, which satisfy the conditions (82), that the respective solutions of the considered CGE models satisfy the conditions (83) for the given values of time and maximize the utility function (84).

Coordination of the objectives on the base of the formulated two-criteria optimization problem can be done in the following way.

By means of interaction with PMD during the process of solving the formulated two-criteria problem, the optimal (in sense of the criterion (84) with the values of weights (85)) point $A(K_1^m, K_2^m) \in \Pi$ andthe respective values of shares of the additional investments $O_j^k[t]$ are determined.

As an example, we mark point $A(6.27, 4.30) \in \Pi$ at Figure 13. This point represents the solution of second stage of the considered optimization problem for the case when $a_{1,2} = 1$, $(W_1 = W_2 = 0.5)$. The inclined line that passes the point A is the equipotential line of the utility function (84) for the considered case corresponding to the maximum value of this function in the set Π.

Analysis of the estimate Π of the Pareto set presented at Figure 13 shows that for different values of the index $a_{1,2}$ chosen by PMD only the following situations are possible:

1. If $a_{1,2} < 0.217$, then the point $B(5.89, 4.36) \in \Pi$ corresponds to the solution of Problem 1.
2. If $a_{1,2} \approx 0.217$, then Problem 1 has solutions corresponding to the points A and B.
3. If $0.217 < a_{1,2} < 1.373$, then the point A corresponds to the solution of Problem1.
4. If $a_{1,2} \approx 1.373$, then Problem 1 has two solutions that corresponds to the points A and $C(6.28, 4.29) \in \Pi$.
5. If $a_{1,2} > 1.373$, then the point C corresponds to the solution of Problem 1.

3. Macroeconomic Analysis and Parametric Control of Economic Growth Processes Based on Dynamic Stochastic General Equilibrium Models

3.1. Linear DSGE Model of Kazakhstan Republic Economy

The nonlinear DSGE model was proposed in [12] on the base of given composition and behaviour of the agents, their interaction in the stochastic conditions and accepting of the principal of rational expectations. This nonlinear DSGE model of the economy consists of the following:

- aggregated first-order conditions of optimization problems [15] of the agents (households and producers of intermediate goods),
- descriptions of rules for activity of the State and
- rules for defining shocks as first-order autoregressions or Gaussian white noises.

The aggregated first-order conditions include the conditions of equilibrium at the markets of labour, capital, intermediate and final goods. The rule of Taylor [16] including the variables of potential GDP Y_t^* is used as the rule of State activity for determining the stock yield. Here and after, the sign "*" marks the endogenous variables that correspond to the potential economy. The derived nonlinear DSGE model includes the both real and potential economy models. The model of the potential economy in its content is similar to the model of real economy except for the potential economy works with flexible prices and wages (in the model of real economy these prices are not flexible), as well as in absence of "mark-up" shocks (which presented in the model of real economy).

The considered nonlinear DSGE model in vector form is given by

$$\mathrm{E}_t F^{\theta}\ X_{t-1}, X_t, X_{t+1}, H_t^{\Sigma_H}\ = 0. \tag{86}$$

Here E_t denotes for conditional expectation for information available in the moment t ($t = 1,2,\ldots$); F^{θ} is the known vector function; θ is the set of parameters consisting of structural model parameters and parameters of shock autoregressions; X_t is the vector consisying of endogenous variables and shocks given by means of first-order autoregressions; X_0 is given; $H_t^{\Sigma_H}$ is the vector that consists of Gaussian white noises; Σ_H is the respective diagonal covariance matrix.

According to methods adopted for DSGE model [12], we derive the linear approximation of the nonlinear DSGE model (86) in the neighbourhood of its stationary point X. The said point is found by solving the vector equation (87) derived from (86) by truncation of time indexes and nullification of white noises:

$$F^{\theta}\ X, X, X, 0\ = 0\,. \tag{87}$$

In the derived linear model, the behaviour of the sector of households is described by the following equations.

a) The equation of consumption in the real economy:

$$C_t = \frac{h}{1+h} C_{t-1} + \frac{1}{1+h} \mathrm{E}_t C_{t+1} - \frac{1-h}{1+h\ \sigma_C}\ R_t - \mathrm{E}_t \Pi_{t+1}\ + \frac{1-h}{1+h\ \sigma_C}\ \varepsilon_t^B - \mathrm{E}_t \varepsilon_{t+1}^B\ . \tag{88}$$

The equation of consumption in the potential economy:

$$C_t^* = \frac{h}{1+h} C_{t-1}^* + \frac{1}{1+h} \mathrm{E}_t C_{t+1}^* - \frac{1-h}{1+h\ \sigma_C}\ R_t^* - \mathrm{E}_t \pi_{t+1}^*\ + \frac{1-h}{1+h\ \sigma_C}\ \varepsilon_t^B - \mathrm{E}_t \varepsilon_{t+1}^B\ . \tag{89}$$

Here C_t is the consumption; R_t is the yield of state bonds; Π_t is the inflation; $h$is the coefficient of habits; σ_C is the value inverse to the elasticity of intertemporal substitution of consumption of households; ε_t^B is the shock of preferences.

b) The equation of investments in the real economy:

$$I_t = \frac{1}{1+\beta} I_{t-1} + \frac{\beta}{1+\beta} \mathrm{E}_t I_{t+1} + \frac{1\ S''(1)}{1+\beta} Q_t - \frac{\beta \mathrm{E}_t \varepsilon_{t+1}^I - \varepsilon_t^I}{1+\beta}, \tag{90}$$

and equation of investments in the potential economy

$$I_t^* = \frac{1}{1+\beta} I_{t-1}^* + \frac{\beta}{1+\beta} \mathrm{E}_t I_{t+1}^* + \frac{1\ S''(1)}{1+\beta} Q_t^* - \frac{\beta \mathrm{E}_t \varepsilon_{t+1}^I - \varepsilon_t^I}{1+\beta}. \tag{91}$$

Here I_t denotes investments; Q_t is the capital price; β is the discount factor, ε_t^I is the investment shock; $S(.)$ is the function of expenses during accumulation of capital (positive, convex and satisfies the conditions $S\ 1\ = S'\ 1\ = 0$ and $S''\ 1\ > 0$).

c) The equation of the capital price in the real economy:

$$Q_t = -\ R_t - \mathrm{E}_t \Pi_{t+1}\ + \frac{1-\tau}{1-\tau+R^K} \mathrm{E}_t Q_{t+1} + \frac{R^K}{1-\tau+R^K} \mathrm{E}_t R_{t+1}^K + \eta_t^Q, \tag{92}$$

The equation of the capital price in the potential economy:

$$Q_t^* = -R_t^* + \frac{1-\tau}{1-\tau+R^K} \mathrm{E}_t Q_{t+1}^* + \frac{R^K}{1-\tau+R^K} \mathrm{E}_t R_{t+1}^{K\,*}, \tag{93}$$

Here R_t^K is the rate of the capital lease; τis the rate of amortization, $R^K = 1 \ \beta + \tau - 1$ is the balanced rate of the capital lease; η_t^Q is the shock of mark-up on the capital.

d) The equation of accumulation of the capital in the real economy:

$$K_t = (1-\tau)K_{t-1} + \tau I_{t-1}, \tag{94}$$

The equation of accumulation of the potential capital:

$$K_t^* = \ 1-\tau \ K_{t-1}^* + \tau I_{t-1}^*. \tag{95}$$

Here K_t is the accumulated volume of capital.

e) The equation of wages in the real economy:

$$W_t = \frac{\beta}{1+\beta} \mathrm{E}_t W_{t+1} + \frac{1}{1+\beta} W_{t-1} + \frac{\beta}{1+\beta} \mathrm{E}_t \Pi_{t+1} - \frac{1+\beta\gamma_W}{1+\beta} \Pi_t + \frac{\gamma_W}{1+\beta} \Pi_{t-1} - \Omega \ W_t - \sigma_L L_t - \frac{\sigma_C}{1-h} \ C_t - hC_{t-1} \ - \varepsilon_t^L - \eta_t^W \ , \tag{96}$$

where

$$\Omega = \frac{1}{1+\beta} \frac{1-\beta\xi_W \ 1-\xi_W}{1+\frac{1+\lambda_W \ \sigma_L}{\lambda_W} \ \xi_W}$$

In the potential economy there is the following equation for wages:

$$W_t^* = \sigma_L L_t^* + \frac{\sigma_C}{1-h} \ C_t^* - hC_{t-1}^* \ + \varepsilon_t^L. \tag{97}$$

Here W_t is the real wage; L_t is the quantity of labour; ξ_W, γ_W are the probability of change of the wage by household and level of indexation of wage according to the Calvo model, σ_L is the inverse value of elasticity of the labour quantity relative to the wage; ε_t^L is the shock of the labour offer, η_t^W is the shock of mark-up on the wage.

f) The balance equation of the sector of households in the real economy:

$$Y_t = C_Y C_t + I_Y I_t + (1 - I_Y - C_Y) G_t, \tag{98}$$

and the respective equation in the potential economy:

$$Y_t^* = C_Y C_t^* + I_Y I_t^* + \ 1 - I_Y - C_Y \ G_t^*. \tag{99}$$

Here C_Y, I_Y are the equilibrium share of consumption and investment expenses in GDP; Y_t isGDP; G_t is the public expenses.

The behaviours of the producing sectors of intermediate and final products is described by the following equations:

g) The equation corresponding to the condition of minimization of expenses in the real economy:

$$L_t = -W_t + \ 1+\psi' \ 1 \quad \psi"(1) \ R_t^K + K_{t-1}. \tag{100}$$

The similar equation in the potential economy is given by

$$L_t^* = -W_t^* + \ 1+\psi' \ 1 \quad \psi"(1) \ R_t^{K\,*} + K_{t-1}^*. \tag{101}$$

Here $\psi \ .$ is the function of additional charges in the case of capital leasing (convex increasing function, $\psi \ 1 \ = 0$).

h) The productions function of the sector of intermediate products in the real economy:

$$Y_t = \phi \ \ \varepsilon_t^A + \alpha K_{t-1} + \frac{\psi' \ 1 \ \alpha R_t^K}{\psi"(1)} + \ 1-\alpha \ L_t \ . \tag{102}$$

The respective productions function in the potential economy:

$$Y_t^* = \phi(\varepsilon_t^A + \alpha K_{t-1}^* + \frac{\psi' \ 1 \ \alpha R_t^{K*}}{\psi"(1)} + \ 1-\alpha \ L_t^*). \tag{103}$$

Here $\phi = (Y + \Phi) \ Y$; α, Φ are the parameters of elasticity of production by capital and fixed charges in the formula of Kobb-Douglas; Y is the equilibrium GDP; ε_t^A is the shock of production.

i. The equation of inflation in the real economy

$$\Pi_t = \frac{\beta}{1+\beta\gamma_P} \mathrm{E}_t \Pi_{t+1} + \frac{\gamma_P}{1+\beta\gamma_P} \Pi_{t-1} + \frac{1}{1+\beta\gamma_P} \frac{(1-\beta\xi_P)(1-\xi_P)}{\xi_P} \ \alpha R_t^K + \ 1-\alpha \ W_t - \varepsilon_t^A + \eta_t^P \ . \tag{104}$$

Here ξ_P, γ_P are the probability of the price change and level of indexation according to Calvo model; η_t^P is the shock of over charge of products.

It is assumed that the expenses are constant in the potential economy:

$$\Pi_t^* = 0, \alpha R_t^K + \ 1-\alpha \ W_t - \varepsilon_t^A = 0. \tag{105}$$

j) The equation of employment in the real economy:

$$E_t = \beta \mathrm{E}_t E_{t+1} + \frac{1-\beta\gamma_E \ \ 1-\xi_E}{\xi_E} (L_t - E_t) \tag{106}$$

and the equation of employment in the potential economy:

$$E_t^* = \beta E_t E_{t+1}^* + \frac{1-\beta\gamma_E \ \ 1-\xi_E}{\xi_E} \ L_t^* - E_t^* \ . \tag{107}$$

Here E_t is the employment; ξ_E is the probability of the employment change according to Calvo model.

The behaviour of the State sector is described as follows:

k) The rule of Taylor defining the state bond yield in the real economy:

$$R_t = \rho R_{t-1} + \ 1-\rho \ \ \pi_t + r_\pi \ \Pi_{t-1} - \pi_t \ + r_Y \ Y_t - Y_t^* \ \ + r_{\Delta\pi} \ \Pi_t - \Pi_{t-1} \ + r_{\Delta Y} \ \ Y_t - Y_t^* - \ \ Y_{t-1} - Y_{t-1}^* \ \ + \eta_t^R. \tag{108}$$

Here $\rho, r_\pi, r_Y, r_{\Delta\pi}, r_{\Delta Y}$ are the parameters; π_t is the inflation shock, η_t^R is the interest rate shock.

The state bonds yield in the potential economy is defined from the equilibrium with constant inflation.

l) The rules for the public expenses are given as the first-order autoregression:

$$G_t = \rho_G G_{t-1} + \eta_t^G, \tag{109}$$

$$G_t^* = \rho_G G_{t-1}^* + \eta_t^G. \tag{110}$$

Here η_t^G is the public expenses shock.

The following shocks of the model (the shock of preferences, shock of labour offer, shock of investments, shock of productivity and shock of inflation) are given by the first-order autoregressions:

$$\varepsilon_t^B = \rho_B \varepsilon_{t-1}^B + \eta_t^B, \tag{111}$$

$$\varepsilon_t^L = \rho_L \varepsilon_{t-1}^L + \eta_t^L, \tag{112}$$

$$\varepsilon_t^I = \rho_I \varepsilon_{t-1}^I + \eta_t^I, \tag{113}$$

$$\varepsilon_t^A = \rho_B \varepsilon_{t-1}^A + \eta_t^A, \tag{114}$$

$$\pi_t = \rho_\pi \pi_{t-1} + \eta_t^\pi. \tag{115}$$

Here $\eta_t^B, \eta_t^L, \eta_t^I, \eta_t^A, \eta_t^\pi$ are the Gaussian white noises.

The shocks η_t^R (shock of interest rate), η_t^G (shock of public expenses), η_t^P (shock of mark-up for products), η_t^W (shock of mark-up to wages), η_t^Q (shock of mark-up to capital) are given as the Gaussian white noises.

The derived linear model of the Kazakhstan Republic economy (88) - (115) in matrix form is given by

$$A^\theta \begin{pmatrix} X_{t-1} \\ R_{t-1} \\ G_{t-1} \end{pmatrix} + B^\theta \begin{pmatrix} X_t \\ R_t \\ G_t \end{pmatrix} + C^\theta \mathrm{E}_t \begin{pmatrix} X_{t+1} \\ R_{t+1} \\ G_{t+1} \end{pmatrix} + D^\theta H_t^{\Sigma_H} = 0 \tag{116}$$

Here $X_t = (C_t, I_t, Y_t, K_t, Q_t, \Pi_t, W_t, L_t, E_t, R_{t+1}^K, C_t^*, I_t^*, Y_t^*, K_t^*, Q_t^*, \Pi_t^*, W_t^*, L_t^*, E_t^*, R_t^{K*}, R_t^*, G_t^*, \varepsilon_t^B, \varepsilon_t^L, \varepsilon_t^I, \varepsilon_t^A, \pi_t)^T$

is the column vector consisting of all endogenous variables of the model (including the shocks given by autoregressions) except for the instruments of the state policy of the bond yield R_t and the volume of public expenses G_t; X_0, R_0, G_0 are given; $H_t^{\Sigma_H} = (\eta_t^A, \eta_t^B, \eta_t^G, \eta_t^L, \eta_t^I, \eta_t^R, \eta_t^\pi, \eta_t^Q, \eta_t^P, \eta_t^W)^T$ is the column vector consisting of the Gaussian white noises;

$$\Sigma_H = \{st.d.\eta_t^A, st.d.\eta_t^B, st.d.\eta_t^G, st.d.\eta_t^L, st.d.\eta_t^I, st.d.\eta_t^R,, st.d.\eta_t^\pi, st.d.\eta_t^Q, st.d.\eta_t^P, st.d.\eta_t^W\}$$

is the set consisting of the standard deviations of the white noises $H_t^{\Sigma_H}$; $A^\theta, B^\theta, C^\theta, D^\theta$ are the matrices with appropriate dimensions; the set consisting of the structural parameters of the model

$$\{\beta, \alpha, C_Y, I_Y, \tau, \lambda_W, R^K, S'', \sigma_C, h, \xi_W, \sigma_L, \xi_P, \xi_E, \gamma_W, \gamma_P, \psi'' \ \psi', \phi, r_\pi, r_{\Delta\pi}, \rho, r_Y, r_{\Delta Y}\}$$

and parameters of the autoregression shocks $\rho_A, \rho_B, \rho_G, \rho_L, \rho_I, \rho_\pi$.

3.2. Estimation of Parameters of Linear DSGE Model Based on Statistics of Kazakhstan Republic Economy

The solution of the model (116) was derived by application of Blanchard-Kahn algorithm [17], [18]. The solution is represented by the vector first-order autoregression:

$$\begin{pmatrix} X_t \\ R_t \\ G_t \end{pmatrix} = Q^\theta \begin{pmatrix} X_{t-1} \\ R_{t-1} \\ G_{t-1} \end{pmatrix} + F^\theta H_t^{\Sigma_H}, t = 1,2,3, \ldots \tag{117}$$

Here X_0, R_0, G_0 are given; Q^θ, F^θ are the matrices of appropriate dimensions.

Similar to approach of [12], the values of the parameters $\beta, \alpha, C_Y, I_Y, \tau, \lambda_W, R^K$ of the model were defined resulted from specificity of the Kazakhstan Republic economy for 2002-2011 years [19]. From the condition that the mean quarterly rate of the public bonds has been about 2.00%, we choose $\beta = 0.9800$; $\alpha = 0.3125$ is the mean share of capital in production, $C_Y = 0.5700$ is the mean share of consumption in GDP structure; $I_Y = 0.2600$ is the mean share of investments in GDP structure; $\tau = 0.0250$ is the amortization rate and $\lambda_W = 0.5000$ is the mark-up on the wage, $R^K = 1\ \beta + \tau - 1 = 0.0454$ is the mean rate of the capital lease.

To estimate the rest of the parameters ($\theta' = \theta \backslash \{\beta, \alpha, C_Y, I_Y, \tau, \lambda_W, R^K\}$ and Σ_H) of the model, we apply the Bayesian approach [20] where the vector $S_t = (C_t, I_t, Y_t, \pi_t, W_t, E_t, R_t)^T$

composed of the said coordinates of the vector X_t, R_t, G_t ^{T} was used as the vector of observed variables.

To apply the Kalman filter in the context of the Bayesian approach, the model (117) was extended by the vector equation of measurement

$$S_t = M \begin{matrix} X_t \\ R_t \\ G_t \end{matrix} . \qquad (118)$$

Here M is the matrix, where each raw contains one unit, all other elements equal to zero.

The logarithms of deviations from the respective linear trend of the values of macroeconomic indicators (Consumption, Investments, GDP, Inflation, average wage, Employment, Rate of refinancing) corresponding to the observed variables were accepted as the results of measurements of the observed variables S_t. For this purpose, we use the statistical data of the Kazakhstan Republic from the first quarter of 2002 year to the third quarter of 2011 year [19].

To apply the Bayesian approach, we define the apriori density of distribution $p = p_0(\theta', \Sigma_H)$ of the parameters θ', Σ_H. In this work we use the form and probabilistic characteristics from [12] except for the mathematical expectations of the apriori distributions of the parameters Σ_H. These mathematical expectations were enlarged by 2.5 times in comparison with the respective values from [12] in connection with the higher selected standard deviations of economic indicators of Kazakhstan in comparison with the zone of Euro.

In accordance with the method of Bayesian approach [20] we use the likelihood function derived on the base of the model (117), (118) with application of the Kalman filter as well as apriori distribution of the parameters $p_0(\theta', \Sigma_H)$ to find a posteriori joint density of distribution of the estimates of the parameters $p = p_1(\theta', \Sigma_H)$ to be found. Then by means of the algorithm of Metropolis - Hastings with use of the density $p_1(\theta', \Sigma_H)$ we generate the sample consisting of 4000000 sets of the parameters θ', Σ_H. Finally, we accepted the respective sample meanings (see Table 17) as the estimates of the parameters to be found.

The quality of the applied method for searching the estimates of the parameters was verified via retrospective prediction. For this purpose, we compute the predictions of the said observed economic indicators for 4 periods from 4-th quarter of 2010 year to 3-rd quarter of 2011 year. The mean square deviations of the obtained expected prediction values of the economic indicators from the respective statistical data equal about 3%.

3.3. Macroeconomic Analysis of Effect of Shocks on Economic Indicators of Kazakhstan Republic

In this section we give some results of the macroeconomic analysis of the results of parametric identification of the model, influence of internal shocks of economy onto the dynamics of GDP and inflation on the base of the computational experiments carried out in the context of Smets and Wouters DSGE model estimated by the statistical data of the Kazakhstan Republic.

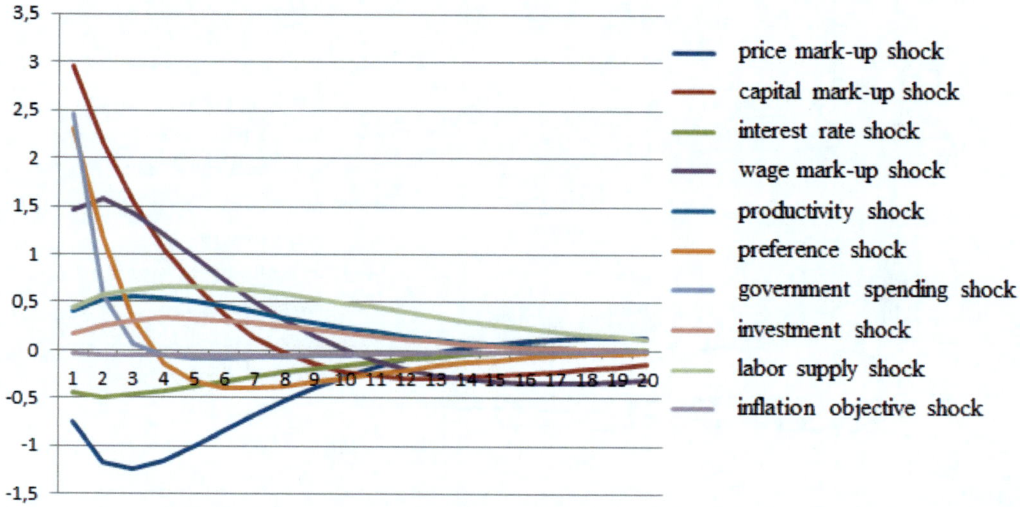

Figure 14. Impulse response of real GDP on unity positive shocks.

Table 17. Estimates of parameters of DSGE model of Kazakhstan Republic

Parameter	Estimated value	Parameter	Estimated value
S''	3.9297	ρ_A	0.4044
σ_C	0.8738	ρ_B	0.4001
h	0.2558	ρ_G	0.2975
ξ_W	0.9256	ρ_L	0.7262
σ_L	1.8452	ρ_I	0.6311
ξ_P	0.8670	ρ_π	0.6731
ξ_E	0.3781	$st.d.\eta_t^A$	1.9318
γ_W	0.3784	$st.d.\eta_t^B$	3.7493
γ_P	0.3309	$st.d.\eta_t^G$	2.6658
$\psi''\ \psi'$	0.4748	$st.d.\eta_t^L$	3.1822
ϕ	2.6152	$st.d.\eta_t^I$	0.2260
r_π	0.8579	$st.d.\eta_t^R$	0.0825
$r_{\Delta\pi}$	0.0116	$st.d.\eta_t^\pi$	0.0502
ρ	0.8967	$st.d.\eta_t^Q$	7.9159
r_Y	0.1198	$st.d.\eta_t^P$	0.9303
$r_{\Delta Y}$	0.0069	$st.d.\eta_t^W$	1.4088

3.3.1. Analysis of Results of Parametric Identification of Smets and Wouters DSGE Model

The computed estimates of the model parameters (Table 16) show the following.

The mean duration of the labour agreement in the Kazakhstan Republic equals to 3.36 years, and of the business agreement - 1.88 years. At that within the period of the labour agreement and business agreement the mean level of indexation relative to inflation equal to 0.38 and 0.33, respectively.

The labour elasticity relative to the wage equals to about 0.54.

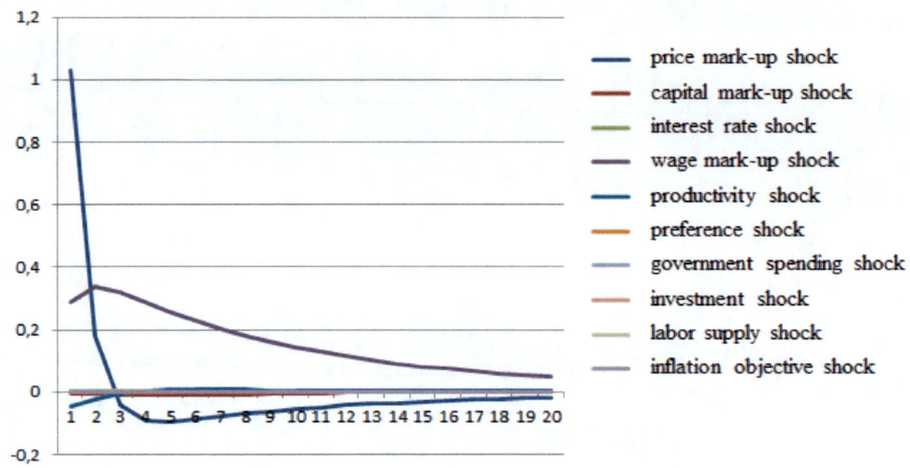

Figure 15. Impulse response of inflation on unity positive shocks.

Habit of consumption equals about 26% of the preceding level of consumption.

The most inertia is demonstrated by the shocks of labour offer, investments, objective level of inflation, productivity and preferences.

The smoothing coefficient 0.90 in the Taylor rule turns out to be much higher than the coefficient published by the National bank of the Kazakhstan Republic obtained in the context of KMOD model (0.12 and 0.24 for profitableness of rates on notes and rate of refinancing, respectively), but closer to the most of the coefficients published in the literature (0.74 for Chili, 0.93 for the zone of Euro, 0.81 for the USA, etc.).

3.3.2. Analysis of Influence of Shocks on GDP and Inflation by Means of Estimation of Impulse Characteristics on Disturbances in Context of Internal Shocks of Kazakhstan Republic Economy

Figures 14 and 15 present the impulse responses of the real GDP and inflation on (unity positive) shocks. Each of the plots presented at the figures is obtained by means of computation of the linear model (117) for the zero initial conditions for the zero period equal to its standard deviation. At that, all values of this shock for nonzero values of time, as well as the values of all other shocks of the model for all values of time are made equal to zero.

Analysis of the plots of impulse responses of the real GDP presented at Figure 14 shows the following:

- The offers of labour force and GDP investments increase with positive shocks of productivity.
- The positive shocks of preferences and public expenses also result in increase of GDP (since these shocks increase the consumption and public expenses, respectively).
- The positive shock of over charge on the price of the products leads to decrease of GDP. The shock of mark-up on the wage results in GDP increase.
- The positive monetary shock leads to fall of production (because of growth of the interest rate).

Analysis of the plots of impulse responses of the inflation presented at Figure 15 shows the following:

- For positive shocks, the mark-up on products and wages, the inflation increases.
- For positive shock of productivity, the inflation slightly decreases.

The rest of shocks do not practically influence the inflation.

These model responses on shocks correspond to the theoretical assumptions.

3.3.3. Decomposition of Dynamics of Indicators into Shares of Influence of Shocks in Retrospective Period

The decomposition of the indicators in the retrospective periods presented at Figures 16 and 17 shows the contributions (in percent) of each shock of the estimated model (117) in the deviations of actual values of GDP and inflation indicators from the respective equilibrium values for the period of time from the first quarter of 2002 year to the first quarter of 2012 year.

These plots indicate how the deviations of GDP and inflation from their respective trends in the retrospective period (from the 1-st quarter of 2002 year to the 3-rd quarter of 2011 year) were formed in accordance with the positive and negative influence of the considered shocks.

For example, the deviation of GDP from the trend of the 3-rd quarter of 2009 year equal to-4.93% is a sum of the positive items:

1. Influences of the shock of labour offer equal to 3.02% of deviation of GDP from the trend.
2. Influences of the shock of the interest rate equal to 1.51% of deviation of GDP from the trend.
3. Influences of the shock of the mark-up on the products equal to 0.85% of deviation of GDP from the trend.
4. Influences of the shock of preferences equal to -3.88% of deviation of GDP from the trend.
5. Influences of the shock of mark-up on the wage equal to -3.63% of deviation of GDP from the trend.
6. Influences of the shock of mark-up on the capital equal to -1.16% of deviation of GDP from the trend.
7. Influences of the shock of public expenses equal to -1.12% of deviation of GDP from the trend.
8. Influences of the shock of productivity equal to -0.69% of deviation of GDP from the trend (other influences are negligibly small).

i.e., the deviation of GDP from the trend equal to -4.93% is a sum of all influences of all shocks of the model for the considered period.

For example, the deviation of inflation from the trend for the 4-th quarter of 2007 year equal to 6.32% is a sum of only positive items (all of the influences are negligibly small):

1. Influences of the shock of mark-up on the products equal to 4.87% of the inflation deviation from the trend.
2. Influences of the shock of mark-up on the wage equal to 1.43% of the inflation deviation from the trend.

Analysis of the diagrams at Figure 16 also shows that the rapid growth of GDP in the interval of 2004-2007 is explained mainly by the shocks of mark-ups (on the wage, products, capital). The decrease of the GDP growth in 2008-2011 years is mainly explained by the negative influence of the shocks of preferences, mark-up on the capital and wage.

Analysis of the diagrams at Figure 17 shows that the deviation of inflation from the equilibrium level in the period of 2002-2011 is almost completely explained by the shocks of mark-up on the products and wages. In other words, all of the shocks of the model did not practically influence on the values of inflation in the considered period of time.

3.3.4. Prediction and Decomposition of Dispersions of Prediction of Indicators on Shares of Shock Influences

In Table 18 below we present the mathematical expectations, standard deviations and decompositions of the dispersion of the predictions of GDP and inflation by the internal shocks of the model for the ten year period. In this table we do not present the shocks that influence on the dispersion less than 1%.

Analysis of Table 18 shows that the main share (about 60%) into dispersion of the prediction of GDP is given by the shocks of mark-up on the capital and preferences. Practically all dispersion of the prediction of inflation (more than 90%) is explained by the shock of mark-ups on the products and wages.

3.4. Parametric Control of Economic Growth Based on Smets-Wouters DSGE Model

To implement the state policy in the sphere of the economic growth, the additive items η_t^R, η_t^G in the expressions of the profitability of the state bonds (108) and rule of public expenses (109) are chosen as the instruments of the state policy. The values η_t^R, η_t^G to be found are searched in form of deterministic instead of the respective shocks. Then the mathematical statement of the parametric control problem is as follows.

On the base of the model (117), find the deterministic instruments $\eta_{T+1}^R, \dots, \eta_{T+40}^R, \eta_{T+1}^G, \dots, \eta_{T+40}^G$, which maximizes the criterion

$$\max_{\eta_{T+1}^R,\dots,\eta_{T+40}^R,\eta_{T+1}^G,\dots,\eta_{T+40}^G} \mathrm{E}_T \sum_{i=1}^{40} \beta^i Y_{T+i} \qquad (119)$$

under the following additional constraints on the endogenous variables of the model: prediction of inflation and interest rate of the bonds for this period must not deviate from their equilibrium values more than 0.5%, and the volume of public expenses more than 5.0% from its equilibrium value. Here T is the number of quarter when the regulation started corresponding to the 3-rd quarter of 2011 year; Y_{T+i} is the real GDP in the period $[T + i - 1, T + i]$; β is the discount coefficient.

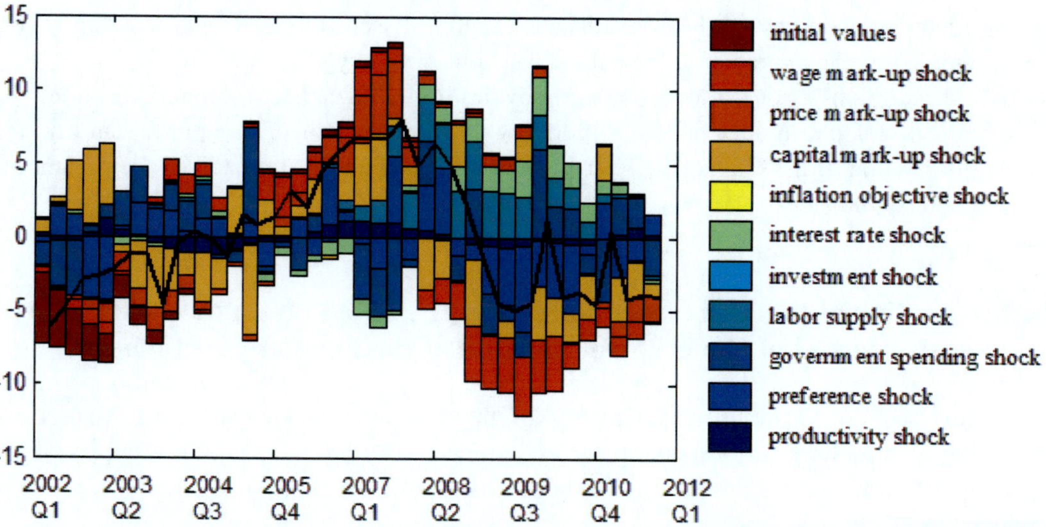

Figure 16. Influence of shocks on dynamics of GDP.

This optimization problem relates to the class of the variational calculus problems. The solution of the stated problem ensure increase of the mean value of GDP in the considered interval by 6.7% in comparison with the prediction for the base scenario derived by means of the model (117) (Figure 18).

4. Parametric Control of Characteristics of Short-term Cycles and Effect of Shocks on National Economy Evolution Based on Several Macroeconomic Models

The provisions about the short-term cycles, which widely consider the factors giving them birth [24], [25] and offering various mathematical models for their study, consist the important part [23], [6] of the modern macroeconomic theory.

Attenuation of the short-term cycles is the most important sphere of the stabilizing policy [26], [27].

In this section we present the following results:

1. Estimation of the hypothesis about generation of the short-term cycle by linear dependence of consumer expenses of the households from the income and linear dependence (with delay) between investments and income growth on the base of CGE model of the economy branches [11], as well as the results of attenuation of the short term arising under these assumptions by the approach of parametric control theory.
2. Estimations of influences of the shocks on the economic indicators on the base of linear approximation of the nonlinear Smets-Wouters DSGE model [12] estimated by the statistical data of evolution of the Kazakhstan Republic economy, as well as the results of attenuation of influences of the internal shocks onto the indicators of the

national economy of the Kazakhstan Republic based on the Smets-Wouters DSGE model by the approach of the parametric control theory.

3. Descriptions of one of the short-term cycles in the Kazakhstan Republic economy by means of one model of the Kondratiev cycle. Research of structural stability and regulation of the characteristics of this business cycle by approach of the parametric control theory.

4.1. Estimation of the Macroeconomic Theory Provisions on the Base of the Computable Model of General Equilibrium of the Economy Branches

In this section we describe the results of the computational experiments focused on estimation of some provisions of the macroeconomic theory on the cyclic oscillations of the macroeconomic indicators as a result of shock changes of demands of final and investment products.

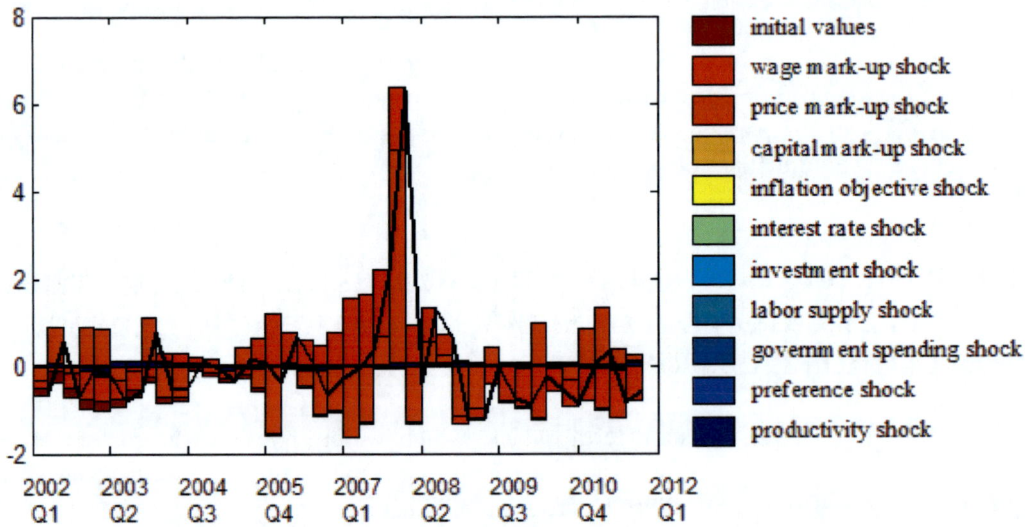

Figure 17. Influence of shocks on quarterly inflation.

Within the limits of the macroeconomic theory, the cyclic oscillations of economic processes can appear in presence of [26]

- linear dependence between the volume of consumer expenses and current income;
- linear dependence between investments and increase of income.

To verify these provisions, we carried out a number of computational experiments aimed at computation of the following scenarios of variation of the said demands of final and investment products:

a) $O_1\ t\ = O_1\ t-1\ + a\ Yg\ t\ - Yg\ t-1\ ;$ (120)

b) $O_{2i}\ t\ = O_{2i}\ t-1\ + b(Yg\ t-3\ - 2Yg\ t-2\ + Yg\ t-1\ ), i = 1, \ldots, 16;$ (121)

c) Joint application of scenarios a) and b);

d) Increase of the shares $O_1\ t$ by k times in comparison with the base variant;
e) Increase of the shares $O_{2i}\ t\ (i = 1, \ldots, 16)$ by l times in comparison with the base variant;
f) Joint application of scenarios d) and e)

Here $t = 2010, \ldots, 2015$ is the time in years; $O_1\ t\ = CO\_p\_17c\ t$ is the share of budget of the households spent for purchasing the final products (exogenous function); $O_{2i}\ t$ is the share of budget of i-th branch spent for purchasing the investment products ($t = 2010, \ldots, 2015$) (exogenous function); $Yg\ t\ = VY\_g\ t$ is the sum of branches' GVA in constant prices (endogenous variable); a, b, l, k are some positive constants ($l > 1$, $k > 1$).

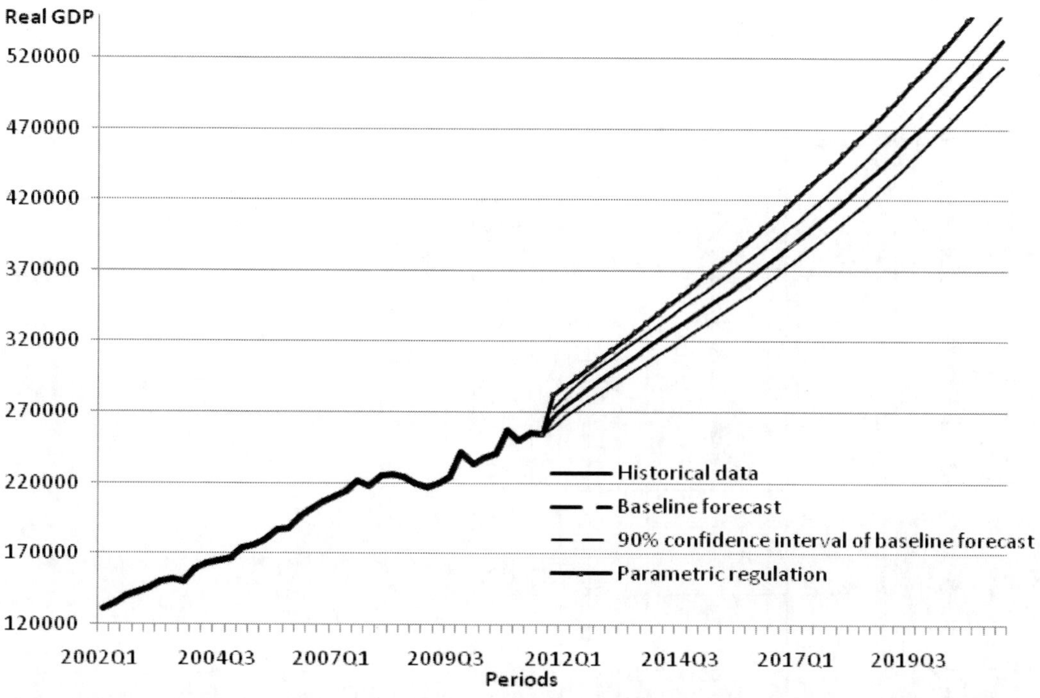

Figure 18. Gross domestic product (in million tenge, in prices of 1994 year).

Application of scenarios a) and b) means the model computation with the values of the parameters O_1 and O_{2i} defined by formulas (120) and (121), respectively, starting from $t = 2010$. At that, the values of all other exogenous model parameters for scenarios a), b), c), d), e), f) corresponds to its base variant.

As a result of computation experiments on the base of the model aimed at realization of scenarios a), b), c), d), e), f) one can observe the cyclic oscillations of the variable $P(t)$ (Consumer price index) (see Figures 19-24). At that, we did not reveal the oscillation phenomenon for the values of the real indicators (in particular, $Yg\ t$).

The experiments results presented in Figure 19-24 prove the respective provisions of the macroeconomic theory [27].

Table 18. Characteristics of prediction of GDP and inflation

Horizon of prediction (quarter)	Prediction of GDP (billion tenge, 1994)		Decomposition of GDP prediction dispersion (%)								Prediction of inflation (%)		Decomposition of dispersion of prediction of inflation (%)	
	Mathematical expectation	Standard devi-ation	Shock of produc-tvity	Shock of preferences	Shock of public expenses	Shock of labour offer	Shock of interest rates	Shock of mark-up on capital	Shock of mark-up on products	Shock of mark-up on wages	Mathematical expectation	Standard deviation	Shock of mark-up on products	Shock of mark-up on wages
1	271.41	8.17	2.25	22.90	17.57	1.47	1.07	45.32	6.13	3.19	1.53%	0.75%	92.24	7.47
4	292.16	10.96	4.04	21.73	15.30	3.25	1.56	41.99	8.27	3.71	1.63%	0.78%	72.31	27.33
10	327.03	13.22	4.04	21.73	15.30	3.25	1.56	41.99	8.27	3.71	1.84%	0.80%	62.65	36.92
20	385.81	16.06	4.09	20.69	14.20	3.70	1.51	39.84	7.90	7.92	1.98%	0.81%	60.71	38.85
30	456.45	19.13	4.03	20.39	13.98	3.67	1.49	39.47	8.01	8.82	2.02%	0.81%	60.55	39.00
40	542.39	22.75	4.02	20.37	13.97	3.67	1.49	39.45	8.03	8.86	2.03%	0.81%	60.53	39.01

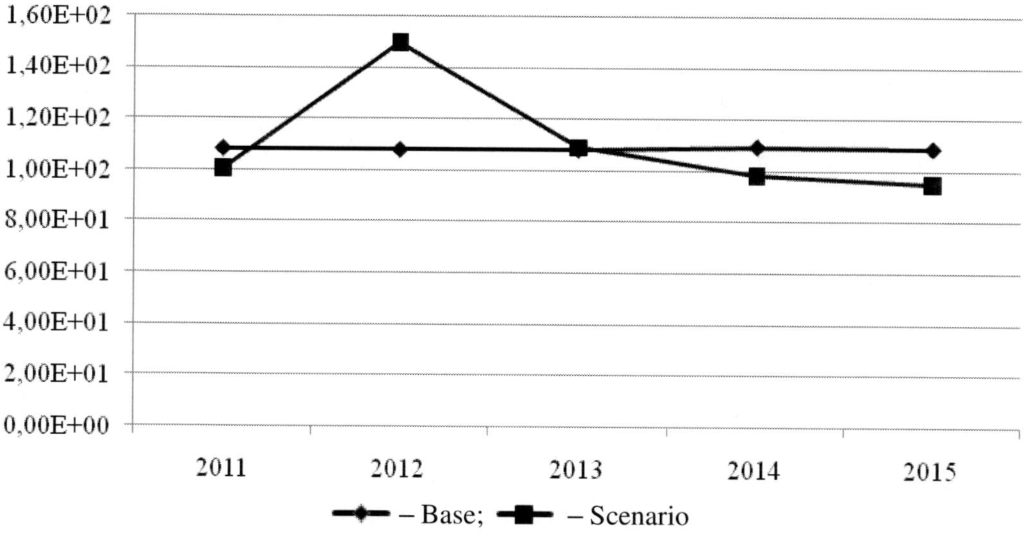

Figure 19. Values of the consumer price index with application of scenario a) with $a = 3\cdot10^{-13}$ and for the base variant.

4.2. Attenuation of the Cyclic Oscillations of the Macroeconomic Indicators by the Parametric Control Methods

In the computational experiments we consider the following problem of attenuation of the cyclic oscillations of the consumer price level arising in application of scenario c) of the economic system development taking into account linear dependence between the volume of consumer expenses and current income, as well as linear dependence between investments and income growth.

On the base of the computable model of the economy branches with application of scenario c) (where $a = 3\cdot10^{-13}$, $b = 10^{-13}$), find the values of the shares $O_i^j(t) = CO\_pj\_iz(t)$ of budgets of j-th producing agent spent for purchasing goods and services produced and rendered by i-th producing agent in 2010-2015 years ($i, j = 1,\ldots,16$), which provide the lower bound of the following criterion K_P which characterizes the deviations of the computed values of the consumer price index $VP[t]$ from the respective desired values of $P[t]$:

$$K_P = \sum_{t=2010}^{2015} \left(\frac{VP[t]-P[t]}{P[t]}\right)^2. \qquad (122)$$

Here we use the computed base values of the model consumer price index without parametric control as the desired values of $P[t]$:.

The constraints on the adjusted parameters are as follows:

$$0.5 \le O_i^j[t]/O_i^j \le 1.5; \quad \sum_{i=1}^{16} O_i^j[t] \le 1; \text{ where } i, j = 1, \ldots, 16; t = 2010, \ldots, 2015. \qquad (123)$$

Here O_i^j are the base values of the said shares obtained as a result of solving the parametric identification problem by data of 2000-2008 years.

The constraints on the growth of the macroeconomic indicator: $Yr[t] \geq 0.95Yc[t]$. Here $Yc[t]$ are the computed scenario values of the sum of GVA of the branches without parametric control; $Yr[t]$ are the computed values of the sum of GVA of the branches with parametric control.

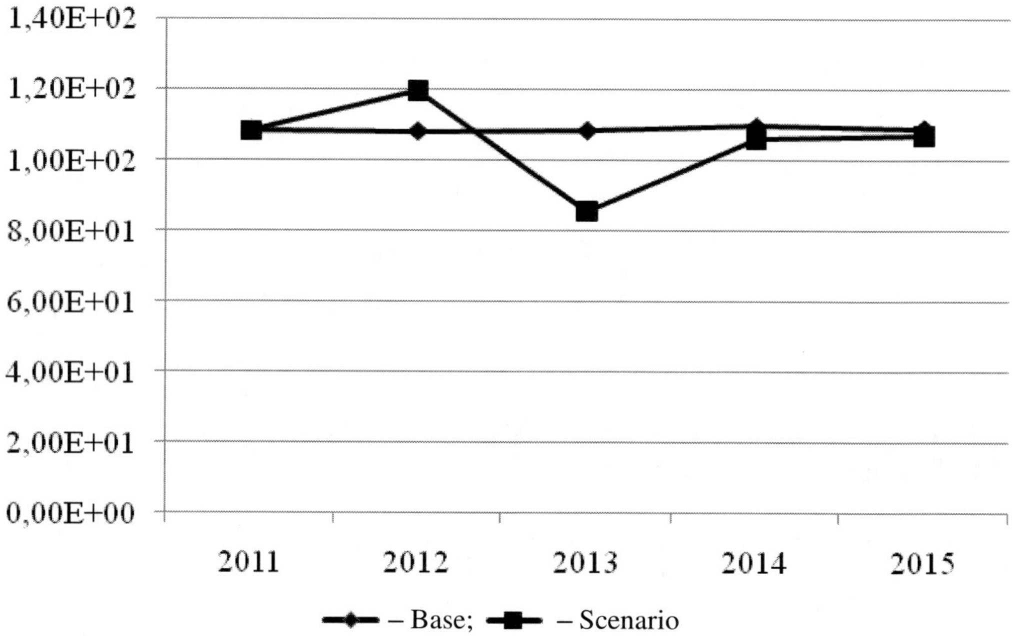

Figure 20. Values of the consumer price index with application of scenario b) with $b = 10^{13}$ and for the base variant.

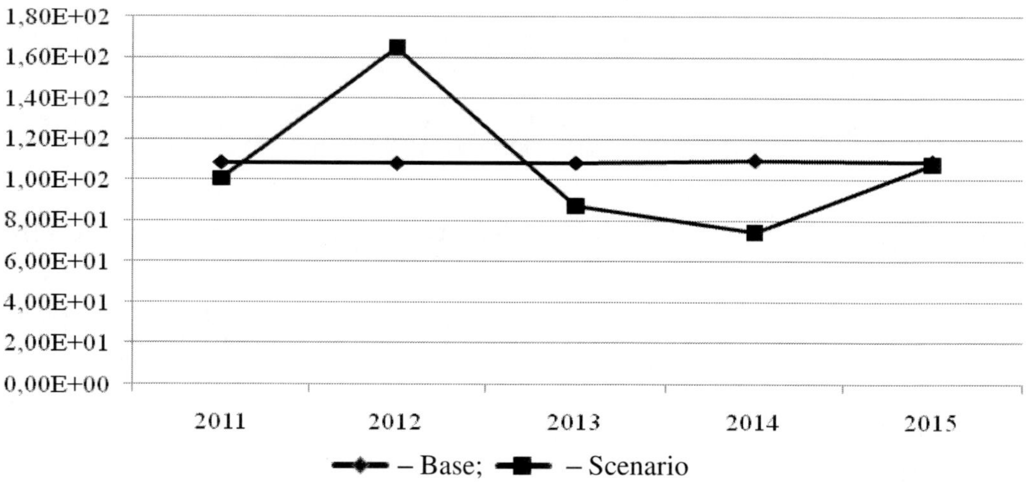

Figure 21. Values of the consumer price index with application of scenario c) with $a = 3{\cdot}10^{-13}$, $b = 10^{-13}$ and for the base variant.

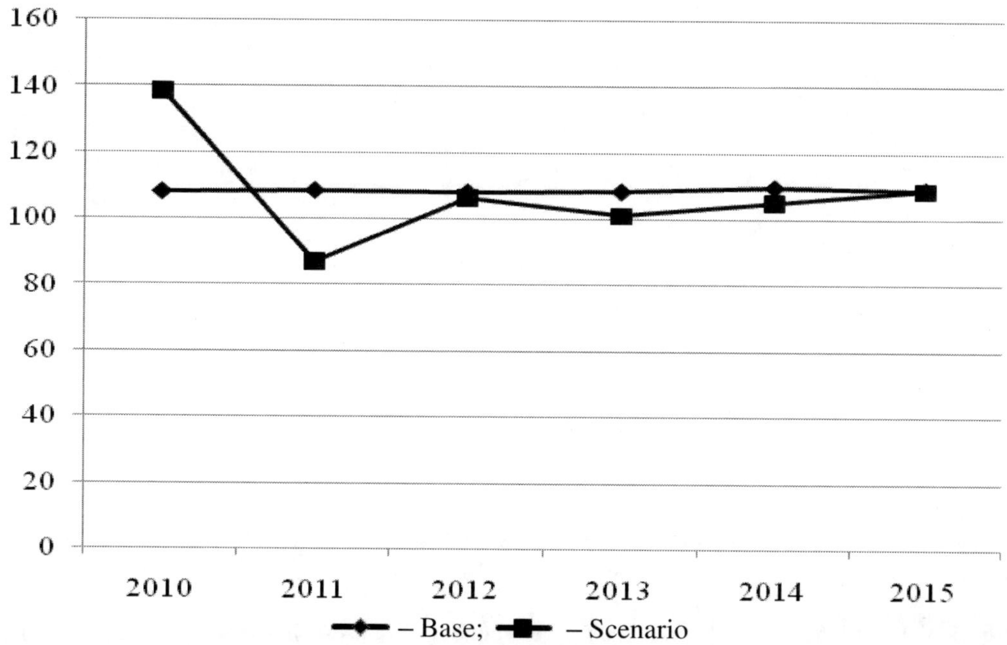

Figure 22. Values of the consumer price index with application of scenario d) with $k = 1.2$ and for the base variant.

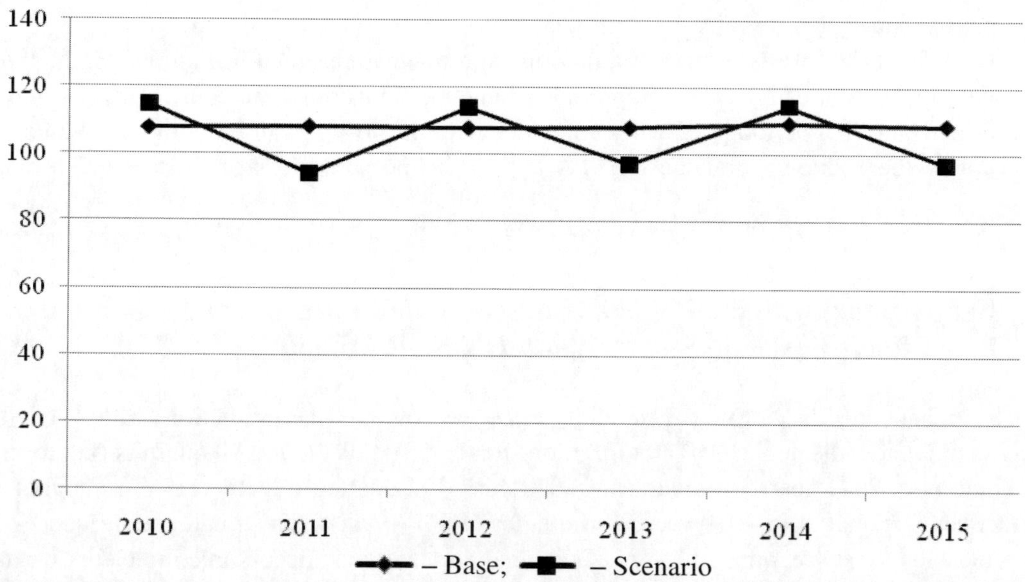

Figure 23. Values of the consumer price index with application of scenario e) with $l = 1.4$ and for the base variant.

The value of the criterion K_P without parametric control is $K_P = 0.424$. The optimal values of the criterion K_P with application of the parametric control law appears to be equal to $K_P = 0.000844$.

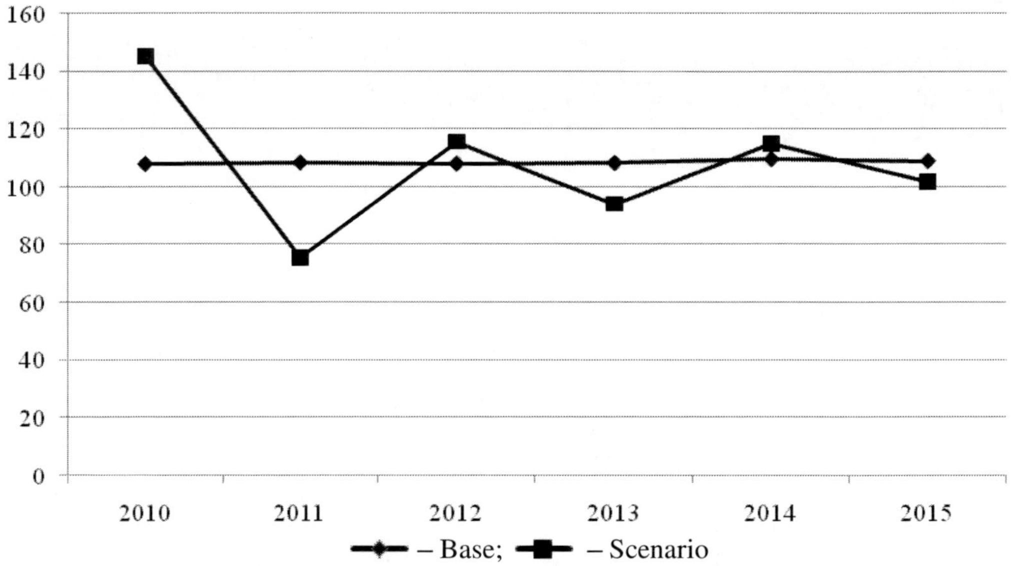

Figure 24. Values of the consumer price index with application of scenario f) with $k = 1.2$, $l = 1.4$ and for the base variant.

The base values of the macroeconomic indicators $Y(t)$ and $P(t)$ values obtained with application of scenario c) and values obtained with application of the optimal law of parametric control are presented in Figure 25.

Analysis of the results of the computation experiments presented in Figures 25, 26 show that with application of the derived optimal parametric control law the consumer price index within the controlled period practically coincides with the desired values, at that the values of the sum of the economy branches' GVAs appears to be some less of the respective base values excepting 2015 year.

4.3. Estimation of Effect of Shocks on Economic Indicators and Attenuation of Their Effects Based on Smets-Wouters DSGE Model

By means of the estimated model (117), we obtain the prediction values of the macroeconomic indicators (GDP and inflation) for 1, 4, 10, 20, 30 and 40 quarters (i.e. for the 4-th quarter of 2011 year, 3-rd quarter of 2012 year, 1-st quarter of 2014 year, 3-rd quarter of 2016 year, 1-st quarter of 2019 year, 3-rd quarter of 2021 year). To estimate the influences of the shocks on the dispersions of errors of the predictions of macroeconomic indicators we use the standard methods for finding the decompositions of the dispersion of the prediction errors for the models of vector autoregressions [27]. The results derived by means of the software Dynare Matlab Toolbox [21] are presented in Tables 18 and 19. These tables do not include the shocks influencing the dispersion less than 0.01%.

Analysis of Tables 19, 20 shows the following. The dispersions of GDP prediction errors are mainly determined by the shocks of preferences, shocks of public expenses and shocks of mark-up on the capital price. The dispersions of the inflation prediction errors are mainly defined by the shocks of mark-up on the products and wages.

Table 19. Prediction (billion tenge in prices of 1994) and decomposition of dispersions of prediction of quarterly GDP

Prediction horizon	Prediction		Decomposition of dispersion (%)								
	Mathematical expectation	Standard deviation	η_t^A	η_t^B	η_t^G	η_t^L	η_t^I	η_t^R	η_t^Q	η_t^P	η_t^W
1	271.413	8.17	2.25	22.90	17.57	1.47	0.09	1.07	45.32	6.13	3.19
4	292.166	10.960	4.04	21.73	15.30	3.25	0.15	1.56	41.99	8.27	3.71
10	327.034	13.229	4.04	21.73	15.30	3.25	0.15	1.56	41.99	8.27	3.71
20	385.810	16.067	4.09	20.69	14.20	3.70	0.14	1.51	39.84	7.90	7.92
30	456.450	19.136	4.03	20.39	13.98	3.67	0.14	1.49	39.47	8.01	8.82
40	542.395	22.751	4.02	20.37	13.97	3.67	0.14	1.49	39.45	8.03	8.86

Table 20. Prediction (%) and decomposition of dispersions of prediction of quarterly inflation

Prediction horizon	Prediction		Decomposition of dispersion (%)			
	Mathematical expectation	Mathematical expectation	η_t^A	η_t^Q	η_t^P	η_t^W
1	1.22%	0.75%	0.29	0.01	92.24	7.47
4	1.36%	0.78%	0.30	0.06	72.31	27.33
10	1.62%	0.80%	0.28	0.15	62.65	36.92
20	1.78%	0.81%	0.27	0.16	60.71	38.85
30	1.83%	0.81%	0.28	0.17	60.55	39.00
40	1.85%	0.81%	0.28	0.17	60.53	39.01

To realize the state policy in the sphere of minimization of influences of the shocks onto the economic indicators we choose the additive components η_t^R, η_t^G (just as in Section 3.4) in the expressions (108), (109) as its instruments. The values to be found are searched as the deterministic values instead of the respective shocks.

The approach of the parametric control to minimization of influence of the shocks consists in the following. Let T be the number of quarter, starting from which the state carries out the policy of parametric control of minimization of influence of the shocks. In each moment of time $t = T, T+1, T+2, \ldots$ on the base of the estimated model (117) written in the form

$$\begin{pmatrix} X_{t+i} \\ R_{t+i} \\ G_{t+i} \end{pmatrix} = Q^{\theta} \begin{pmatrix} X_{t+i-1} \\ R_{t+i-1} \\ G_{t+i-1} \end{pmatrix} + F^{\theta} H_{t+i}^{\Sigma_H}, i = 1,2,3,\ldots,40 \qquad (124)$$

we find the deterministic values of the instruments $\eta_{t+1}^R, \ldots, \eta_{t+40}^R, \eta_{t+1}^G, \ldots, \eta_{t+40}^G$, minimizing the criterion

$$L_t = \mathrm{E}_t \left(\sum_{i=1}^{40} \beta^i \left(\pi_{t+i}^2 + \lambda_Y Y_{t+i}^2 \right) \right), \min_{(\eta_{t+1}^R, \ldots, \eta_{t+40}^R, \eta_{t+1}^G, \ldots, \eta_{t+40}^G)} L_t \qquad (125)$$

(that characterizes the expected discounted summary deviation of the values of GDP and inflation from the respective equilibrium values (trend)) under the following constraints on the endogenous variables of the model. The mathematical expectation of inflation and bond rate must not deviate from the respective equilibrium values more than by 0.5% in this time horizon:

$$\left| \mathrm{E}_t \pi_{t+i} \right| \leq 0.005, i = 1,2,3,\ldots,40, \qquad (126)$$

$$\left| \mathrm{E}_t R_{t+i} \right| \leq 0.005, i = 1,2,3,\ldots,40, \qquad (127)$$

and the mathematical expectation of the public expenses volume - by 5.0% from their trend values:

$$\mathrm{E}_t G_{t+i} \leq 0.05, i = 1,2,3, \dots ,40. \tag{128}$$

At that, in each moment of time t, the value of the current state of economy is known for the moment of time t - (X_t, R_t, G_t). After obtaining new information (the values of variables $(X_{t+1}, R_{t+1}, G_{t+1})$ at the next moment of time $(t + 1)$), the values of instruments $\eta_{t+2}^R, \dots, \eta_{t+41}^R, \eta_{t+2}^G, \dots, \eta_{t+41}^G$ are computed anew by means of solving the problem (124)-(128) for the respective time interval. Here β is the discount coefficient, $\lambda_Y$is some weighting coefficient.

Introduce the new minimization criterion:

$$L_t = \sum_{i=0}^{40} \beta^i \left((\mathrm{E}_t \pi_{t+i})^2 + \lambda_Y (\mathrm{E}_t Y_{t+i})^2 \right), \min_{(\eta_{t+1}^R, \dots, \eta_{t+40}^R, \eta_{t+1}^G, \dots, \eta_{t+40}^G)} L_t. \tag{129}$$

It is not hard to verify that the criterion L_t differs from L_t by the value that does not depend on the variables $(\eta_{t+1}^R, \dots, \eta_{t+40}^R, \eta_{t+1}^G, \dots, \eta_{t+40}^G)$.

From the relation (124), taking its mathematical expectation from both of its parts, we obtain

$$\mathrm{E}_t \begin{pmatrix} X_{t+i} \\ R_{t+i} \\ G_{t+i} \end{pmatrix} = Q^{\theta} \mathrm{E}_t \begin{pmatrix} X_{t+i-1} \\ R_{t+i-1} \\ G_{t+i-1} \end{pmatrix} + F^{\theta''} \begin{pmatrix} \eta_{t+i}^R \\ \eta_{t+i}^G \end{pmatrix}, i = 1,2,3, \dots ,40 , \tag{130}$$

Here $F^{\theta''}$ is the matrix composed of two respective columns of the matrix F^{θ}. As a result, the optimal values of the variables of the problems (124)-(128) and (126)-(130) coincide. The obtained optimization problem (126)-(130) belongs to the classic (deterministic) kind of the variational calculus problems that is solved by the method of linear-quadratic programming (with application of Matlab) for each moment of time t.

Further we estimate effectiveness of application of the parametric regulation formulated above under assumption that the state applies this policy for 30 years (120 periods). I.e., it is assumed that at each period of time t ($t = T, T + 1, T + 2, \dots, T + 119$; T is the number of quarter corresponding to the 3-rd quarter of 2011 year) the state establishes the values of the parameters $\eta_{t+1}^R, \eta_{t+1}^G$ by means of application of the above optimization problem (124)-(128) (or, equivalently, (126)-(130)). In computational experiments we assume that the economy is precisely described by the estimated model:

$$\begin{pmatrix} X_{T+i} \\ R_{T+i} \\ G_{T+i} \end{pmatrix} = Q^{\theta} \begin{pmatrix} X_{T+i-1} \\ R_{T+i-1} \\ G_{T+i-1} \end{pmatrix} + F^{\theta} H_{T+i}^{\Sigma_H}, i = 1, 2, 3, \dots ,160,$$

where X_T is known.

To estimate effectiveness of application of the parametric control approach formulated above, we use the Monte-Carlo method with estimation of 100 scenarios of the economy development. Let us present the enlarged algorithm for estimation of application of the parametric control approach.

1. Generation of the sample consisting of 100 elements, the sets of the values of the Gaussian random values (white noises) $\{H_{T+1}^{'\Sigma_H}, H_{T+2}^{'\Sigma_H}, \dots, H_{T+120}^{'\Sigma_H}\}_j$ where $j = 1, \dots, 100$, $H_{T+i}^{\Sigma_H} = [H_{T+i}^{'\Sigma_H}, \eta_{T+i}^{R}, \eta_{T+i}^{G}]^T$ with known probabilistic characteristics of the noises $\Sigma_{H'}$.
2. For each element of the sample $\{H_{T+1}^{'\Sigma_H}, H_{T+2}^{'\Sigma_H}, \dots, H_{T+120}^{'\Sigma_H}\}_1$, $\{H_{T+1}^{'\Sigma_H}, H_{T+2}^{'\Sigma_H}, \dots, H_{T+120}^{'\Sigma_H}\}_2$, …, $\{H_{T+1}^{'\Sigma_H}, H_{T+2}^{'\Sigma_H}, \dots, H_{T+120}^{'\Sigma_H}\}_{100}$:

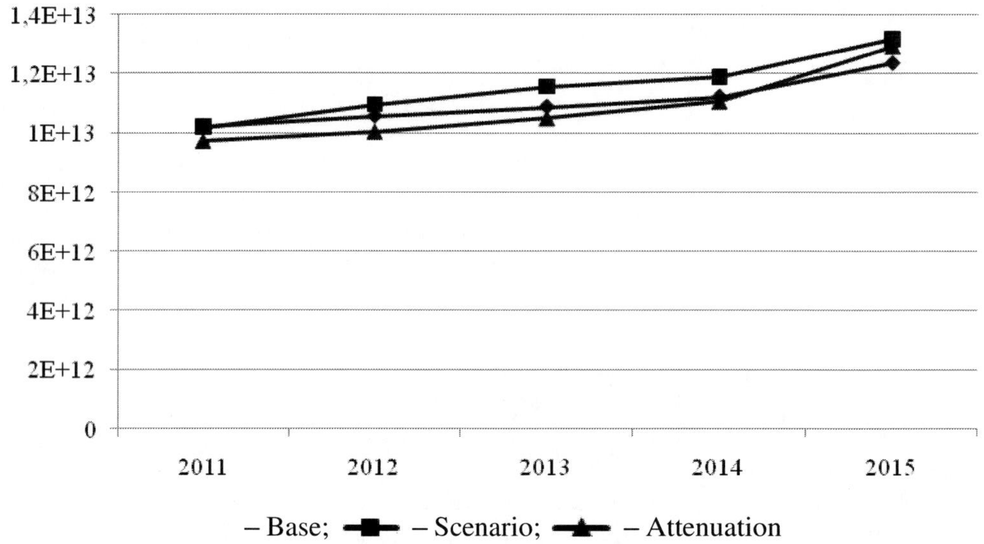

Figure 25. Computed values of sums of GVA of economy branches.

2.1. Construct Computation of the Model with Parametric Control:

Solve the optimization problem (126)-(130) for the period of time $t = T$, i.e. find the respective $\eta_{T+1}^{R}, \dots, \eta_{T+40}^{R}, \eta_{T+1}^{G}, \dots, \eta_{T+40}^{G}$; from the derived set of the values, take $\eta_{T+1}^{R}, \eta_{T+1}^{G}$ (the other values $\eta_{T+2}^{R}, \dots, \eta_{T+40}^{R}, \eta_{T+2}^{G}, \dots, \eta_{T+40}^{G}$ are truncated).

2.2. Compute the model for one step with the values of shocks $H_{T+1}^{\Sigma_H}$, which consist of the values of instruments $\eta_{T+1}^{R}, \eta_{T+1}^{G}$ defined by the state and values of shocks $H_{T+1}^{'\Sigma_H}$ that have been realized by the economy independently (exogenously) from the state policy.

$$\begin{matrix} X_{T+1} \\ R_{T+1} \\ G_{T+1} \end{matrix} = Q^{\theta} \begin{matrix} X_{T} \\ R_{T} \\ G_{T} \end{matrix} + F^{\theta} H_{T+1}^{\Sigma_H},$$

2.3. The steps 2.1. and 2.2 are repeated for the values $t = T + 1, T + 2, \dots, T + 120$.

3. By the derived 100 trajectories of GDP and inflation, find the mean trajectory (expected value of prediction) and standard deviation of the prediction for the time interval $t = T, T + 1, T + 2, \dots, T + 120$. The computed values are compared with the base prediction.

The results of realization of the proposed algorithm show that for the used sample of the shocks, the parametric regulation of attenuation of influences of the shocks provides decrease of the predicted standard deviation of GDP by 58.3% in mean in the prediction horizon from the 3-rd quarter of 2011 year to the 3-rd quarter of 2011 year (see Figure 27).

The parametric regulation of attenuation of influences of the shocks ensures decrease of the predicted standard deviations of inflation by 32.0% in mean in the predicted horizon from the 3-rd quarter of 2011 year to the 3-rd quarter of 2021 year and decrease of the selective standard deviation of inflation by 47.8% in comparison with actual data in the period of the 1-st quarter of 2002 year to the 3-rd quarter of 2011 year (see Figure 28).

4.4. Parametric Control of Characteristics of Short-term Cycles and Effect of Shocks on National Economy Evolution Based on Several Macroeconomic Models

4.4.1. Model Description

This model [28] combines the descriptions of non-equilibrium economic growth and non-uniform scientific and technological advance. The model is described by the following system of equations, including two differential and one algebraic equations:

$$\begin{cases} n(t) = Ay(t)^{a}, \\ dx/dt = x(t)(x(t)-1)(y_0 n_0 - y(t)n(t)), \\ dy/dt = n(t)(1-n(t))y(t)^2(x(t)-2+\dfrac{\mu + l_0}{n_0 y_0}), \\ n_0 = Ay_0{}^{\alpha}. \end{cases} \tag{131}$$

Here t is the time (in months); x is the efficiency of innovations; y is the capital productivity ratio; y_0 is the capital productivity ratio corresponding to the equilibrium trajectory; n is the rate of saving; n_0 is the rate of saving corresponding to the equilibrium trajectory; μ is the coefficient of withdrawal of funds; l_0 is the job growth rate corresponding to the equilibrium trajectory; A and a are some model constants.

The preliminary estimation of the model parameters is carried out based on the statistical information of the Republic of Kazakhstan for the years 2001-2005 [29]. The deviations of the observed statistical data from the calculated ones do not exceed 1.9% within the considered period.

As a result of solving the problem of preliminary estimation of the parametric identification, the following values of the exogenous parameters are obtained: $\alpha = -0.0046235$, $y_0 = 0.081173$, $n_0 = 0.29317$, $\mu = 0.00070886$, $l_0 = 0.00032161$, $x(0) = 1.91114$.

The preliminary prediction for 2006 and 2007 is characterized by the errors 6.1% and 12.1%, respectively, for the capital productivity ratio, and 2.3% and 11%, respectively, for the rate of saving.

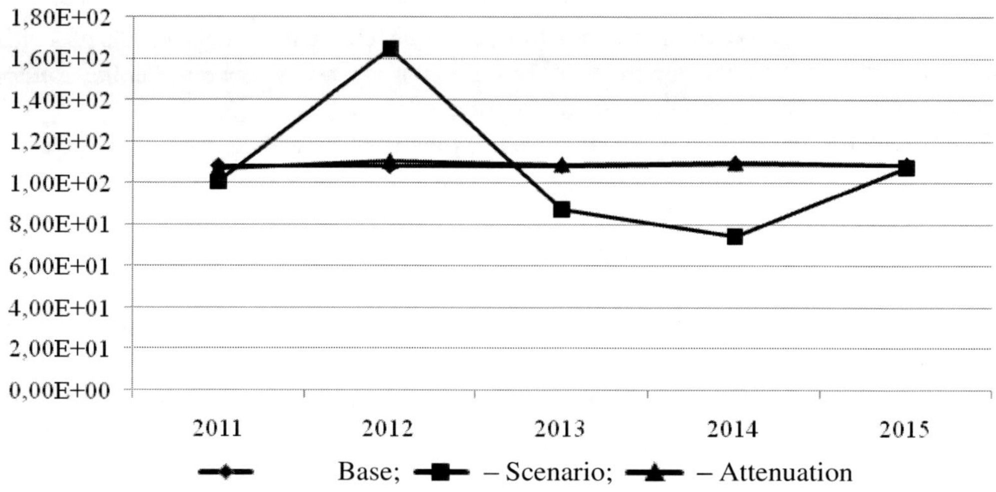

Figure 26. Computed values of consumer price index.

The respective cyclic phase trajectory of the Kondratiev cycle model is presented in Figure 29.

The estimation of the period of the cyclic trajectory corresponding to the statistical information of the Republic of Kazakhstan for the said years equals to 232 months.

4.4.2. Estimating Robustness of the Kondratiev Cycle Model without Parametric Control

According to the component 4 of the parametric control theory (Section 1), we made estimation of structural stability (robustness) of the mathematical model in the selected compact of the state space of the model. In this work we use the results of application of one computational algorithm of estimation of structural stability of the mathematical model (in form of continuous-time dynamic system) of the economic system of the country on the base of Robinson theorem [30]. In particular, this theorem states that if the chain-recurrent set $R(f,N)$ of the flow f in some compact N of its state space is empty, then this flow is weakly structurally stable in N.

For the specific mathematical model of economic system, we can define e.g. the parallelepiped of its state space including all possible trajectories of evolution of the economic system for the considered time interval as the compact N.

Figure 30 presents the estimation of the chain-recurrent set $R(f, N)$ obtained as a result of the application of the chain-recurrent set estimation algorithm for the region $N = [1.7, 2.3]\times[0.066, 0.098]$ of the phase plane Oxy of system (131). Since the set $R(f, N)$ is not empty, then one cannot make a conclusion about the weak structural stability of the Kondratiev cycle model in N on the basis of the Robinson theorem. However, since there is a non-hyperbolic singular point in N, namely, the center $(x_0 = 2 - \frac{\mu+l_0}{n_0 l_0},\ y_0)$ [31], then system (131) is not weakly structurally stable in N.

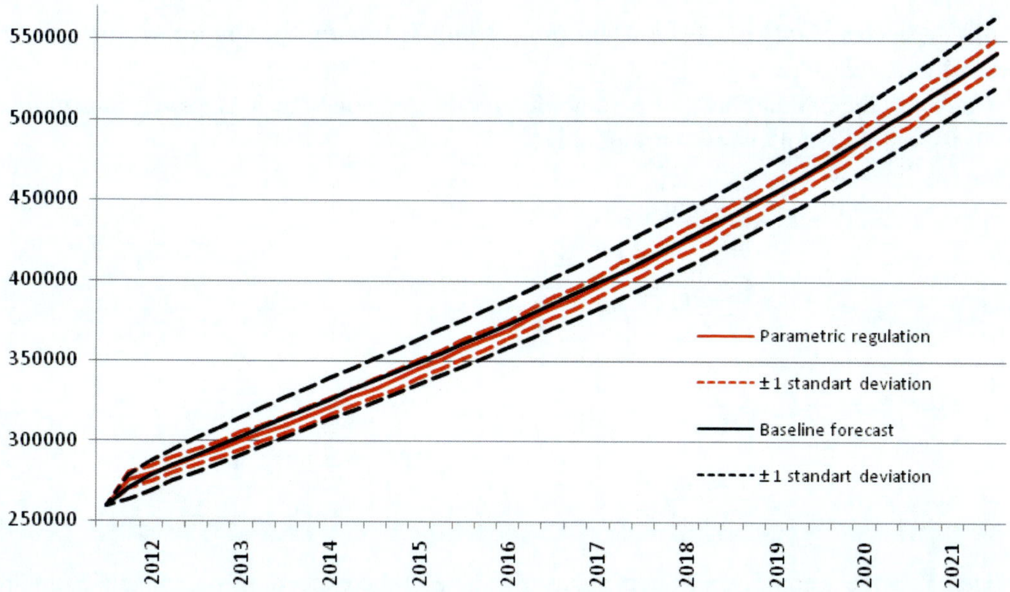

Figure 27. Predicted values of real GDP for the base scenario and approach of parametric control.

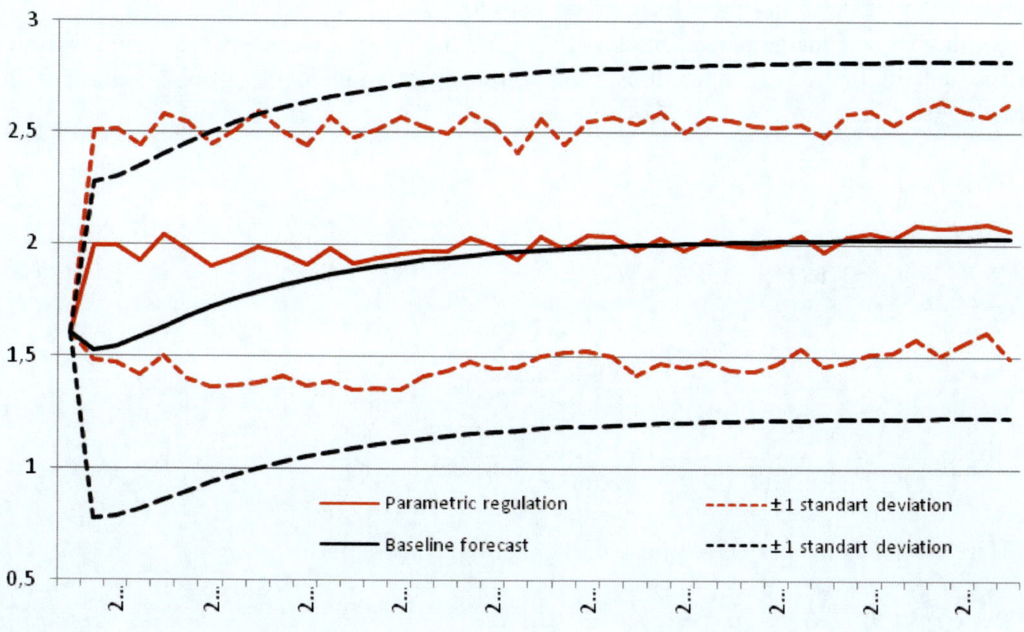

Figure 28. Predicted values of inflation for base scenario and approach of parametric control (%).

4.4.3. Parametric Control of the Evolution of the Economic System on the Basis of the Kondratiev Cycle Model

Choosing the optimal laws of parametric control is carried out in the environment of the set consisting of the following four relations:

$$1)\ n_0\ t\ = n_0^* + k_1 \frac{y\ t\ -y(0)}{y(0)};$$

$$2)\ n_0\ t\ = n_0^* - k_2 \frac{y\ t\ -y(0)}{y(0)}; \qquad (132)$$

$$3)\ n_0\ t\ = n_0^* + k_3 \frac{x\ t\ -x(0)}{x(0)};$$

$$4)\ n_0\ t\ = n_0^* - k_4 \frac{x\ t\ -x(0)}{x(0)}.$$

Here k_i is the scenario coefficient; n_0^* is the value of the exogenous parameter n_0 obtained as a result of the preliminary estimation of the parameters.

The problem of choosing the optimal law of parametric control at the level of the econometric parameter n_0 can be formulated as follows.

On the basis of mathematical model (131), find the optimal parametric control law in the environment of the set of algorithms (132) ensuring attainment of optimal values of the following criteria:

$$1)\ K_1 = \frac{1}{36}\ {}_{t=1}^{36} y(t),\ \ \max_{(132)} K_1;$$

$$2)\ K_2 = \frac{1}{36}\ {}_{t=1}^{36} x(t),\ \ \max_{(132)} K_2; \qquad (133)$$

$$3)\ K_3 = \frac{1}{36}\ \frac{{}_{t=1}^{36} x(t)}{x(0)} + \frac{{}_{t=1}^{36} y(t)}{y(0)}\ ,\ \ \max_{(132)} K_3;$$

$$4)\ K_4 = \frac{1}{T}\ {}_{t=1}^{36}\ \frac{x\ t\ -x(0)}{x(0)}^{\ 2} + \frac{x\ t\ -x(0)}{x(0)}^{\ 2}\ ,\ \ \min_{(132)} K_4.$$

Here T =232 is the period of one cycle) under the constraints.

$$0 \leq y(t) \leq 1,\ 0 \leq n(t) \leq 1,\ 0 \leq x(t). \qquad (134)$$

The base values of the criteria (without parametric control) are as follows:

$K_1 = 0.06848$; $K_2 = 2.05489$; $K_3 = 2.08782$; $K_4 = 0.0307$.

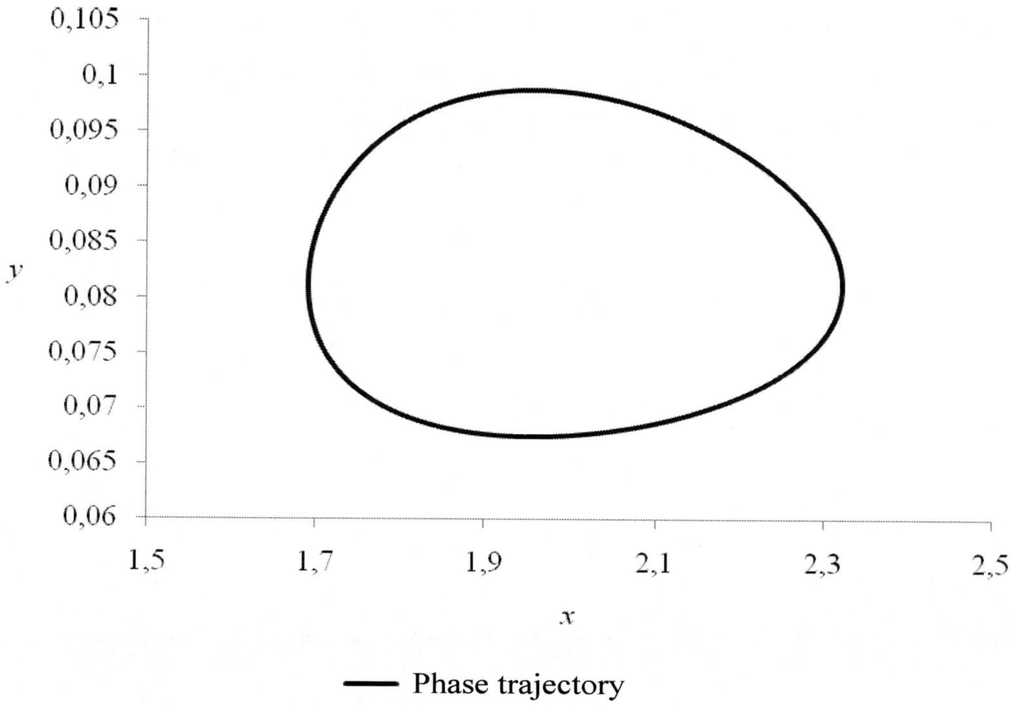

Figure 29. Cyclic phase trajectory of the Kondratiev cycle model.

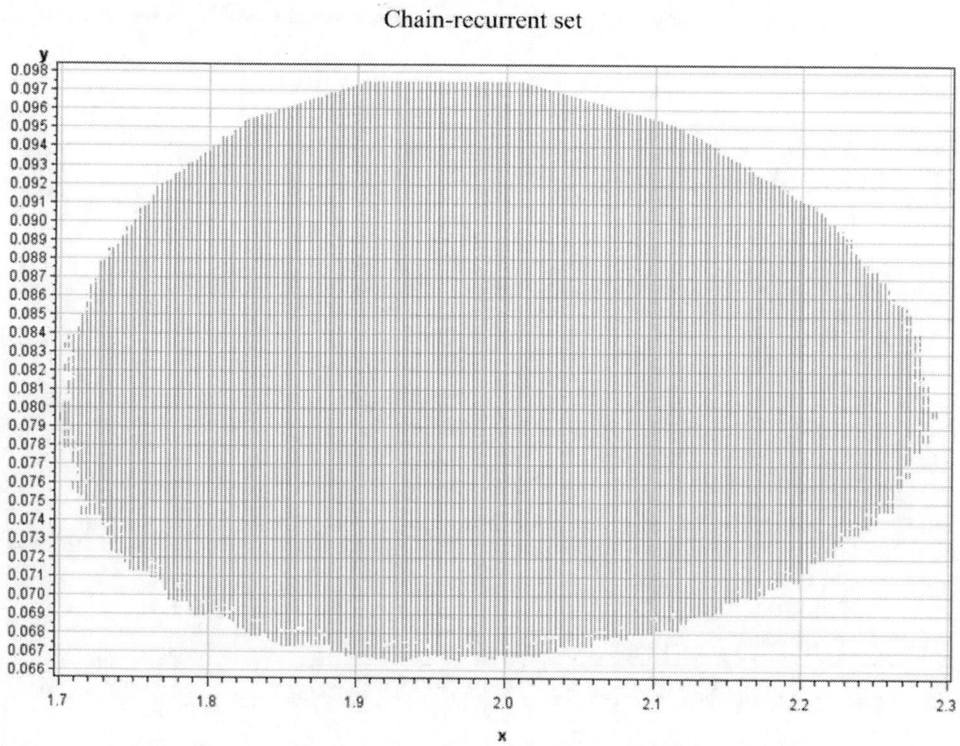

Figure 30. Chain-recurrent set for the Kondratiev cycle model.

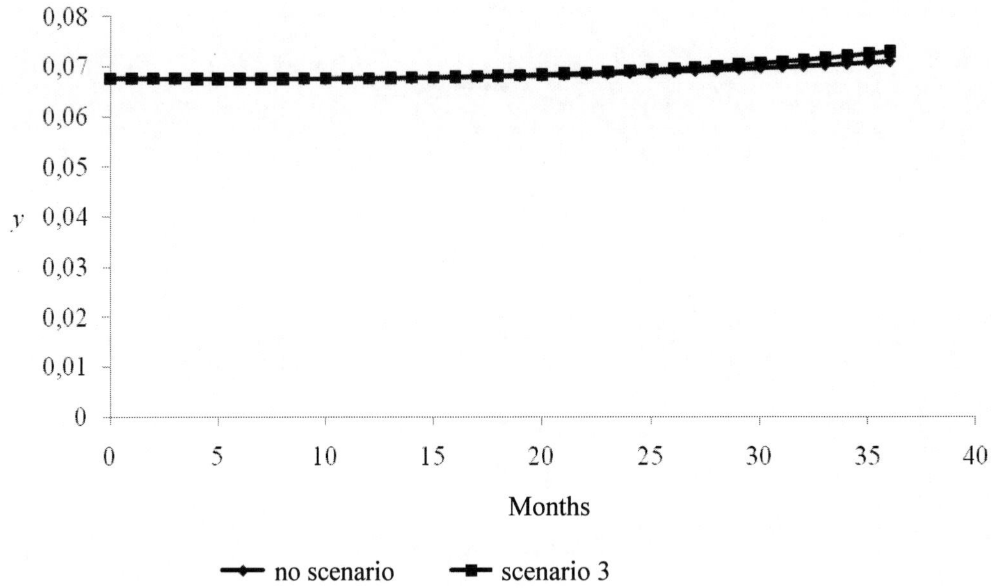

Figure 31. Capital productivity ratio without parametric control and with use of law 3 optimal in a sense of criterion 1.

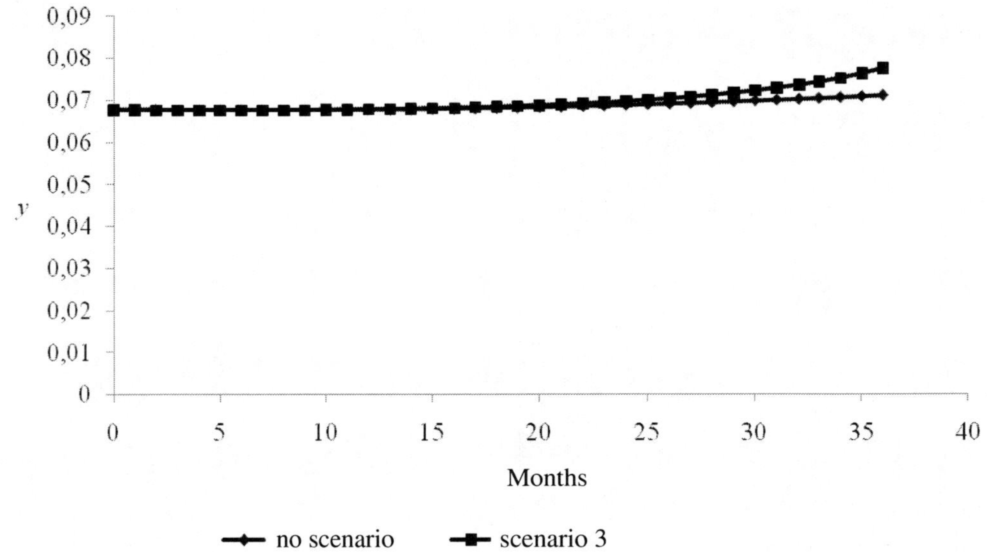

Figure 32. Capital productivity ratio without parametric control and with use of law 3 optimal in a sense of criterion 2.

Table 21. Values of coefficients and criteria for optimal laws

Criterion	Optimal law	Coefficient value	Criterion value
1	3	0.2404966	0.06889
2	3	0.47668	2.230337
3	4	0.071862	2.19674
4	4	0.300519	0.007273

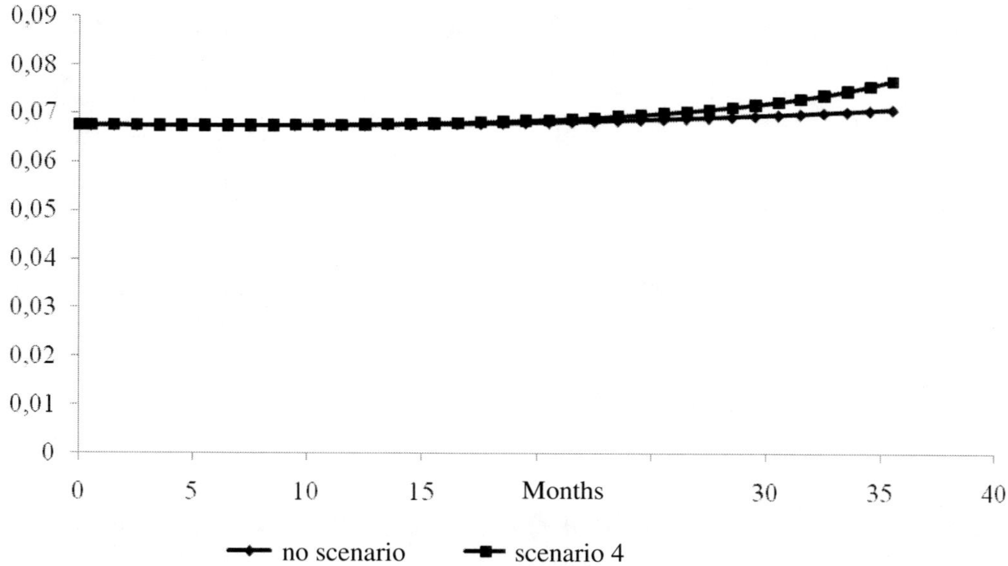

Figure 33. Capital productivity ratio without parametric control and with use of law 4 optimal in a sense of criterion 3.

The values of all criteria for the control law, which is optimal in a sense of the respcctive criterion, from set (132) represented before are obrained as a result of solving the problems formulated above on the application of the parametric control approach to the evolution of the economic system. The results are presented in Table 21.

The values of the model endogenous variables without applying the parametric control as well as with use of the optimal parametric control laws for each criterion are presented below in graphic form (Figures 31-35).

4.4.4. Estimating Structural Stability of Kondratiev Cycle Mathematical Model with Parametric Control

To carry out this analysis, the expressions for optimal parametric control laws (132) with the found values of the adjusted coefficients are substituted to the right-hand part of the second and third equations of system (131) for the parameter n_0. Then, by use of the numerical algorithm for estimating the weak structural stability of the discrete-time dynamic system for the chosen compact set N determined by the inequalities $1.7 \leq x \leq 2.3$, $0.066 \leq y \leq 0.098$ in the state space of the variables (x, y), the estimation of the chain-recurrent set $R(f, N)$ as the empty (or one-point) set is obtained. This means that the Kondratiev cycle mathematical model with the optimal parametric control law is estimated as weakly structurally stable in the said compact set N.

4.4.5. Analysis of Dependence of Optimal Value of Criterion K1 on Parameter for Variational Calculus Problem Based on the Kondratiev Cycle Mathematical Model

Let us analyze the dependence of the optimal value of criterion K_1 on the exogenous parameters μ (share of withdrawal capital production assets per month) and a for parametric control laws (132) with the found optimal values of the adjusted coefficients k_i, where the values of the parameters (μ, a) belong to the rectangle A = [0.00063, 0.00147]×[-0.01, 0.71] on the plane.

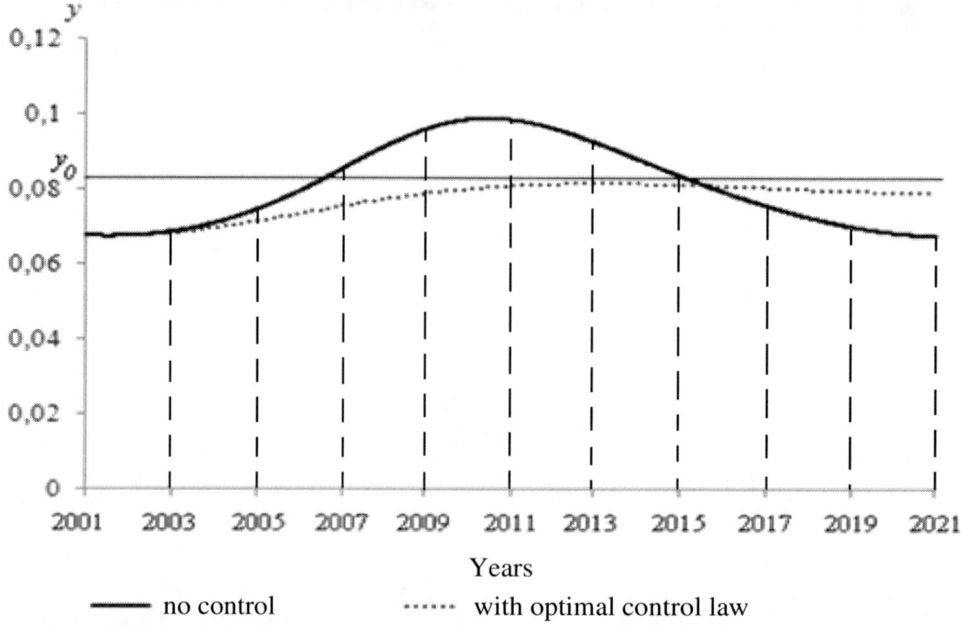

Figure 34. Capital productivity ratio without parametric control and with use of law 4 optimal in a sense of criterion K_4.

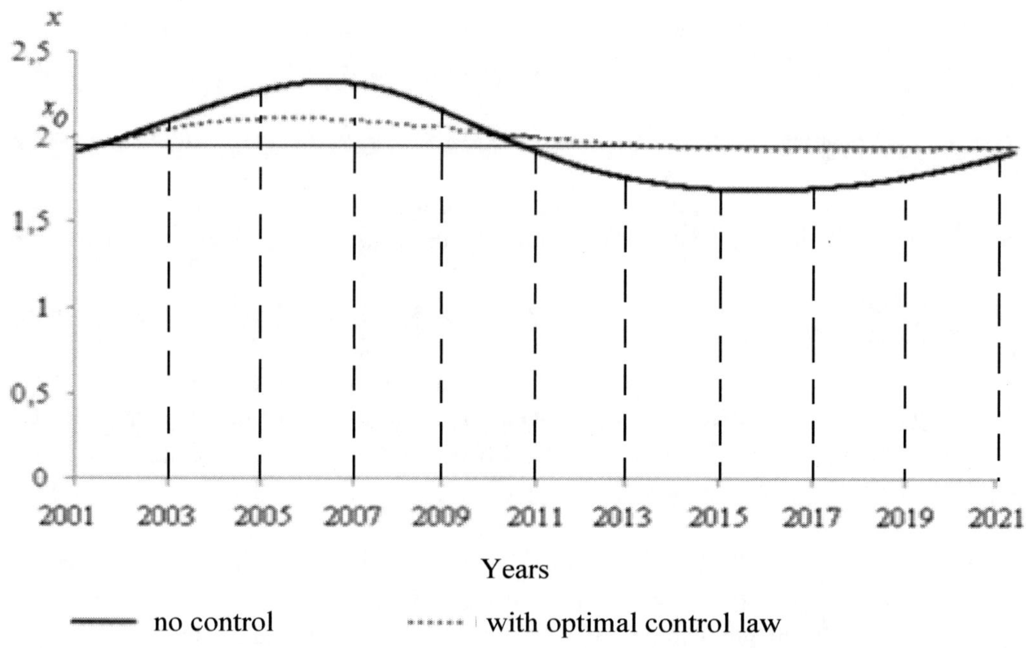

Figure 35. Efficiency of innovations without parametric control and with use of law 4 optimal in a sense of criterion K_4.

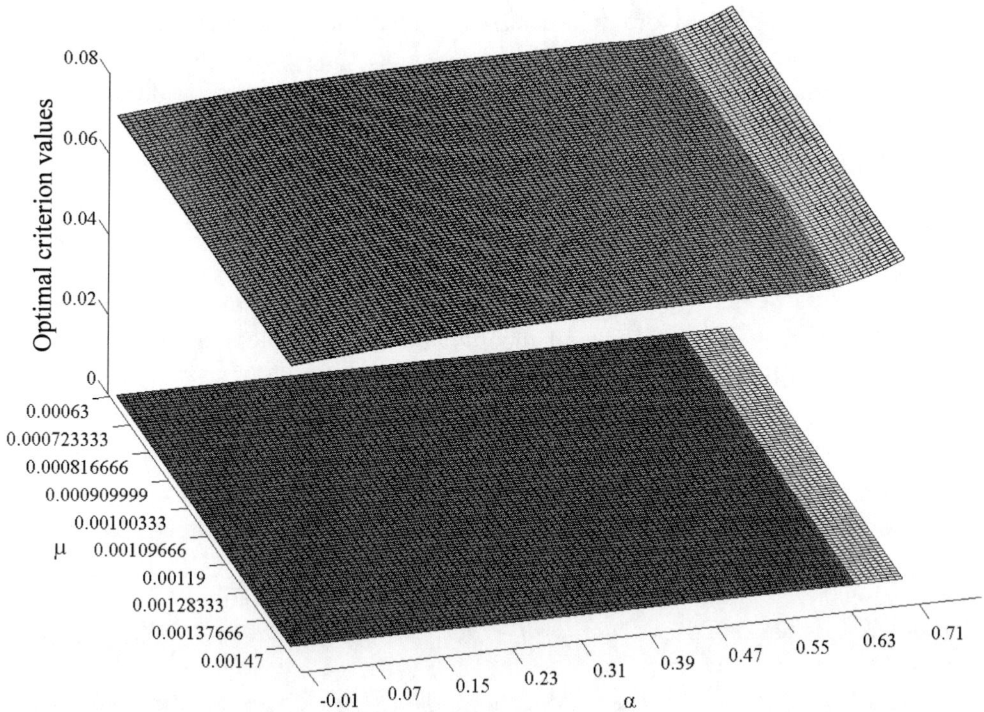

Figure 36. Plots of the dependencies of optimal value of criterion K on exogenous parameters μ, a. Here the colours corresponds to the numbers of laws of parametric control (132) as follows: ▇ - 3, ▇ - 1.

The plots of dependencies of the optimal values of criterion K_1 (for parametric control laws 1 and 3 yielding the maximum criterion values) on the uncontrolled parameters (see Figure 36) were obtained as a result of the computational experiment. The projection of the intersection line of the two surfaces on the plane (μ, α) consists of the bifurcation points of the extremals of the considered variational calculus problem.

CONCLUSION

The chapter illustrates the possibility of offering alternative policy rules based on the application of the parametrical regulation theory by example of the following models: CGE model of economic sectors, couple of CGE models with common economic optimization tools, Smets-Wouters DSGE model and the model of Kondratieff cycle.

Within CGE (or DSGE) model the approach of choosing the optimal law of parametrical regulation which provides extreme value (agreed with the decision maker) for the efficiency criterion of economic growth or criterion characterizing the degree of impact of various shocks on macroeconomic indicators can be considered as a guideline for political rules of economic policy.

Within the framework of making recommendations on the selection of effective policy rules using the set (library of macroeconomic models) by example of two models (CGE model of economic sectors and CGE model with a sector of knowledge) the approach to finding agreed solutions based on two criteria (representing real GDP and real GVA of innovative sector of the economy, respectively) of the correspondingly formulated optimization problems with general economic optimization tools based on these models can be considered as recommendations on the political rules of economic policy.

As part of the Kondratieff cycle model the approach to selecting the rules based on the study of dependence of criterion optimal values (agreed with the decision maker) on values of uncontrollable external factors can be considered as recommendations on the political rules for stabilization policy.

Thus, the results of the research presented in this paper represent the elements of the modern paradigm for macro analysis and recommendations development on the choice of alternative economic rules.

REFERENCES

[1] Kydland, F. E., Prescott, E. C. (1977).Rules rather than Discretion: the Inconsistency of Optimal Planes. *Journal of Political Economy*, Vol. 85, No. 3, 473-490.

[2] Turnovsky, S. J. (2000). *Methods of Macroeconomic Dynamics.* Cambridge: MIT Press.

[3] Dornbusch, R., Fischer, S. (1987). *Macroeconomics.* Singapore: McGraw-Hill.

[4] Ashimov, A. A., Sagadiyev, K. A., Borovskiy, Yu. V., Iskakov, N. A., and Ashimov, As. A. (2008). On the market economy development parametrical regulation theory. *Kybernetes,* Vol. 3, No. 5, 623-636.

[5] Ashimov, A. A., Sultanov, B. T., Adilov, Zh. M., Borovskiy, Yu. V., Novikov, D. A., Nizhegorodtsev, R. M., and Ashimov, As. A. (2012). *Macroeconomic Analysis and Economic Policy Based on Parametric Control*. New York: Springer.

[6] Tarasevich, L. S., Grebennikov, P. I., Leusskiy A. I. (2006). *Macroeconomics.* Moscow: Higher education. (in Russian).

[7] Acemoglu, D. *Introduction to modern economic growth.* (2009). Princeton: Princeton Univ. Press.

[8] de la Croix, D., Michel, P. (2002). *A theory of economic growth – Dynamics and policy in overlapping Generations*. Cambridge, MA: Cambridge Univ. Press.

[9] Leontief, W. (1966). *Essays in Economics: Theories and Theorizing*. London: Oxford University Press.

[10] Sharaev, Yu. V. (2006). *Theory of economic growth*. Moscow: Publishing house of SU HSE. (in Russian).

[11] Makarov, V. L., Bakhtizin, A. R., Sulashkin, S. S.(2007). *The Use of Computable Models in Public Administration*. Moscow: Scientific Expert. (in Russian).

[12] Smets. F., Wouters, R. (2003). An Estimated Dynamic Stochastic General Equilibrium Model of the Euro Area. *Journal of the European Economic Association*, Vol. 1, No. 5, 1123–1175.

[13] Nelder, J. A., Mead, R., (1965). A simplex method for function minimization. *The Computer Journal*, No. 7, 308-313.

[14] Orlov, A. I. (2002). *Econometrics.* A Textbook. Moscow: Publishing house "Ekzamen". (in Russian).

[15] Sims, C. (2004). *Random Lagrange Multipliers and Transversality* [PDF]. http://sims.princeton.edu/yftpb/Macro2004/rlg.pdf.

[16] Orphanides, A., (2007). Taylor Rules - Federal Reserve [PDF]. www.federalreserve.gov/pubs/feds/2007/200718/200718pap.pdf.

[17] Fernández-de-Córdoba, G., Torres, J. L. (2009). Forecasting the Spanish economy with an Augmented VAR-DSGE model. *Working Papers*, No 2009-1. Malaga: Malaga Economic Theory Research Centre.

[18] Blanchard, O. J., Kahn, C. M. (1980). The Solution of Linear Difference Models under Rational Expectation. *Econometrica*, Vol. 48, No. 5, 1305-1312.

[19] The Agency of Statistics of the Republic of Kazakhstan. [VEB page]. http://www.stat.kz.

[20] Fernández-Villaverde, J. (2010). The econometrics of DSGE models. *International Review of Economics*. Vol. 1, No. 1, 3-49.

[21] Hamilton, J. D. (1994). *Time Series Analysis*. Princeton: Princeton University Press.

[22] *Dynare.* [VEB page]. http://www.dynare.org.

[23] Turnovsky, S. J. *Methods of Macroeconomic Dynamics*. Cambridge: MIT Press 2000.

[24] Kolemaev, V. A. (2002). *Mathematical economics*. Moscow: Unity. (in Russian).

[25] Kydland, F. E., Prescott, E. C. (1977) Rules rather than Discretion: the Inconsistency of Optimal Planes. *Journal of Political Economy*, Vol. 85, 473-490.

[26] Dornbusch, R., Fischer S. (1990). *Macroeconomics*. New York: McGraw-Hill.

[27] Tumanova, E. S., Shagas, N. A. (2004). *Macroeconomics.* Moscow: Infra-M. (in Russian).

[28] Dubovskiy, S. V. (1995). Kondratiev cycle as an object of modelling. *Mathematical modelling,* Vol.7, No.6, 65-74. (in Russian).

[29] *Statistical year-book of Kazakhstan*. Ed. K. S. Abdiev (2001-2010). Almaty: Statistical Agency of the Kazakhstan Republic.

[30] Robinson, C. (1980). Structural Stability on Manifolds with Boundary. *Journal of differential equations*, No. 3, 1-11.

[31] Bautin, N. N., Leontovich E. A. (1990). *Methods and techniques of qualitative analysis of dynamic systems on plane*. Moscow: Nauka. (in Russian).

In: Economic Developments and Emerging Markets ... ISBN: 978-1-62417-986-0
Editors: Guy Boscarino and Derek Notte

Chapter 3

DEVELOPMENT OF THE WORLD ECONOMY IN THE SYSTEM OF CONTINENTAL-OCEANIC POLARIZATION

Leonid A. Bezrukov*
V.B. Sochava Institute of Geography SB RAS, Irkutsk, Russia

ABSTRACT

This study examines the main economic mechanisms as reported in scientific literature that are responsible for the performance of the system of "core-periphery" relations in the world economy. It is pointed out that there is a need for a more comprehensive validation of the "core-periphery" model through the use of alternative and complementary methodological approaches.

Such an additional substantiation is suggested in the form of the economic-geographical concept of the continental-oceanic dichotomy to furnish a means of adding immeasurably to our fundamental knowledge of the reasons behind the current structure of the world economy. The aforementioned dichotomy is considered to mean a successive and fundamental divarication of the integral world economy into two opposite and, at the same time, mutually complementing types of economies: continental and oceanic (maritime).

A geoeconomic typology has been performed for a number of key countries of the world by categorizing them as continental and oceanic (maritime), based on assessing the degree of their "transport-geographical continentality". It was possible to unravel the workings of the redistribution mechanism for income (the mass of surplus product) between landlocked and maritime countries in the process of international division of labor. The study highlights the main negative socio-economic implications of increased transport expenditures for landlocked commodity producers and regions.

This study has ascertained a correlation of the degree of transport-geographical continentality of the countries with the level of their socioeconomic development which has been assessed by using the indicators of per capita GNP, per capita export, and life expectancy.

---

* Corresponding author: Leonid A. Bezrukov. Doctor of Geographical Science, Head of the Laboratory of Georesource Studies and Political Geography, V.B. Sochava Institute of Geography SB RAS (Russia, Irkutsk), Professor of Politology and National History, History Department, Irkutsk State University. E-mail: bezrukov@irigs.irk.ru.

Based on estimating the changes in the volumes of world production and export as regards products of different transportability depending on the remoteness of producers and exporters from the sea, this study revealed the most general regularities of territorial organization of the world economy in the system of continental-oceanic polarization. It is established that the maritime countries and maritime zones play a leading role in the shaping of the world economy and price formation in world commodity markets.

It is suggested that there is a need for a radical change in the international commercial rules in order to eliminate the one-sided economic advantages of the maritime (seaside) countries to the detriment of the continental (landlocked) countries.

1. INTRODUCTION

Organizing the main kinds of economic activity across the globe, the ongoing global processes are transforming the world economy to a unified market of commodities, services, capitals, labor force, information, technologies, etc. This is not to say, however, that the world is becoming more homogeneous and prosperous. Not only does globalization open up new avenues for an acceleration of scientific and technological, economic, social and cultural progress, but it also creates for many countries and regions additional costs and threats. Many researchers are quite justified in claiming that the acceleration of globalization and accumulation of the bulk mass of production and services exacerbates the problem of world inequality and is accompanied by an increase in socioeconomic disproportions, and by emergence and conservation of zones of backwardness and stagnation.

The reasons behind a dramatic unevenness of development of countries and regions are usually recognized through various theoretical constructs. The most advanced of them are the "core-periphery" concept (model) and the theoretical constructs deriving from them: the concept of "world systems", the theory of "peripheral dependent development", the theory of "internal colonialism", and others. According to the "core-periphery" model, central to structuring of the world and the world economy, countries and regions and their economic systems is a constant redistribution of the resulting surplus product from peripheral and semi peripheral zones to corresponding "prospering" centers [1–4].

The functioning of the core-periphery system of relations in the most general form is explained in the scientific literature by the operation of several mechanisms, typically having a strictly economic character. The case in point is primarily the "innovation" mechanism which is considered to mean a qualitative transformation of the center (core) through generation, introduction and diffusion of innovations, the specialization of the center in effective sectors of the economy, and the attainment of the highest productivity of labor to receive an "intellectual" or "technological" rent [1–6]. Another mechanism for the "core-periphery" model is provided by nonequivalence of the international exchange of products with a different degree of processing: since the prices of finished products grow faster when compared with primary goods, exporters of raw materials and semi finished products come off worse — according to "trading conditions" the ratio of the index of export prices to the index of import prices) [6–11]. Furthermore, in accordance with the so-called "law of first price, the original (starting) prices of raw materials from colonial and dependent countries were fixed historically at the lowest level of advantage to buyers from "mother countries" and other advanced states [12]. The inequitable relations between the core and the periphery are also cause by the operation of the mechanism (law) of the "average rate of profit" where some of the surplus value of the sectors with a low organic composition of capital is appropriated by the sectors with a high composition thereof [13]. An additional mechanism that is responsible for worldwide inequality also includes the fundamental differences between the advanced system of relation of private property in the West which furnishes a means of transforming fixed assets to the source of capital, and the absence of such a system in the countries of the third world and of the former socialist camp [14, 15].

Besides, it is a rather widespread opinion that not only do the purely economic mechanisms have a determining influence upon the unevenness of world development, but also a steady current decline in state, distance and other barriers as well as geographical conditions and factors involved in these processes. Thus, according to the eminent Russian economist O.T. Bogomolov, "the influence of the natural, resource and climatic differences and of the geographical location of countries in the international division of labor is gradually dwindling giving way to scientific and technological progress, and, as a result of the unprecedented successes of the transport in the latter half of the 20th century, the "global economic space" appears to become more compact and more accessible for development and profitable exploitation" [10, p. 163]. Pointing out some independency of the level of contemporary socioeconomic development of separate countries from the possession of natural resources, one of the leading American economic futurologists L.C. Thurow mentions that the availability of natural resources dropped out of the competition equation [16]. L.C. Thurow explains these changes by invoking a reduction in cargo and passenger transportation costs during the last several decades, because the current transportation costs have created such a world in which the resources can be transported cheaply where they are needed [16].

Thanks to progress in the transport-communication sphere, an ever increasing proportion of products manufactured on the globe are involved in the international trade turnover: the growth rates of worldwide exportation of goods in the latter half of the 20th century were outstripping, on the average, the growth rates of industrial production by a factor of 1.5, and the growth rates of GDP by a factor of 1.5 [13]. In this connection, some authors argue that the transport factor no longer has any role. For instance, an eminent Russian geographer A.D. Armand [17] is of the opinion that "the principle of payment for the distance" is becoming devoid of its determining significance". His statement is reiterated by the German researcher F.J. Radermacher, who claims that, because of the revolution in transport and information technologies, the distance no longer has an important role [18]. But do these tendencies operate in an identical manner for all regions of the globe, to what extent can we claim that there is a decline in the role played by the geographical and transport factors in the formation of contemporary world-economy pattern and, generally speaking, can the aforementioned factors be used in ascertaining the reasons behind the extreme heterogeneity of the world economy?

A number of researchers [7, 19], recognizing some sketchy character of the "core-periphery" concept and explaining its economic mechanisms, are justified in pointing to a need to seek new or at least complementary methodological approaches and substantiations. It is quite clear that when analyzing the processes of international and regional development accompanied by an enhancement in the territorial differentiation of society, it will obviously not suffice to confine attention to the framework of economic theory only.

To accomplish this, it is necessary to take into consideration, comprehensively and in all its entirety, the influence of basic geographical factors underlying today's heterogeneity of the world, which calls for the development of appropriate synthesizing economic-geographical concepts.

2. The Continental-Oceanic Dichotomy Concept, and the Income Redistribution Mechanism between Continental and Oceanic Countries

Such a concept that supplements the economic cause-and-effect workings responsible for the emergence and functioning of the core-periphery model of the world economy can be represented, we believe [20], by the economic-geographical concept of the "continental-oceanic dichotomy". This dichotomy is considered to mean a consistent and fundamental divarication of an integral world economy into two opposite and, at the same time, mutually complementing types of economy — continental (landlocked) and oceanic (maritime) which differ drastically, depending on their macrolocation relative to the sea, in the organization and effectiveness of the economy, the specific character of the interaction with the external world, and in the pathways of development.

The suggested economic-geographical concept of the continental-oceanic dichotomy is built upon the radical differences in effectiveness of more expensive land and economical sea transportation, on the one hand, and on the characteristic features inherent in the macrolocation of the countries and regions relative to the sea.

The point is that the transport component of the end price of many commodities is still characterized by considerable amounts, reaching several tens of percent for many kinds of raw materials and semifinished products. Transportation costs will therefore have an important significance, as before, in the establishment of international and interregional economic ties, and it is likely that they still remain a crucial factor that governs the processes of economic exchange and, in the final analysis, of production itself.

A quantitative assessment of the former initial cause for the continental-oceanic dichotomy, i.e. the differences in effectiveness of land and sea transportation, was made by correlating the average rates of return from cargo transportation by rail transport with the average freight rates for sea transportation. Calculations showed that the freight rates for sea transportation as of the end of the 20th–beginning of the 21st century are lower than the average rates of return for freight railroad transportation of West Europe and Japan by a factor of 70–80, North America — by a factor of 20–25, and Russia — by a factor of 5–10 [20].

In view of the obvious technical and economic advantage of sea transportation over the more costly land transportation, the continental countries and regions differ from maritime ones by an increased level of costs for transportation of a unit of identical products.

A quantitative assessment of the latter cause of the dichotomy under consideration, i.e. the macrolocation of the countries relative to the sea, was carried out by using this author's technique, based on determining the degree of "transport-geographical continentality" (TGC) of the countries as the measure of remoteness of most of their demoeconomic potential from sea and ocean routes with year-round navigation [20].

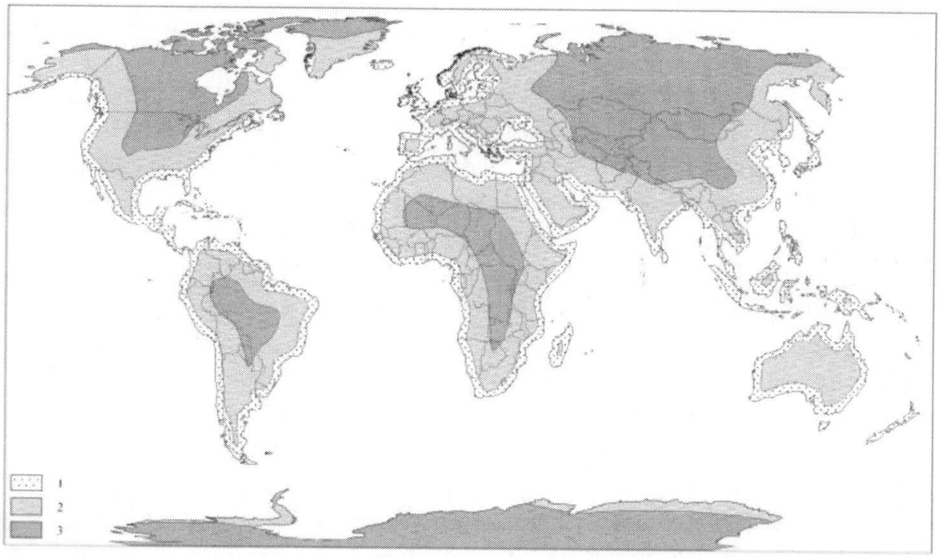

Figure 1. Transport-geographical remoteness of the territories of the Earth's continents and islands from oceans, seas and sea ports with year-round navigation Zones of transport-geographical remoteness (distance from oceans, seas and sea ports, km): 1 – marine (0–200); 2 – continental (200–1000); 3 – ultra-continental (over 1000).

The "maritime" and the "continental" zones incorporated a strip as wide as 200 and 200–1000 km from the shore, respectively, and the "ultracontinental" zone included the territories at a distance of more than 100 km from the shores (Figure 1). The indices of the degree of continentality of the countries were updated with due for their having inland water ways accessible for sea vessels.

The values of the degree of TGC lie within 1 to 100 points. The minimum value equal to 1 point shows that the entire population of a given country lives within the coastal strip not further than 50 km from the shore, while the maximum value 100 points indicates that the entire population is concentrated in the zone at a distance of more than 2000 km from the water edge. The larger the index of the degree of a country's continentality, the more unfavorable the macrolocation of a given country and, primarily, of its main demoeconomic centers relative to the sea and ocean routes.

The selection of the countries to be assessed took into consideration, in the first place, their significance in the world economy, second, the diversity of macrolocation relative to the sea, and, third, the location in all habitable parts of the world.

The representativeness of the sample is supported by the fact that the 37 countries used in this analysis constitute more than 70% of the entire habitable land surface where, as of the early 2000s, about 72% of the world population lived and nearly 80% of the world GNP was produced.

The characteristic of the degree of continentality of the countries of the world is formalized according to six classes (points): very high (60.1–100), high (35.1–60), increased (20.1–35), moderate (10.1–20), low (5.1–10), and very low (1.0–5). Results of calculations for the 37 countries of the world intimate that a very high degree of continentality corresponds to Kyrgyzstan, Kazakhstan and Mongolia; high — Chad, Afghanistan, Congo (DRC), and Russia; increased — Iran and Paraguay; moderate — Belarus, China, India,

Nigeria, Ukraine, Switzerland, Poland, Hungary, m1exico, and the US; low — the Republic of South Africa, Argentina, Canada, Brazil, Germany, Saudi Arabia, France, Thailand, and Egypt, and very low — Turkey, Italy, Australia, Indonesia, Great Britain, Sweden, Japan, Island, and Singapore.

Based on the resulting indices of the degree of TGC and their respective specified classes, all countries of the world are subdivided, to a first approximation, into two geoeconomic types: continental, and oceanic (maritime) [20]. The former type includes countries with a very high and a high degree of continentality (Kazakhstan, Mongolia, Chad, Afghanistan, Russia, and others), and the latter type involves countries with a very low and a low degree (Japan, Great Britain, Italy, Australia, Brazil, Germany, Turkey, and others). The intermediate position corresponds to the countries with a moderate and increased degree of TGC, gravitating (according to their economic, political and geographical location, the sectoral structure of transport, and to other factors) towards either the continental type (Ukraine, Iran, etc.) or the oceanic type (US, China, Poland, etc.).

The logic of economic-geographical approach provides a general insight into the mechanism of income redistribution (the mass of surplus product) between continental and oceanic countries which stems from the inequality of their transportation costs and which takes place in the process of the international division of labor. In accordance with the rules of today's international commercial practice, the distribution of transportation costs connected with the delivery of goods between the seller (producer) and the buyer (consumer) is regulated by the so-called "basic delivery conditions" which, along with the payment conditions, constitute one of the fundamental (obligatory) conditions of any foreign trade contract [21].

The export value is taken into consideration in most cases on uniform basic FOB[1] terms. According to these terms, the seller/exporter shall bear all expenses connected with transportation of goods to the exporting sea port, and with loading on board ship. As a result, it transpires that in the case of the delivery of products to the world market from continental countries, most of the aggregate transportation costs shall be covered by their sellers (producers). It is they who should pay for the most expensive section of the route, i.e. the expenses connected with overland transportation of cargo within their territory to sea ports and the borders.

Imports are usually registered on the basic condition CIF[2] terms. According to these terms, importers (or, more exactly, consumers of imported commodities) are in a disadvantageous position — in this case, too, again because of continental countries. They have to cover the expenses connected with the most expensive section of the route, i.e. the overland transportation costs of cargoes from sea ports to inland regions of their respective countries, which increases substantially the end prices of imported goods.

Consequently, when interacting with the world market, landlocked producers and exporters, having to compensate for their own increased transportation costs, derive much smaller revenues compared to maritime produces and exporters. Only in this case, however, commodities from a continental country can compete, as regards prices, with goods from a maritime country. In the event of goods imported from the world market, however, landlocked consumers will make payment on prices far in excess of the prices of consumers

[1] FOB – Free on Board.
[2] CIF – Cost, Insurance, Freight.

from oceanic countries. Thus the drastic differences in the degree of continentality are responsible for the fundamentally unequal levels of transportation costs in continental and oceanic countries which generate a perceptible difference in prices of commodities of respective producers and consumers, which, in turn, leads to a constant redistribution of the mass of surplus product in favor of oceanic countries.

The main economic consequences at the international level from the operation of the mechanism of the continental-oceanic dichotomy are revealed in a general form by comparing the price competitiveness in the world market of products from continental (landlocked commodity producers) and oceanic (maritime commodity producers) countries. Given an identical labor intensity of the landlocked and maritime commodity producers, for the former producer the amount of transportation costs and, hence, the share of the transportation component in the product cost will be considerably higher than for the latter producer. Given the equality of the main expenses (of course, with the exception of transportation costs), the prime cost and, hence, the value and price of identical products in continental countries and regions — because of the increased transportation component — will be obviously higher than in oceanic countries, which actually signifies price non-competitiveness of products from landlocked producers in the world market.

If, however, the value and price of identical products are the same in the world market, the structure of their prime cost (or expenses) in continental countries must transform to some extent. In this case, the relative price competitiveness of products from landlocked producers can be achieved primarily through lower remuneration of labor and lower depreciation allocations (and, where possible, through cuts in expenses on energy, fuels, raw materials, etc.) as well as through some profit squeeze. Such deformation of the entire system of financial and economic indicators of the performance of landlocked producers — cutdown in remuneration of labor, depreciation allocations, profits, etc. — is automatically responsible for a decline in investment possibilities, production efficiency, budget revenues, purchasing capacity of the population, consumer market capacity, etc. The final consequences of these negative changes include a slackening of the rates of economic growth, and a deterioration in the living standards of the population in continental countries when compared with oceanic ones. The mechanism considered above is, we believe, one of the leading mechanisms that are responsible for the differentiation in the world economy pattern. "Not only does the continental-oceanic income redistribution mechanism have a strong direct and indirect influence on contemporary international and regional development, but also it serves (or served in the past) as the primary cause for the emergence of other mechanisms that are shaping the core-periphery relations in the world economy. Thus, by virtue of their centuries-long supremacy in effective sea navigation which was then consolidated by the historical inertia, the maritime West-European countries adjusted "for themselves" the entire system of the international division of labor by monopolizing long-distance intercontinental trade and building it upon nonequivalent exchange. Largely because of these advantages, the advanced oceanic countries of the West, the core of which was composed of West-European countries and peoples, are now the "prospering" center of the world economy (the "golden billion") which differs drastically from less developed and semiperipheral zones. On the whole, however, the mechanism of redistribution of the mass of surplus product from continental to oceanic countries makes it possible to gain insight into the most general primary traits of formation and structuring of the world economy irrespective of its particular bloc-specific, formational, regional and other groups and subgroups, largely having a secondary character.

Hence it is clear that the views maintained by a number of economists and economic geographers concerning a decline in the role played by the geographical and transport factors call for serious revision. The values of transportation costs are highly differentiated in regions and countries of the world; they are determined primarily by the differences in effectiveness of transportation between the dominant modes of transport. The most dynamical mode of transport is the sea transport that ensures an unprecedented (for the other kinds of conveyance) cheapness of intercontinental transportation. For the last century the proportion of the transport component in the price of products for the consumer has decreased appreciably, on the average, for developed countries, which was due largely to widespread use of the economic advantages of sea transportation.

However, a drastic reduction in sea transportation costs led to a still greater transport-economic polarization of regions and countries of the world within the continental-oceanic context, primarily to an enhancement of the pre-existing marked "squeeze" of landlocked regions when compared to maritime ones. Thus, according to data reported by H. Scheer [19], the transportation of a comparable amount of cargo by railroad within Germany from Passau (Bavaria) to Bremen to 700 km is now more costly when compared with the sea transportation of the same cargo to Bremen from Australia to tens of thousands of kilometers. According to L. Annovazzi-Jarab [22], the transportation costs for a standard container from Baltimore (US) by sea to the coast of Africa (Cameroon) to a distance of 9 thou km amount to about 3 thou US dollars, whereas the delivery of this same container from the coast to the Central African Republic by rail and motor road transport to 950 km will now amount to 13 thou US dollars. The American economists of authority J.L. Gallup, J.D. Sachs and A.D. Mellenger A.D. [23] established a consistent regularity of an increase in transportation costs with the distance from the sea coasts to inland regions, a feedback of the amount of such costs with values of per capita GNP for countries of the world, and with competitiveness of products. According to calculations made by S. Radelet and J.D. Sachs [24], the shipping cost for landlocked developing countries is, on the average, by 50% larger than for developing countries having the entry to the sea.

For most landlocked regions and countries the high transportation costs are still an important obstacle to obtain an effective entrance to the world market and to ensure an equitable competition with other regions and countries. Hence, the concepts of cheapness of transportation costs refer largely to maritime regions with sea transportation, whereas in the heart of the continents the transport-economic situation remains fundamentally different. It turns out that in the latter half of the 20th century the factor of shipping cheapness involving the main competitive advantages corresponded to maritime regions, primarily leading oceanic countries which are the creators of today's world economy and the key "players" in the world market.

3. Correlation between the Degree of Continentality of Countries and the Level of Their Development

The assumption that the degree of transport-geographical continentality of the countries of the world must have a definite and rather strong influence upon their key socioeconomic indicators will now be verified by an attempt to reveal the closeness of correlation between the desired degree of TGC and the level of development of particular countries. In our calculations, we use the set of 37 countries supported by the estimates of the degree of continentality made above. This sample is quite representative, because the countries included in it encompass the entire possible range of change and the degree of TGC as well as of the level of socioeconomic development, and are situated in all parts of the world; many of them play a leading role in the world economy and politics. An important point also is that most of the countries under consideration have existed independently (as sovereign states) for a rather long period of time (over the course of the last 80–100 years at least). Theoretically, for them the correlation between the degree of TGC and the socioeconomic indicators must be the most natural and organic.

Some exception is provided by the former colonies, especially those which achieved independence in the mid- and latter half of the 20th century. The consequences of colonial enslavement are severely detrimental to the contemporary level of socioeconomic development of a number of countries, such as India, Indonesia, Nigeria, Chad, and Congo (DRC). Another exception is provided by the CIS countries which have been for centuries integral parts of a unified state — initially, the Russian Empire, and then the Soviet Union. In a unified state, such as the USSR in particular, the difference in the level of socioeconomic development of the individual parts of the gigantic state smoothed over to a certain extent, including the difference caused by the characteristics of the macrolocation of those parts relative to the sea. Historically, the period of sovereign existence of the CIS states is still too short for these characteristics — after the disintegration of the USSR — to have a marked influence on the level of their development. That is the reason why Kazakhstan, for example, stands out among the other states with a similar degree of continentality as having relatively better socioeconomic indicators. Nevertheless, there is good reason to include in the sample the former colonies as well as the CIS countries in order to thoroughly cover and analyze the largest states of the world, albeit this would lead to some decrease in closeness of correlation between the degree of TGC of the countries of the world and the level of their development.

As the most important indicators of the level of development of the countries of the world, we used the following per capita GNP having regard to the parity of the purchasing capacity (PPC) of the currencies in 2008 and life expectancy in 2010 according to data reported in [25], and per capita export in 2010 according to [26]. The indicators of per capita GNP for PPC and life expectancy form part of the calculated indicators as used by the UN technique in determining the "Human Development Index (HDI). This compound indicator of economic development attempts to get away from purely monetary measurements by combining per capita GNP with life expectancy and literacy in a weighted average. The indicator of "per capita export" is a measurement of the currently important globalization of the degree of a country's involvement in international trade and is indicative primarily of the competitiveness of a national economy and of its commodities and services — their quality and value, the degree of processing, and the "mobility" of products, etc.

To determine the correlation between the degree of continentality of countries and their most important socioeconomic indicators, all the data series were ranked from 1 to 37 in order to avoid any large spread in absolute values. In accordance with the ideology of our investigation, the lowest rank for each of the socioeconomic indicators was assigned to the country with the "worse" (minimum) value of this indicator. The lowest rank of the indicator of the degree of TGC was assigned to the country with its smallest value, i.e. the location of this country is "better" (nearer) relative to the sea when compared to the other countries.

The degree of transport-geographical continentality of the countries is related to their most important socioeconomic indicators by a linear inversely proportional dependence. The closeness of rank correlation between the indicators under consideration turned out to be largely moderate. The coefficient of pair correlation between the degree of TGC and per capita GNP is –0.63, the degree of TGC and per capita export — –0.57, and the degree of ЕПС and life expectancy — –0.69. The resulting correlation coefficients signify that the factor of the degree of ЕПС accounts for 32–48% of the variation in the most important socioeconomic indicators. In the general case, with an increase in the degree of continentality of the countries of the world, there is a regular deterioration in their key socioeconomic indicators. In spite of the moderate closeness of the correlation, our calculations clearly show statistically that there are dramatic basic differences in the economic effectiveness of countries of the continental and oceanic types.

The macrolocation of the countries relative to sea routes is still of important significance in the contemporary global economy. Remoteness from the sea deprives the corresponding countries (primarily those with a very high and high degree of continentality) of the potential of the necessary economic dynamism, impairs the possibilities of participating in world economy relations, hinders and rises the cost of access to the ever increasing world flows — to traditional flows of goods as well as to finance, innovation and information flows.

On the other hand, it would be inappropriate to draw any straightforward conclusions about a direct decisive influence of transport-geographical continentality on the level of contemporary socioeconomic development of the countries; incidentally, this was demonstrated by results of correlation analysis. The reasons why the states of the world now show a different level of development — in spite of their similar macrolocation relative to the sea — are rooted in the past. The point is that the history of world civilization has accumulated a host of economic, political, cultural, religious, institutional and other differences between countries, the action of which often "overrides" the direct influence of the degree of their continentality. There would be no need for a retrospective journey into the history, if the factor of the past, initial level of development, the belonging of a country to the "core" or the "periphery" of the world economy had no property of cumulativeness and if it were not largely determining as regards the subsequent trajectory of national economic growth, "concealing" in this case the action of purely geographical factors.

The dominant role of the maritime countries of West Europe in the exploration of the globe is universally known, combined with the global proliferation of western culture, and the formation of the world market and the international division of labor. Most of West-European countries boast a favorable location relative to the sea (a very low and low degree of continentality).

Hence, it comes as no surprise that they have retained their centuries-long leadership in the key socioeconomic indicators. However, the level of development of the landlocked countries in West Europe with a moderate degree of TGC (Switzerland, and others) in no way differs from the level of neighboring maritime states, and is accounted for by their comparative proximity to the major world centers and markets, their relatively short distance from the sea, many years of close economic ties with maritime countries, adaptation to the specific character of the landlocked location of the entire economic structure (specialization in tertiary industries, transition to the manufacture of solely transportable finished products, etc.). According to the main socioeconomic indicators, the West-European countries are similar to the US and the so-called countries of "resettlement capitalism": Canada and

Australia, as well as to Japan, the major centers of which have an advantageous maritime (or, arbitrarily, maritime) location.

The other "pole" as regards the advantageous location relative to sea routes (a high degree of continentality) and the level of development includes the backward countries of "Black" Africa and Afghanistan. According to the most important socioeconomic indicators, they are closely followed by a group of countries with a different degree of TGC, including with a very low degree, with a common trait of their colonial or semicolonial past (Nigeria, India, Indonesia, and Mongolia). This example is a demonstration that the factor of initial common backwardness "outweighs" considerably the action of the factor of location relative to seas. Thus, both the landlocked Chad and Nigeria with its access to the sea have similar, low socioeconomic indicators. Such indicators of the countries of Africa are also due to the complicated natural conditions as well as to the heritage of the period of colonial enslavement, and the insufficiently considered, often destructive, modernization of the economy over the last several decades. An advantageous location relative to the sea creates solely some possibilities for economic growth, while the way in which they are realized by a particular country is entirely determined by the social-historical conditions. An example of the start of successful realization of the advantages from its maritime location is provided by a rapid economic growth of China — primarily in its regions in the neighborhoods of the sea. Another instance of this type is the rapid pace of development (over the last several decades) of the so-called "new industrial countries" of East Asia: Taiwan, Hong Kong, South Korea, and Singapore which are distinguished for their insular or maritime location (a very low degree of TGC). One further example is seen in a relative prosperity characteristic for a number of small (so-called offshore) insular countries which are the centers of financial, touristic, trade and other services of international significance (Bahama Islands, Brunei, Barbados, etc.). At the same time, not a single developing landlocked country performs such functions and has high socioeconomic indicators. Consequently, with a successful combination of the internal and external conditions, the factor of advantageous location relative to the sea may well play an important role in an acceleration of economic development of corresponding countries. Thus the geoeconomic continental-oceanic typology of the countries of the world, constructed through the use of rigorous economic criteria, provides a more penetrating insight into the transport-geographical possibilities of their involvement in the functioning of the world economy, which is an important precondition for universal socioeconomic development. The suggested typology of the countries is useful in explaining the most characteristic traits of their multiform specific character, reflecting the influence of the macrolocation relative to the sea on the level of economic development and the living standard of the population, on the degree of involvement in the international division of labor, etc.

4. The Regularities of Territorial Organization of the World Economy in the System of Continental-Oceanic Polarization

The influence of transportation costs manifests itself in a very much differentiated manner not only according to the remoteness of regions and countries from sea routes but also according to the groups of products (commodities or cargoes). The analysis of the differentiation of transportation costs for the kinds and groups of commodities was made on the basis of their classification according to "relative transportability". This is considered to man the relative possibilities of overcoming the distance by a particular kind of product (goods) expressed in terms of the price (value) of a unit of its mass.

Typical kinds of products circulating in the world market were originally combined into groups. Next, these groups were distributed according to the degree of processing of products into three aggregated classes: "raw materials", "semifinished products", and "finished products". The relative transportability of most of the commodities was obtained directly from data on their world prices, usually signifying the value of a unit of mass of a particular commodity. The groups of the kinds of products under consideration were then distributed among the six (adopted by us) categories of relative transportability encompassing the entire actual range of this indicator (Table 1). Finished products have a relative transportability ranging from very high (the most expensive goods) to moderate (less expensive but heavier goods), while semifinished products and raw materials varied from moderate to low (because of the smaller value of a unit of their mass).

Hence, the general regularity is thus: the higher the degree of processing and readiness of products for the end consumer, the better the economic possibilities for their transportation. While virtually all known modes of transport can be used to deliver goods of the first two categories of relative transportability, including the air transport, the transfer of products of decreased and low transportability to long distances is possible mainly by the most economical modes of transport, namely the sea transport, pipelines and, partly, the rail transport. Hence it becomes clear that the main competitive advantages during the transportation of mass raw-material cargoes correspond primarily to maritime regions.

The magnitude of the economic payoff they obtain when compared with the landlocked regions can be judged, on the one hand, from the role of raw materials and semifinished products in international operations and, on the other, from the proportion of transportation costs in the end prices of goods (i.e. from the amount of the eventual cost saving through the involvement of the sea transport). It is the raw materials and semifinished products of decreased and low transportability in the form of mass fluid, loose, bulk and timber cargoes, have an important role in international transportations, and in the volume of cargoes shipped by the sea transport (in tonnage) they exceed 75–80% [27]. As regards transportation costs, we emphasize that in Russia, in the first half of the 2000s, their share in the end prices of semifinished products was estimated [20] at 10–30%, while in the prices of raw materials it was in the range from 20–40 to 70–80%.

It must also be emphasized that, on the one hand, raw materials and semifinished products serve as the basis for the entire technological and reproductive "pyramid" of the world economy and, on the other, exports of raw materials and semifinished products make up the framework for the economies of most continental countries, including Russia. In view of the unfavorable transport-geographical location of these countries relative the sea and ocean routes, the issues related to transportability of products and transportation costs still remain extremely challenging problems.

In order to understand the most general regularities in the territorial organization of the world economy in the system of continental-oceanic polarization, an assessment was made of the intensity (volumes) of worldwide production and export of products of different transportability depending on the distance of producers and exporters from the sea [20]. The analysis used seven well-known kinds of products representing four categories of relative transportability: diamonds and gold (very high transportability), aluminum (moderate), oil and bauxites (decreased), and coal and iron ore (low). Our findings concerning the distribution of world reserves, production (extraction) and export of the kinds of products under consideration in the zones of different remoteness from the sea (sea routes): maritime, continental and ultracontinental, were obtained on the basis of systematizing an extensive body of factual material characterizing these indicators for all countries (and, where necessary, for regions inside large countries) as of the end of the 20th–beginning of the 21st century (Table 2). The data used in the analysis were taken from numerous reference books, statistical bulletins, publications and electronic sources of information. The results obtained were analyzed to reveal the following regularities of development.

Table 1. Relative transportability of separate groups of products (commodities) depending on the degree of their readiness for direct consumption (from the stage of processing)

Very high (over 1000)	High (100–1000)	Increased (10–100)	Medium (1–10)	decreased (0.1–1)	Low (less than 0.1)
A. FINISHED PRODUCTS					
Precious metals and stones, and articles made of them (gold, platinum, palladium, diamonds) Military aircraft Electronic equipment (supercomputers)	Precious metals and stones, and articles made of them (silver) Electronic equipment (liquid crystal display computers) Clothing and footwear Pharmaceutical drugs	Pharmaceutical drugs Passenger cars (of elite classes) Transport vehicles (special-purpose) Electronic equipment (computers) Household equipment	Passenger cars Autotrucks Products of heavy engineering (steam boilers, wires) Wood and paper-based products (paper, board)		
B. SEMIFINISHED PRODUCTS					
			Unprocessed (primary) metals (aluminum, nickel, copper, tin, zinc, titanium, chromium, magnesium) Semifinished products of chemistry and petrochemistry (chemical fibers)	Unprocessed (primary) metals (lead) Oil products (gasoline, fuel oil, diesel fuel) Semiproducts of chemistry and petrochemistry (benzene, methanol, ethylene, polyvinyl chloride, polyethylene, carbamide, ammonia Semiproducts of ferrous metallurgy (steel, pig iron) Wood and paper-based products (cellulose pulp, lumber)	Semiproducts of chemistry and petrochemistry (sulfuric acid)

Table 1. (Continued)

Very high (over 1000)	High (100–1000)	Increased (10–100)	Medium (1–10)	decreased (0.1–1)	Low (less than 0.1)
C. RAW MATERIALS					
			Food products (pepper, coffee, tea, beef, butter)	Fuel (raw petroleum) Ores and concentrates (bauxites, zinc ores) Food products (sugar, wheat, rice, corn, fruits) Chemical raw materials (propane, butane) Agricultural and wood raw material (cotton, round timber) Nonmetal mineral resources (graphite, magnesite)	Fuel (coal) Ores and concentrates (iron) Chemical raw materials (sulfur, phosphates) Nonmetal raw materials (sodium salt, barites) Mineral-building materials (gravel, crashed stone, clay, sand, cement)

Wholesale prices of mass unit of commodities in the main world centers of production and sales in 2000, thou US dollars/t.

Table 2. Distribution of world reserves, production (extraction) and export of separate major kinds of products with different relative transportability according to the zones of remoteness from the sea (sea routes) in the late 20th–early 21st century, %

Kind of product	Relative transportability	Items	Unit of measurement	World total	Zones of remoteness from the sea (km)		
					Maritime (0–200)	Continental (200–1000)	Ultracontinental (over 1000)
Diamonds (1998)	Very high	Reserves	t	728	5.8	40.7	53.5
		Production	t	25.9	18.3	53.6	28.1
		Export	t	20.7	18.3	53.6	28.1
Gold (1998)	Very high	Reserves	thou t	51.8	25.0	60.9	14.1
		Production	t	2516	28.4	59.1	12.5
		Export	t	1735	29.0	57.2	13.8
Aluminum (1999)	Medium	Reserves	-	-	-	-	-
		Production	thou t	23670	65.2	19.5	15.3
		Export	thou t	13480	67.1	10.8	22.1
Petroleum (1998)	Decreased	Reserves	billion t	153.0	71.8	17.8	10.4
		Production	mln. t	3535	72.3	17.3	10.4
		Export	mln. t	1870	80.6	12.5	6.9
Bauxites (1998)	Decreased	Reserves	billion t	23.0	75.6	18.7	5.7
		Production	mln. t	126.3	85.3	9.4	5.3
		Export	mln. t	37.9	95.3	2.6	2.1
Coal (2000)	Low	Reserves	billion t	1028.5	27.5	45.3	27.2
		Production	mln. t	4535.7	36.4	45.6	18.0
		Export	mln. t	599.3	69.1	20.1	10.8
Iron ore (2003)	Low	Reserves	billion t	170.1	30.9	53.4	15.7
		Production	mln. t	1200	49.9	45.5	4.6
		Export	mln. t	580	59.9	38.1	2.0

First, at the transition from world reserves to production and from production to export, the contribution from the maritime zones as regards products of decreased and low transportability (oil, bauxites, coal, and iron ore) steadily increases, and the contributions from the continental and ultracontinental zones decrease (most dramatically for the latter zones), whereas no such tendency is observed in the case of products with very high transportability (diamonds and gold). Consequently, the increased role of the maritime zones in international trade in raw materials and semifinished products is objectively conditioned by the price-raising and limiting effect of considerable transportation costs arising in the event of overland shipping of bulky products of decreased and low transportability to long distances.

Second, the distribution of the world export of the most important kinds of products having a decreased and low transportability is strictly ordered according to the zones of remoteness from the sea: the proportions of the volumes of export increase dramatically from the ultracontinental to maritime zones. While for the kinds of products under consideration (oil, bauxites, coal, and iron ore) the proportion of the ultracontinental countries in the world export constitutes a mere 2–11%, the contribution from the maritime zones reaches 60–95%. This indicates once again that the world market of raw materials and, partly, semifinished products is the market primarily of the maritime zones and oceanic countries dominating in international trade because of large-scale use of economical sea transportation.

It is readily apparent from the regularities we have ascertained above that the maritime zones are the least affected by the limitations imposed by rises in prices associated with transportation costs. The principal method of reducing the transportation costs of raw materials is provided by the sea transport, delivering to the world market more than 90% of oil, iron ore, coal and bauxites (Table 3). The maritime zones of the world are interconnected by economical, stable and powerful "oil", "coal", "iron ore" and other "bridges" which minimize the transportation costs between the regions of production and consumption of the most important kinds of raw materials.

On the other hand, the landlocked regions on the globe, as is evident from the resulting correlations, continue to be devoid of such favorable transport possibilities and have to remain in some "economic isolation" from the world market for a great variety of kinds of products of decreased and low transportability.

Immense masses of varied raw materials of decreased, low and, partly, moderate transportability continuously arriving via oceanic "bridges" are handled in the port-industrial complexes of the developed countries of West Europe and North America, and the new industrial countries of east Asia.

Table 3. Proportion (%) of exports in world production (extraction) of separate major kinds of products of decreased and low relative transportability, and proportion (%) of sea transportation in such exports in the late 20th–early 21st century

Kind of product (as of the year)	Production (extraction)	Exports		Proportion	
		total	includes by sea	of exports in production	of sea transportation in exports
	mln. t	mln. t	млн т	%	%
Petroleum (1998)	3535	1870	1685	52.9	90.1
Bauxites (1998)	126.3	37.9	37.1	30.0	97.9
Coal (2000)	4535.7	599.3	559.0	13.2	93.3
Iron ore (2003)	1200	580.0	528.0	48.3	91.0

The inland (landlocked) regions of these continents and countries are encroached by predominantly ready ("beneficiated") products of very high, increased and, partly, moderate transportability. Such a division of labor yields a huge economic effect to the developed oceanic countries through a reduction in transportation costs and a minimization of the total production costs and prices of products. A different situation is observed in Russia, for example; the major centers of production and export of bulky kinds of raw materials and semifinished products emerged primarily in the ultracontinental zone, i.e. in Siberia, in the Ural region, and in the Volga region.

That the oceanic countries and maritime zones dominate in the world economy is also determined by the influence of the trade-economic reasons and factors. The very historically established differentiation of a unified world market of commodities into two main subdivisions, namely the market of raw materials and semifinished products, and the market of finished products, appear to be largely caused by the drastic differences of commodities in these subdivisions from one another as regards relative transportability, i.e. the conditions of their delivery in the process of international trade, and the significance of the transport factor as a whole. While a leading role in competitiveness of raw materials and semifinished products is played by the costs of their production and transportation which are heavily dependent on the relatively static natural-resource and transport-geographical factors, the basis for competitiveness of finished products is formed by their consumer properties conditioned by the dynamical competitive advantages, namely scientific and technological progress, innovations, development of the intellectual potential, etc. [4, 10].

The first trade-economic reason for the advantages of the maritime regions in the world market of raw materials and semifinished products has to do with the characteristics of exchange business, the most widespread form of international trade in these commodities. Exchange trade constitutes a constantly operating market of large-scale, qualitatively homogeneous and substitutional goods controlled by the largest and, usually, maritime trade-organizational centers of the oceanic countries only. Since the producers of raw materials supply to the world market the same standardized products, their markets establish monopolies of major buyers imposing advantageous (for themselves) lower prices to isolated and competing (with one another) exporters [6, 10, 11]. Since the control over the markets of raw materials and semifinished products (and corresponding pricing) is usually exercised by powerful corporations from the developed oceanic countries, it is quite natural that the interests of the producers and exporters of the countries and regions remote from the sea are ignored. In particular, the world prices in the commodity markets of raw materials that are dictated by the buyers from the developed oceanic countries are in no way meant to compensate for the increased (compared with the average international level) transportation costs of suppliers of raw materials from the continental countries. In connection with this situation with the disadvantageous location of the countries and regions at a great distance from the sea, we can refer to a neat remark made by L.C. Thurow in reference to the heart of the capitalist system implying a maximization of profits, namely that a sentimental adherence to some geographical part of the world is not involved in this system [16].

The second reason behind the advantages of maritime exporters over landlocked exported reveals itself in the consequences of the creation (during the last several decades of the 20th century) of the general world logistic system of deliveries of raw materials which ensured significant economies in transportation costs [28]. A decrease in transportation costs has had the highest beneficial effect in the maritime zones which are the home for numerous warehouses of exchange products, whereas the sellers and buyers from the landlocked zones were scarcely affected by it.

The third reason has to do with the formation of prices in world commodity markets of raw materials and semifinished products by their leading exporters. The matter is that the international value (and, hence, the price) of a commodity is comprised not of the weighted-mean national production costs but is formed under the determining influence of the expenses in the countries which are the main suppliers of a given commodity to the world market [10, 11]. Since, however (as established above), the maritime zones account for most of the world exports of the most important raw materials, the key role of the oceanic countries in pricing in the world commodity markets of raw materials and semifinished products appears quite explicable.

Let us illustrate what has been said above with an example of international trade in coal. More than half of the world exports of coal comes now from Australia, Indonesia, SAR and Colombia, i.e. the oceanic countries where coal is extracted solely in maritime zones. The lower limit of fluctuations in world prices of coal will be obviously represented by the weighted-mean expenses connected with its production and by transportation of the aforementioned countries as the leading “players” in the world coal market. Intermittent drops of world prices of coal retain a rather high profit of its export from the aforementioned and other oceanic countries, whereas exports of coal from continental Russia characterized by increased rail transportation costs are at the profitability threshold [29].

Unlike the situation with raw materials and semifinished products, the transport factor in international trade in finished products is essentially unimportant. However, today’s world market of finished products, more specifically unique, hi-tech and sophisticated commodities, is a market where the main suppliers and “creators” of products and, accordingly, the legitimate masters of the market are represented by powerful corporation of the developed and new industrial countries [6, 9, 10, 30], being, with rare exception, of the oceanic type. It is precisely these “actors” that dictate to the buyers of the entire world their price terms, whereas the continental countries occupy here a purely subordinate position.

It will be emphasized in closing that because of the operation of the factors and causes considered above, the world markets of raw materials as well as of finished products are, above all, the markets of the oceanic countries and maritime zones. The suppliers and buyers of these countries exert a determining influence upon pricing in the world commodity markets. Therefore, the world economy and the world market necessarily pattern after the developed maritime countries and act in their interests, i.e. adhering to the principles of a relatively distance-independent oceanic exchange where the world prices are formed having regard to their own minimal (with respect to the oceanic countries) shipping costs of goods.

The fact that the oceanic countries and maritime zones are dominant in the world economy is a logical result of the combined operation of the preconditions, terms and factors which are mainly based on the radical differences in the effectiveness of sea and overland transportation and, accordingly, on the difference in economic possibilities of the maritime and landlocked regions. This salient feature of today’s world economy structure within the context of the continental-oceanic dichotomy is also supported by the unity (continentality) of the World Ocean and the dominant political-economic role of the leading states belonging to some or other extent to the oceanic type (with the exception of Russia). A relatively low level of transport capacity of the economy in the oceanic countries is favorable to their primordial orientation to an intense economic interaction with the external world where, in the general case, they can take advantage of the processes of globalization, internationalization and transnationalization of the economy; furthermore, it dictates a need for a poorly pronounced (compared to the continental countries) requirement for a firm establishment of the territorial division of labor and the organization of the economy on a district basis.

On the other hand, a strong dependence of the countries of the continental (landlocked) type on transportation distances and increased shipping costs requires a constant reduction in their costly burden and dictates, therefore, a fundamentally different (adaptive) style of economic activity and communication with the external world. For Russia is the largest continental country, it is important to focus on establishing, above all, short-distance intraregional and, partly, intradistrict economic ties, with a certain limitation of the less efficient long-distance relations, maintaining a relatively low level of transportation tariffs and on magistralization of the transport, ensuring a general self-sufficient development and forming a receptive domestic market, achieving a firm territorial division of labor, organizing the economy on a district basis and the principles of economic zoning, etc.

Conclusion

In summary it may be said that the geographical factor in dividing our planet's surface into land and the World Ocean continues to play a very important role in economic processes. This factor is responsible for a fundamental dualism of continental and oceanic countries and, in the long run, determines their unequal economic possibilities, differences in the territorial organization of society, the economic life style and in the character of interaction with the external world. Although it may appear that, with the transition to industrialized society, a direct dependence of the economy on geographical conditions and factors is decreasing in general, extensive use of the latest information technologies as such is in no way accompanied by a reduction in transported amounts of matter and energy. In spite of the general tendency for a reduction in transportation costs, there is neither an appreciable decrease in the role played by the transport-communication factor in the economy, nor the disappearance of the difference in the effectiveness of sea and land transportation, nor the loss by the sea transport of its leading role in ensuring the uninterrupted operation of the world economy.

The globalization process leads to the shaping of the world's political-economic configuration in the form of the "oceanic" model for the international division of labor that ensures constant access to specific key advantages (i.e. to world income), largely for the most advanced seaside countries. The continental-oceanic dichotomy is one of the main lines which are responsible for the geoeconomic differentiation of the planet that reflects the unevenness of development of states and regions in the general system of world-economy core-periphery relations. Some independency of the level of socioeconomic development of individual advanced countries on geographical conditions and factors, which is sometimes pointed out in publications, must not be absolutized, because a decrease in the role of natural resources, rather than of the factor of economic-geographical location is implied in most cases. We would be justified in conjecturing that in the foreseeable future the division of the world economy into two types of economies — continental and oceanic — and the ensuing consequences will largely retain their significance.

Hence, mankind's global challenges should incorporate the issue related to elimination and a substantial attenuation of the influence from the continental-oceanic dichotomy on the world's socioeconomic polarization. The policies and practices of organization of economic activities adaptive to the burden of overland distances as used in landlocked countries are capable of resolving it only partially. Its fundamental resolution, however, implies a radical restructuring of the very foundations of operation of the world economy by introducing the more equitable commercial-economic relations between continental and oceanic countries.

From the concrete perspective, the case in point must be reformation of those international commercial rules and norms active in the interests of developed oceanic countries which determine the "basic delivery conditions" of goods. L.C. Thurow arrived, for example, to a well-grounded conclusion that the rules of world trade were written by those who exercise control over the conditions of access to the world's largest market. For that reason, Great Britain wrote the rule of world trade in the nineteenth century, and the United States — in the twentieth century [16].

It should only be noted that Great Britain — the universally recognized lawgiver in the realm of international trade — "wrote" these rules in the 18th – 19th century, relying largely on the customs existing at that time (in the 16th – 17th century and earlier) in Holland, in the Italian state-cities of Genoa and Venice, and in the cities of the Hanseatic League. It is quite natural that the trade and economic relations with other countries, including with landlocked ones, were built by Italian and Hanseatic port cities and by Holland and England according to their own models. Yet, those models disregarded the partners' overland transportation costs across the territories of their countries — the partners had to cover them for their own account. Upon occupying the dominant position in the world economy and in the world market, the West-European seaside countries, with Great Britain at the head, fixed the (inherent and favorable to them) logic of trade relations in the international commercial rules and regulations regulating the "basic delivery conditions" of goods.

Radical reformation of the international commercial rules regulating the "basic delivery conditions" of goods serve as a necessary condition for elimination of the unilateral economic advantages of the oceanic countries in prejudice of the landlocked countries through a more equitable distribution of transportation costs between them with regard to foreign trade transactions.

Only through such a reformation (which will, of course, take a long period of time to complete) will it be possible to achieve a truly mutually beneficial division of labor between the countries of the world, no matter what their macrolocation relative to the sea routes. An enhancement of the processes of internationalization and globalization of the economy will then not yield an increasingly larger unilateral profit to the oceanic countries to the detriment of landlocked ones.

There is no question that the suggested transformation of the trade and economic relations must be initiated by Russia which is on the international scene the most authoritative and competent among the continental countries. The achievement of this goal along the suggested lines, more specifically the solution of the continental-oceanic dichotomy problem at a global level, would be more appropriately geared via international organizations, such as UNO, WTO, and others. The optimal solution of the continental-oceanic dichotomy problem at a global level will eliminate one of the chief causes for the core – periphery polarization of the world economy will orchestrate improved cooperation between all countries across the globe.

REFERENCES

[1] Braudel, F. *Civilisation matérielle, économie et capitalisme, XVe-XVIIIe siècle*. Vol. 3, Le temps du monde, Paris, 1979.

[2] Wallerstein, I. World-Systems Analysis. In: *World System History*, ed. by G. Modelski. In: *Encyclopedia of Life Support Systems (EOLSS)*. Oxford: Eolss Publishers, 2004.

[3] Gritsai, O. V., Ioffe, G. V. and Treivish, A. I. *Core and Periphery in Regional Development*. Moscow: Nauka, 1991 [in Russian].

[4] Mironenko, N. S. Geographical Approach to Core-Periphery Relations in the World Economy. Vest. MGU, Ser. 5. *Geografiya*, 2005, No. 1, pp. 25–38 [in Russian].

[5] Soros, G. *Open Society: Reforming Global Capitalism*. London: Little, Brown and Company, 2000.

[6] Delyagin, M. G. *World Crisis: General Theory of Globalization*. Moscow: INFRA, 2003 [in Russian].

[7] Prebisch, R. Capitalismo periférico. Crisis y transformación. Mexico-City: Fondo de Cultura Económica, 1987.

[8] Todaro, M. P. *Economic Development*. Moscow: MGU–UNITY, 1997 [in Russian].

[9] Inozemtsev, V. L. Split Civilization. *Existing Prerequisites and Eventual Consequences of the Post-Economic Revolution*. Moscow: Academia – Nauka, 1999 [in Russian].

[10] Bogomolov, O. T. *Anatomy of Global Economics*. Moscow: IKTs "Akademkniga", 2004 [in Russian].

[11] Faminsky, I. P. Concerning Price Formation in the World Market. In: *International Economic Relations*, ed. by I. P. Faminsky. Moscow: Ekonomika, 2004, pp. 208–230 [in Russian].

[12] Toffler, A. *The Third Wave*. London: Pan Books Ltd., 1981.

[13] Bolotin, B. M., Glaziev, S. Yu. and Nizhegorodtsev, R. M. Integration of Countries with the Economy in Transition Into the Modern World Economy. In: *Teaching the Market*, ed. by S. Yu. Glaziev. Moscow: Ekonomika, 2004, pp. 480–517 [in Russian].

[14] De Soto, H. *The Mystery of Capital: Why Capitalism Triumphs in the West and Fails Everywhere Else*. New York, NY: Basic Books, 2000.

[15] Chernyshev, S. *Sovereign Russia: How to Earn one's Living Together with the Country.* Moscow: Izd-vo Yevropa [in Russian].

[16] Thurow, L. C. *The Future of Capitalism: How Today's Economic Forces Shape Tomorrow's World*. New York, NY: William Morrow, 1996.

[17] Armand, A. D., *Geography of the Information Century*, Izv. RAN. Ser. geogr., 2002, No. 1, pp. 10–14 [in Russian].

[18] *Radermacher, F. J. Balance or Destruction*: Eco-Social Economy as the Key to Global Sustainable Development. 2004,Vienna: Ecosocial Forum Europe.

[19] Scheer, H. Solare Weltwirtschaft. Strategie für die ökologische Moderne. München: Kunstmann Antje Gmbh, 1999.

[20] Bezrukov, L. A. *The Continental-Oceanic Dichotomy in International and Regional Development*. Novosibirsk: Geo, 2008 [in Russian].

[21] Plotnikov, A. Yu. The New Incoterms – 2000. *Content, Practice of Use and Commentaries*. Moscow: Izd-vo PIOR, 2002 [in Russian].

[22] Annovazzi-Jarab, L. Landlocked Countries: Opportunities, Challenges and Recommendations. *Trade Facilitation. The Challenges for Growth and Development.* New York and Geneva: United Nations, 2003, pp. 81–136.
[23] Gallup, J. L., Sachs J. D. and Mellenger A. D. Geography and Economic Development. *International Regional Science Review*, 1999, No.22.2, pp.179–232.
[24] Radelet, S. and Sachs, J. D. Shipping Costs, Manufactured Exports, and Economic Growth. (1998) http://www.earth.columbia.edu/sitefiles/File/about/director/pubs/shipcost.pdf
[25] Human Development Report 2010. The Real Wealth of Nations: Pathways to Human Development. New York, NY: UNDP.
[26] http://unctad.org/en/Pages/Statistics.aspx-UNCTAD, 2012.
[27] Mogilevsky, I. M. Transport and Communications: past, Present, Future. Moscow: Nauka, 2005.
[28] Errera, S. and Brown, S. L. *Trading Energy Futures and Options*. US: PennWell Books, 2002.
[29] Lipin, A. S. Estimating the Balanced Transport Tariff of Coal Export. *Region: ekonomika i sotsiologiya*, 2006, No. 2, pp. 93–104 [in Russian].
[30] Castells, M. *The Information Age: Economy, Society and Culture*, Vols. 1–3, Cambridge, MA; Oxford, UK: Blackwell, 1996; 1997, 1998.

In: Economic Developments and Emerging Markets ... ISBN: 978-1-62417-986-0
Editors: Guy Boscarino and Derek Notte

Chapter 4

GREEN ECONOMY IN RURAL AREAS THROUGH GEOTOURISM AND GEOPARKS

***Neda Torabi Farsani**[*,1], **Joana Lima**[1], **Celeste Eusébio**[1], **Carlos Costa**[1] and **Celeste Coelho**[2]*

[1]Department of Economic, Management and Industrial Engineering, Aveiro University, Aveiro, Portugal
[2]Department of Environment and planning, (CESAM: Centre for Environmental and Marine Studies), Aveiro University, Aveiro, Portugal

ABSTRACT

There are a variety of niche tourism markets which develop depending on tourism supply and demand. In recent decades, geotourism as a new niche market has emerged to introduce geo-heritage sites – geology, geomorphology, geography, and geodiversity – as new tourist attractions and geoparks as new geotourism destinations. Geoparks are an innovation for the protection of natural, cultural and geological heritage sites and sustainability. The European Geopark Network and the UNESCO Global Geoparks Network have introduced geotourism as one of the main targets for geoparks in parallel with conservation and education. For that purpose, it is crucial to apply the network concept to geotourism activities. Network activity not only provides an opportunity to benefit from the open world market, but also facilitates knowledge transfer. Geoparks, through geotourism and innovative strategies, strive to promote green economy for local communities. This chapter aims at introducing geotourism as a new green economy for rural development and popularization of sustainability and geo-sciences.

[*] ntorabi@ua.pt.

INTRODUCTION

The world community faces several environmental problems such as global warming, air pollution, etc. The growing concern within the earth and environmental sciences and governments demonstrates the importance of earth preservation. For instance, governments try to minimize the negative impacts of industrial and economic sectors on the environment, as do the earth scientists and NGOs. The emergence of green economics as an economic policy for people and planet is clear evidence for earth conservation (Cato, 2009). The concept of green economy, which focuses on the interaction between environment and the economy, was publicly announced in the 1992 Rio Conference - Rio de Janeiro, Brazil. According to the United Nations (1992), green economy is a strategy for sustainable development and poverty eradication. Bapna and Talberth (2011) noted that green economy is a vision for growth and development and can improve the people's lives with sustainable development. They believe that green economy promotes a triple bottom line: sustaining and advancing economic, environmental and social well-being. The five key components of a green economy knowledge-sharing platform are as follows: promoting network activity, partnerships and connections; matching 'knowledge seekers' with 'knowledge providers' in order to facilitate knowledge exchange; providing an (online data) platform and a geo-referenced resource database for green economy; integrating expert analysis and policy review; and preparing a finance facility to support developing countries in designing and implementing green economy policies (UNCSD Secretariat, 2012).

Nowadays, tourism, like other industries, faces a multitude of significant sustainability-related challenges such as energy and GHG emissions, water consumption, waste management, loss of biological diversity, and effective management of cultural heritage. However, in order to overcome the challenges, tourism strives to change itself to green economy through following sustainability principles.

The term 'sustainable tourism' originated from the general concept of 'sustainable development', which 'meets the needs of the present without compromising the ability of future generations to meet their own needs', was introduced by Brundtl and in 1987 (Beeton *et al.*, 2007). The appearance of the concept of sustainable tourism in 1987 indicates that tourism is a pioneer industry for sustainability and development of green economy. Moreover, ecotourism, rural tourism and, in the most recent decade, geotourism are tourism products that develop the sustainability and are known as green tourism.

This chapter is an initial attempt to introduce a new niche market in tourism – Geotourism – as green economy, especially in rural areas. In addition, this chapter targets two major purposes: (i) to investigate the economic development in geoparks through geotourism and (ii) to identify the strategies applied in several territories for stimulating geotourism as a green economy.

GEOTOURISM: GREEN NICHE MARKET

The new concept of geotourism as a new niche market, with a special interest in the geo– geology, geomorphology, geodiversity, and geography, can add opportunities to cultural sustainability and rural development. Geotourism as a branch of sustainable tourism allows tourists and visitors to travel in a territory in order to experience, learn from, and enjoy earth heritage.

At present, geotourism is a new movement helping travellers to increase their knowledge about natural resources, the cultural identity of host communities, and ways of preserving them. The 'emerging tourism' niche of geotourism is still at an early stage of commercial development in most countries, and geoparks as a sustainable development model for protected areas are pioneers in the development of geotourism marketing (Farsani *et al.*, 2012a).

According to some authors (Tourtellot, 2000: 2; Stokes *et al.*, 2003; Fennell 2003; Buckley, 2003; Dowling and Newsome, 2006) geotourism is an emerging niche market which follows sustainability principles. Since geotourism is a niche market, it tends to be small and specialized; this market involves less impact, and interests visitors who seek a specific experience, activity and place. Therefore, geotourism is less likely to negatively impact the area (Gorman, 2007).

Geotourism, which was introduced publicly in 2002 by the Travel Industry Association of America and National Geographic Traveller Magazine, incorporates sustainability principles (Alan, 2002). They introduced geotourism *as tourism that sustains, or even enhances, the geographical character of a place, such as its culture, environment, heritage, and the well-being of its residents* (Tourtellot, 2000: 2). This concept encompasses more than ecotourism and sustainable tourism. Geotourism differentiates itself from ecotourism through focusing on the geological and geomorphological character of a region while many of ecotourism's definitions limit ecotourism to only occurring in protected or natural areas and mainly related to biodiversity. Geotourism is one of the core activities of geoparks which are under the umbrella of geographic tourism (Dowling and Newsome, 2006; Farsani *et al.*, 2012a). It is obvious that development of the local economy would occur as a result of tourism activities. Moreover, geoparks are pioneers in the development of geotourism; they strive to stimulate socioeconomic activities and sustainable development.

Geoheritage, which draws attention to the geological and geomorphological elements of nature worthy of conservation, has for years been considered less vulnerable than other environmental values. Therefore, it has not received the same amount of attention from the conservationist movement as cultural and ecological heritage (Reynard and Coratza, 2007). Thereupon, the new concept of geopark emerged to eliminate this challenge.

Patzak and Eder (1998: 33-34) argued for the definition of a geopark as a territory comprising a number of protected geological heritage sites of special geological significance, rarity or beauty. These geological features are representative of a region and its geological history, events and processes. Geoparks are of particular value for education, science, culture and socioeconomic development (mainly through tourism) (UNESCO, 2006). It is worth mentioning that geotourism in geoparks includes wider aspects of tourism products than National Geographic's definition and diagram (Farsani *et al*, 2012b). Moreover, geotourism in geoparks strives to integrate tourism products and fun with geoscience educational programs for visitors, locals and schoolchildren (Figure 1).

Since geoparks follow sustainability, it can be said that geoparks are ecoparks with emphasis on geological and geomorphological heritage and which attempt to introduce them as new tourist attractions.

Geoparks not only pursue sustainability principles in their practices, but also stimulate rural development, revival of cultural identities, and geo-knowledge transfer from the professional to public level. Geoparks try to carry out the embodiment of geo-sciences – the transformation of tacit geo-knowledge into processes and practices – for the popularization of earth preservation. Regarding this, geoparks take advantage of network activities at local, national and international level (such as Global Geoparks Network (GGN), European Geoparks Network (EGN), Asia Pacific Geoheritage and Geopark Network (APGGN)) which is one of the key components of a green economy knowledge-sharing platform.

The above definitions illustrate that geotourism as a form of sustainable tourism can be a strategy for green economic development for local economies. The next section tries to conceptualize the theme of greening economies in geoparks through geotourism.

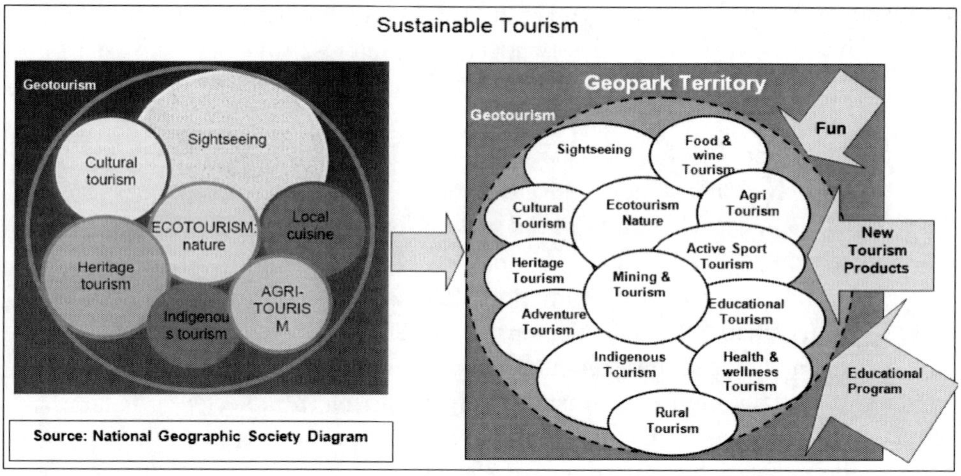

Figure 1. Geotourism Diagram (own construction).

GREEN ECONOMIC DEVELOPMENT IN GEOPARKS THROUGH GEOTOURISM

During the last decade several crises have spread (climate, biodiversity, fuel, food, water, and more recently, in the global financial system) and are impacting the possibility of sustaining prosperity worldwide and reducing extreme poverty. Consequently, there is increasing evidence that the greening of economies may be a new economic paradigm, according to which material wealth is not achieved through growing environmental risks, ecological scarcities and social disparities (UNEP, 2011). However, to attain this objective, new conditions are required to promote investments in the transition to a green economy. According to UNEP (2011:21), the key for a “transition to a green economy is to enable economic growth and investment while increasing environmental quality and social inclusiveness”. In this sense, policy makers have a key role in creating the conditions for public and private investments to incorporate broader environmental and social criteria and several challenges emerge when actual economies start adopting and implementing green strategies to achieve economic development.

Several studies reveal that “greening tourism can lead to broad economic, social and environmental benefits for host countries and their communities” (UNEP, 2011: 419). Additionally, tourism is an available alternative activity in the isolated, and sometimes resource-scarce, rural areas, and one that can help create new jobs, reduce communities' poverty, prevent land abandonment and, consequently, prevent desertification (Anand *et al.*, 2012). However, this activity is a complex mix of business and social activities and their impacts happen with different nature and intensity in different destinations – there is no simple cause-effect relation (Godfrey and Clarke, 2000). So, the degree to which geotourism can be a driving force of economic development depends upon the characteristics of the local economy where the geopark is located, the definition of realistic goals which must be in line with these characteristics and the ability to achieve these goals. Geotourism only contributes to the economic development of local communities, especially in a context of Green Economy, if appropriate conditions in destinations are provided. In this context, the development of strategies to implement green tourism activities in geoparks will provide significant returns both for business and society.

Geotourism can offer several activities to visitors and constitute a challenge for managers and planners, because it combines the components of geoparks with components such as other attractions, accommodation, tours, activities and interpretation (James and Hose, 2008). The use of these spaces by tourists could provide both conservation of geodiversity and economic development of local communities, always adopting a vision of sustainability. Sustainability principles should be applied to all types of tourism activities, operations, establishments and projects developed in geoparks. In this context, geotourism, as a form of sustainable tourism, should invest in energy efficiency, in consuming less water, minimising waste, conserving biodiversity, cultural heritage and traditional values, creating jobs, enhancing the attractiveness of geoparks, generating local income and integrating local communities in planning process, with a view to improving Quality of Life of the host communities. The global awareness of the need to adopt a vision of sustainability in tourism led to the promotion of green types of tourism (for example, ecotourism, rural tourism, community-based tourism and geotourism) as a low-impact, environmentally sensitive way to travel, also responding to the growing demand for tourist activities related to natural areas over recent years (Anand *et al.*, 2012; Bimonte and Faralla, 2012; Law *et al.*, 2012). These tourists are found to prefer small and family-run facilities, as well as local and ''green'' products, and they are also more concerned about their environmental and cultural impacts and stimulate a greater multiplier effect (Ballantyne *et al.*, 2009; UNEP, 2011; Bimonte and Faralla, 2012). Additionally, according to UNEP (2011) there several researches that show that tourists seeking environmental and culturally differentiated destinations are willing to pay more for this experience. This represents a strategic argument for adopting green economy strategies and developing geotourism as a type of tourism that feeds the interest of this growing segment and as an important and effective driver of local economic development.

The greening of tourism involves efficiency improvements in the use of resources, is expected to reinforce the local employment and to create significant opportunities oriented to preserve local culture and natural environment. It should also contribute to the development of the local economy and to poverty reduction, some of the guiding principles for a Green Economy (UNEP, 2011; Anand *et al.*, 2012). Geotourism can be seen as a "greening form" of tourism, as its objectives are related to geo-preservation and conservation as well as to increasing the involvement of local communities in the tourism value chain (Farsani *et al.*, 2012a). This vision for tourism in geoparks makes it easier for this kind of tourism destination to overcome one major challenge identified by the UNEP (2011) for a destination trying to adopt green economy strategies: to incorporate biodiversity and local community involvement concerns in destination planning and investment. However, planning green economy strategies in tourism is a process characterized by complexity and high levels of uncertainty, because limited data is available and because there are few examples of targeted and effective strategies (Law *et al.*, 2012).

Globally, geotourism, as a form of tourism, can have positive or negative impacts (Cooper *et al.*, 1998; Eusébio, 2006), depending on how it is planned, developed and managed. Then, a set of enabling conditions is required for geotourism to become sustainable, generating net benefits for host communities. However, is important to remember that the impacts of geotourism to local vulnerable communities are rarely quantified (UNEP, 2011; Eusébio and Lima, 2012). Green economy-oriented strategies that include monitoring economic, environmental and socio-cultural impacts will be relevant tools to minimize the costs and maximize the benefits of geotourism for local economies.

In this sense, it must be considered that, although during the initial stages of geotourism development this activity can bring significant economic benefits to the geopark area, when tourism development exceeds a certain level (desired level) the economic costs of tourism development will be felt more significantly (Williams and Shaw, 1991). The danger of over-dependence on tourism, the seasonality of production and the low rate of return on investments, the increase in propensity to import and negative externalities (e.g. loss of quality of life of the local communities) (Mathieson and Wall, 1982; Williams and Shaw, 1991; Eusébio, 2006; Lima, 2008) are the economic costs of tourism most cited in the literature for local areas. The agents responsible for green tourism development in geoparks should consider both economic benefits and economic costs of tourism in order to quantify the net economic benefits (benefits minus costs). When geotourism generates positive net economic impacts, it is contributing to improving the economic development of local communities and should be promoted. On the other hand, if geotourism generates negative net economic benefits to local communities the agents responsible for the development of these spaces should identify the factors responsible for that situation and implement strategies, anchored in green economy principles, to simultaneously increase the benefits of geotourism and decrease its costs.

The potential positive impacts of geotourism on local economic development are related to increases in local employment, income, production, diversification of the economy or infra-structures for local communities. According to several authors (e.g. Kadt, 1979; Sharpley and Telfer, 2002; Eusébio, 2006; Lima, 2008) the main reasons that tourism may positively influence the economic development of local communities are: income generation, economic valuation of free goods, modernization of the local economic structure, increased local production, boosting of traditional economic activities, utilization of unused equipment, creation of public infrastructures and employment growth. When adopting a green economy strategy, these effects can be maximized. Geotourism, especially as a branch of sustainable tourism, will bring benefits to local economic sectors related to tourism but also to residents who could benefit from increased employment in this sector. Thus, if the destination has adequate planning for tourism development, the effect of increasing demand through tourism will help to decrease the leakages in order to meet the needs of the additional demand generated (Eusébio and Lima, 2012). The economic impact of geotourism on local development will depend on the tourism business model adopted in geoparks, in particular the quantity and type of products sourced from the local economy and if locals are involved in tourism business operations. In destinations where a large percentage of tourist needs is locally supplied, local contribution and tourism multipliers tend to be higher, resulting in a correspondingly greater economic impact. This effect can vary dramatically between destinations, but there is increasingly convincing evidence that more sustainable tourism (particularly in rural and nature areas) can lead to more positive poverty-reducing effects (Anand *et al.*, 2012; Eusébio and Lima, 2012).

Therefore, economic diversification should be a priority for the "green economic development process" of the destination, especially in a geopark that aims not only to increase the level of geotourism expenditure but also to increase the multiplier effect of that expenditure on local economies. Diversification of the economy means greater benefits that tourism can trigger within that area (greater multiplier effect). In this sense it is important to highlight that tourism development should contribute to diversification within the local economy and never replace that diversification (Inskeep, 1991). The increase in demand generated by geotourism not only causes effects on local production, but also brings the possibility to the local economy of increasing their level of local employment (Kadt, 1979), which meets the goal of greening economies.

Geotourism should also be planned as a promoter of co-operation among local actors if various components associated with tourism are interpreted in a cross-sectorial, integrated and interdisciplinary approach, promoting networking, cooperation and interaction between the actors involved. In this context, as suggested by UNEP (2011:438), programmes which promote green tourism activities "will be more effective if produced by a multi-stakeholder participatory planning process, as well as through the development of partnerships at local, national, regional and international levels". *Geoparks* are places where this type of planning process in order to promote green tourism activities is usually considered (Farsani *et al.*, 2012a). This process will stimulate increases not only in geotourists' expenditure, as explained before, but also in the multiplier effect of geotourism on local economies. Local businesses networking in the tourism value chain enhance local economic opportunity (through employment, local contribution and multiplier effects) while also enhancing local competitiveness among tourists demanding greater local content. This becomes a win-win situation that promotes a greater number and variety of activities in a given destination, a "buy local" movement and the growth of specialised niches (UNEP, 2011). Efforts made by tourism businesses to work together, to include local communities within value creation, as well as public "buy local" policies and the development of infrastructure and supporting industries, create not only the new conditions for business development, but also for a more equitable growth and less leakages (UNEP, 2011; Nijaki and Worrel, 2012).

A geopark with local independent entrepreneurs who are receptive to innovation and cooperation with the public sector and other private entrepreneurs will have greater capacity to identify the skills needed to support the transition to a green economy. In this context, all tourism businesses located in geoparks should be thinking about the generic skills required to use resources efficiently to achieve sustainable growth. Additionally, the majority of tourism businesses located in geoparks are SMEs (Small and Medium Enterprises) with the potential to generate greater income and opportunity from green strategies. However, in this regard, a challenge to greening SMEs appears: their lack of access to capital. In order to overcome this difficulty, all levels of governments should facilitate the financial flow or other kinds of incentives (reduce administrative fees or offer technical, marketing or business administration support) to these green tourism projects with an emphasis on contributions to the local economy and poverty reduction (UNEP, 2011).

If we consider the various types, nature and intensity of the impacts that geotourism brings, destination planning is of particular importance for geoparks that pursue the implementation of green economy strategies. Assessment, diagnosis and planning are essential steps when trying to maximize benefits and minimize the costs of tourism development and when a green strategy is adopted. Local stakeholders need to establish mechanisms for coordinating with institutions responsible for the environment, energy, labour, agriculture, transport, health, finance, security and other relevant areas (UNEP, 2011). A growing body of evidence shows that greening tourism can lead to broad economic, social and environmental benefits for host countries and their communities (Mill and Morrison 2006; World Economic Forum, 2010; UNEP, 2011). Nevertheless, evaluating the economic effects stimulated or increased through geotourism, in addition to the other important goals of geoparks (nature/heritage protection and conservation, and promotion of research and education), assume special importance for geoparks insofar as this evaluation constitutes an important argument for increasing public support and gaining political and public support (Hartling and Meier, 2010), essential for implementing green economy strategies (UNEP, 2011). However, this task seems to be quite complex in practice, due either to the difficulty of obtaining an instrument capable of measuring all impacts of geotourism in geoparks, or to the difficulty of obtaining statistical data to carry out such analysis. As mentioned by Hughes (1994: 404), even for tourism in general "there is also a lack of longitudinal data which monitor the effects over time of an injection of expenditure", so when we are trying to analyse the economic effects of a much younger and specific type of tourism, this problem grows significantly. Without reliable and comparable statistical background information about the development of geotourism in geoparks (number of visitors, local demography, local economy features/indicators, the tourism facilities and services, economic development factors and about the factors that influence the dimension of the tourist multipliers), both the characterization of the local economies where geoparks are located and the economic impacts of geotourism in geopark development are very difficult to assess.

Hartling and Meier (2010) reported that geoparks are very heterogeneous spaces because of their various and varied goals (e.g., education, nature protection, protection of geologic heritage, research), their unclearly defined borders and gateways, as well as their close association between cultural and natural heritage. Additionally, the lack of data about the characteristics of geotourism, and even about green tourists (Dolnicar and Matus, 2008), creates a barrier for destination managers to evaluate that segment's potential value and, in this regard, makes the implementation of green economy strategies difficult. Consequently, the estimation of the economic effects of tourism in geoparks faces several constraints and the basis for greening these economies may be compromised. Therefore, future studies are suggested to collect and analyse this kind of data, over large periods of time and in different geoparks, also producing comparable statistical information. These studies should be an ambitious goal for those responsible for geopark tourism management, in order to guarantee a more coherent and effective definition of green economy strategies to maximize the contribution of geotourism to achieving sustainable growth of these tourism destinations. However, it is important to reinforce that every geopark is unique and therefore each development strategy to implement green tourism activities in this type of tourism destinations must be sensitive to the geopark's assets and challenges.

The next section presents the strategies applied in some geoparks as new geotourism destinations for promoting green economy.

STRATEGIES APPLIED IN GEOPARKS FOR DEVELOPMENT OF GREEN ECONOMY

Although there is profuse literature around destination sustainability, there is little research referring to green economy strategy development for tourism destinations, with a few exceptions suggesting criteria to guide green economy development at a destination level (DeLacy and Lipman, 2010; Law *et al.*, 2012). This section tries to give some contributions in this regard.

Geotourism strives to sustain the region's landscape through perpetuation of local values and attracting visitors who actively promote local values through their travel behaviour. The responsibility of geotourism is to keep things 'as they are' for the resident or develop things 'their way' rather than change for visitors (Nickerson, 2010). Geotourism as a target of geoparks increases the value of local business, local landscape and culture connected to the geopark.

Geoparks on one hand preserve the natural, geological, cultural and environmental values and on the other hand strive for development of green local economy in their territory. Regarding this, they have applied some innovative strategies to create new job opportunities, seasonal or part time jobs, reduce unemployment and migration rate, especially in rural areas. Since most of the abiotic natural objects are located in rural areas, development of geotourism in these areas can be a chance for rural development (Bębenek, 2006). Geotourism development also represents a partnership between the government, local people, and private sectors: local businesses, active tourism companies, tour agencies, restaurants and accommodation among others. This partnership is welcomed because it makes good economic sense and can benefit all partners (Dowling, 2009). Geotourism has, nowadays, been introduced as a best form of sustainable rural tourism marketing.

In China, the focus is the enhancement of employment and new economic activities linked to the specialties of each geopark. The establishment of individual geoparks should favour new orientations in tourism (geotourism). Local governments of Yuntai Geopark in Henan (China) and Jinhu Geopark in Taining County, Fujian Province, introduced geotourism as a new economic growth point which can alleviate the expense of geoheritage preservation, increase local revenues and enhance employment. For example in 2002, the number of visitors of Yuntai Geopark in Henan soared to 940,000 and the income from admission fees increased to 27.2 million Yuan, up 68% and 97%, respectively, compared with the previous year. Moreover, more than 60 hotels were newly built in the county, and about 4,000 jobs were offered by tourism development (Xun and Ting, 2003).

The involvement of local communities in geopark activities and geotourism is a key factor in sustainable tourism which is followed by geopark authorities.

Farsani *et al.* (2011), by using a quantitative method among 25 geoparks around the world, found that the majority (80%) of geopark authorities believe that in geopark territory, involving local businesses in tourism marketing such as tours, is the best way to promote the local economy. Moreover, 68% of geoparks try to link their activities with other local tourism activities such as boating, bird watching and cultural activities. According to questionnaires completed in the aforementioned research, geoparks strive to engage local communities in geopark activities. Regarding this, during the summer time (from May to October), the Adamello Brenta European Geopark (Italy) authorities employ 40 temporary staff who usually work in the 'Visitors Centre', the 'Environmental Educational Office' and on the dates when the park carries out the sustainable mobility projects.

Geoparks and geotourism not only try to engage locals in conservation, educational and geotourism activities but also apply strategies for attracting more visitors to these new territories and strive to promote local products and geo-products which are the symbols of geological and cultural heritage.

It is noteworthy that the private sector or family businesses only undertake innovation such as creating new products or organizing new tourism activity when an organization promises them to be profitable. Branding (certification) of local products will support the local producer. Local producers and customers will benefit from branding advantages which are higher-quality, environmentally friendly and so on (Farsani *et al.*, 2012a). Regarding this, the UNESCO Global Geoparks Network provides opportunities and possibilities, through a common logo, for small firms which are the members of the Geopark Network to boost their creativity and initiatives (Zouros and Mckeever, 2009).

At local level, in some geoparks (such as: Vulkaneifel Geopark (Germany); Cabo de Gata-Nijar Natural Park (Spain); Naturtejo Geopark and Arouca Geopark (Portugal); Nature park TERRA.vita European Geopark (Germany); Parco Naturale Adamello Brenta (Italy);Eisenwurzen European Geopark (Austria)) geotourism markets take advantage of the geopark brand or other eco-labels and certifications (e.g. in festivals, local products, events, publications etc.). In addition, some local producers use the geopark brand for promoting their products (Figure 2).

Figure 2. Promoting local products by using geopark brand in Naturtejo Geopark (Portugal) (Source: Naturtejo Geopark Authorities, 2011).

Eusébio and Lima (2012) noted that geotourism can take advantage of other industries which are or were active in the geopark area and create new tourism products. For example, there are many remains of lead, zinc, copper, tin, silver, and iron mines in Geological, Mining Park of Sardinia (Italy), and at present the tourism sectors represent a strategy for rehabilitating this economy through geotourism. Regarding this, they established geo-mineral museums to present unique and particular examples of mining engineering in the past to visitors; and they have revived the Italian Centre of Coal Culture and Saint Barbara's Cave (Agus and Atzei, 2009). Furthermore, Bohemian Paradise Geopark (Czech Republic) arranged a trip by a small mining train through a mine shaft (260 meters long) for promoting mining tourism as a geotourism product and new means for income generation. In addition, organizing wine tours related to geology, soil, and topography, historic mining operations (even active mines) and mining trails are more examples in this regard.

Making geo-products is another innovative strategy, and has been applied by some geoparks, family businesses, restaurants, bakeries, and local businesses for education and development of the local economy through geotourism. It is worth mentioning that geo-products which are made based on geological elements of geoparks not only introduce local products and local handicrafts to visitors, but increase public awareness and knowledge of visitors about the geology and geomorphology of that territory. Therefore it can be said that geo-products are pedagogic tools in geoparks as well. Making geo-products in Arouca Geopark (Portugal) constitutes a good example for making geo-products (Figure 3).

Preserving landscape and ancient architecture and changing their land use can be the next strategy for sustainability and generating income through geotourism in geoparks. The authorities of geoparks rebuild historic and old places instead of modernizing them. Regarding this, the Réserve Géologique de Haute-Provence European Geopark (France) has rebuilt an old house for the art instalment in Veil Esclangon, and the Naturtejo European Geopark (Portugal), at Portas de Almourão geomonument, and Penha Garcia Ichnological Park, has changed the use of four dilapidated historical buildings; they have been converted to popular tourism facilities like a conference hall, tourism office, and museum to follow the tenets of sustainable tourism in the geopark. It is noteworthy that some geoparks such as Naturtejo Geopark (Portugal) encourage local businesses and small and medium-sized enterprise (SMEs) to rebuild the old houses for touristic objectives (e.g. geo-bakery, family guest house).

Holding conferences, exhibitions, fairs, festivals and events in geoparks is another strategy for attracting more tourists to these territories. For instance, Naturtejo Geopark established a dinosaur exhibition in 2010 and this exhibition received 22041 visitors over 7 months in many educational programs (Naturtejo Geopark Authorities, 2010).

Figure 3.Geo-Break in International Congress of Geotourism – AROUCA Geopark 2011 (Photo by Farsani N.T).

In addition, organizing workshops and integrating fun, sport and comedy with educational programs, especially for schoolchildren, is not only a strategy for popularization of geo-science and gaining sustainable objectives but can create new job opportunities for companies, geotours and businesses (such as rock climbing for schoolchildren, geo-kayaking, hiking, geo-cycling).

Using sustainable energy – for geotourism facilities – to reduce air pollution is a sustainable innovative strategy applied in geoparks. E-bikes in Nature Park TERRA.vita (Germany) are equipped with an electric motor to ease biking in hilly areas. About 25 to 30 recharging stations are installed along the trails. The stations are equipped with a solar panel to provide carbon-free energy for the bike batteries (Escher, 2010).

Geo health & wellness tourism can also be promoted under the geotourism umbrella. Establishment of brine swimming pools and salt baths in Subbetica Geopark (Spain) is a geo-innovation offered to inhabitants and visitors to this geopark (Arroyo and Barquero, 2010). Furthermore, mud therapy, spa therapy, peat therapy, ice therapy and salt therapy are other geo health activities which are applied or can be applied in the future for attracting tourists, especially seniors, to geoparks.

Consequently, geotourism as a niche market is still at an early stage of commercial development, but experiences and examples illustrate that the creation of geoparks and the development of geotourism as sustainable tourism can be an innovative solution for the development of rural and green economies.

CONCLUSION

Geotourism is one of the core activities of geoparks which are under the umbrella of geographic tourism. It is obvious that development of the local economy would occur as a result of tourism activities. Moreover, geoparks are pioneers in the development of geotourism; they strive to stimulate socioeconomic activities and sustainable development by attracting an increasing number of visitors. Regarding this, geoparks have applied innovative strategies and facilities such as making geo-products, organizing geo-sports, establishing rural accommodation, providing geotourism maps, establishing geopark museums, geotours, and geo-sightseeing. Moreover, the experiences in some geoparks demonstrate that authorities of geoparks attempt to create new job opportunities, seasonal and part-time jobs for local communities and businesses. Also, they try to promote local products through strategies such as organizing festivals, events, workshops, and using brands.

Consequently, geoparks are a key factor in geotourism development and they maximize geotourism to the benefit of the local green economy and educate people about the evolution of their local landscape. Geoparks encourage local enterprises and cottage industries to get involved in geotourism and geo-marketing. Geoparks' aims are always related to local agents' participation, conservation and networking. All these aims are important pillars for the achievement of green economies and sustainable development. A geopark promotes these processes and, with local independent entrepreneurs that are receptive to innovation and cooperation with the public sector and other private entrepreneurs, will have a greater capacity to develop as a green tourism destination and as a green local economy. Geoparks through network activities at international, national and local level strive to provide an opportunity to benefit from the open world market. For instance, if a local businessperson or producer joins the geopark local network as an open economy model, she/he can take advantage of the recommendation of geopark authorities for creative innovation, together with the usage of the geopark brand for local products which are related to geopark activities. Besides this there is the possibility of being introduced to other geopark markets, at a national and international level. Furthermore, under the umbrella of the network, local businesspeople become stronger and more profitable. They also become eager to apply innovation. Therefore, special attention to financial slack, policy and management, innovation, competition marketing, open marketing opportunities such as network activities and niche marketing are key issues for future geotourism management in geoparks. Furthermore, the lack of data about the characteristics of geotourism, geotourists and even about green tourists, as mentioned above, creates a barrier for destination managers to evaluate that segment's demands and potential value, making it more difficult to design and implement green economy strategies. Consequently, future studies are suggested to collect and analyse this kind of data, over large periods of time and different geoparks, also producing comparable statistical information, in order to guarantee a more effective definition of green economy strategies to maximize the contribution of geotourism to the local economies of geoparks.

ACKNOWLEDGMENT

The authors acknowledge the Portuguese Science Foundation for the PhD grant to one of the authors (Neda Torabi Farsani); they also wish to thank Margarete Patzak – the Programme Specialist from Global Earth Observation Section, Division of Ecological and Earth Sciences UNESCO – for her contribution in collecting information. The authors appreciate the help of managers and officials for providing the intended information. And thanks to David Tucker for editing the English grammar of this book chapter.

CONTRIBUTORS

Neda Torabi Farsani (ntorabi@ua.pt) is a PhD student in tourism management, at the University of Aveiro, Portugal. She has a Master's degree (2006) in Physical Geography from Azad University, Iran and a Bachelor's Degree in Natural Resources Engineering, Range and Watershed Management of the University of Yazd, Iran (2002). She has written several articles in fields of geotourism, geoparks, and sustainable tourism, gastronomy in geoparks, rural tourism, green spaces, and physical geography in international conferences and published in international journals.

Joana Lima (jisl@ua.pt) has been a Lecturer at the Department of Economics, Management and Industrial Engineering (DEGEI) of University of Aveiro, Portugal, since September 2010 and is a researcher at the GOVCOPP Research Unit. Since October 2008, she has been a Research Assistant at the DEGEI of University of Aveiro and she has been a PhD Student in Tourism, at the University of Aveiro, since January 2010. She holds a Master's degree in Management and Development of Tourism, University of Aveiro (2008), and a Degree in Economics, Faculty of Economics of the University of Coimbra, Portugal (2005). Her main research interests include Tourism Socio-Economic Impacts, Social and Economic Development, Social Tourism and Tourism Experience.

Maria Celeste de Aguiar Eusébio (celeste.eusebio@ua.pt) is an Assistant Professor of Tourism and a researcher at the GOVCOPP Research Unit at the University of Aveiro (Portugal). She holds a Degree in Tourism Management and Planning from the University of Aveiro (1995), a Master in Economics from the University of Coimbra (1998) and a Doctoral Degree in Tourism, also from the University of Aveiro (2006). Her research interests include tourism economics, tourism impacts, tourism forecasts and consumer behavior in tourism and social tourism.

Carlos Manuel Martins da Costa (ccosta@ua.pt) is based at the University of Aveiro, Portugal, where he is Associate Professor, Leader of the PhD Programme and the Tourism Research Unit. He is also Editor of the Portuguese Tourism & Development Journal. He holds a PhD and MSc on Tourism Management (University of Surrey, UK), and a BSc on Urban and Regional Planning (University of Aveiro, Portugal). Carlos is at present involved in research covering areas such as planning, networks, policy, education, internationalisation, SMETs, and innovation. He is also leading a spin-off company in Aveiro University responsible for consultancy with tourism organisations and companies.

Celeste Oliveira Alves Coelho (coelho@ua.pt) Full Professor of Applied Environmental Sciences at the University of Aveiro, Ph.D. in Geography from the University of Aberdeen, U.K. Celeste has more than 30 years of experience in both fundamental and applied research, in the areas of: Applied Physical Geography, Natural Resources Planning and Management, Hydrology and Integrated River Basins Management, Soil Degradation and Desertification Processes (environmental and socio-economic dimension) and Natural and Technological Hazards. She has been involved in teaching in Tourism, Urban and Regional Planning, Environmental Engineering, Coastal Sciences, Territorial Planning and Natural and Technological Hazards, at the Universities of Aveiro, Porto, Coimbra and of Mozambique. She is the coordinator of the Group of Integrated Watershed Management of the Centre of Environmental and Marine Studies (CESAM). She is coordinator of the PhD. on Environmental Sciences and Engineering and Co-coordinator of a joint Ph.D. on Territorial Planning, Risk and Public Policy (Univ. Coimbra, Univ. Aveiro, Univ. Lisbon). She has coordinated several national and international applied and fundamental research projects. She has been the guest editor of international journals and books and has more than 100 papers published in scientific journals and conference proceedings.

References

Agus, D. & Atzei, E. (2009). Geological mining park of Sardinia, geological heritage, mines and its human history. In: N., Zouros, T., Ramsay, P., McKeever, & M., Patzak, (Eds.), *European Geoparks, earth heritage protection and sustainable local development* (pp. 145), The Natural History Museum of the Lesvos Petrified on behalf of the European Geoparks Network, SIGRI LESVOS, Greece GR-81103.

Alan, A. L. (2002). Geotourism and what geographers do. *Tourism Geographies*, 4 (4), 347-348.

Anand, A., Chandan, P. & Singh, R. B. (2012). Homestays at Korzok: Supplementing Rural Livelihoods and Supporting Green Tourism in the Indian Himalayas. *Mountain Research and Development*, 32(2), pp.126-136.

Arroyo, B.M. & Barquero, A.S. (2010). An example of local development in a geopark. *Proceeding of 9th European geoparks conference* (pp.100-101), 1-5 October, Lesvos Island- Greece .

Ballantyne, R., Packer, J., & Hughes, K. (2009). Tourists' support for conservation messages and sustainable management practices in wildlife tourism experiences. *Tourism Management*, 30(5), pp. 658–664.

Bapna, M., & Talberth, J. (2011).economics ecosystem services green economy Rio 2012 sustainable businesssustainable development. Retrieved May 23, 2012, from http://www.wri.org/stories/2011/04/qa-what-green-economy.

Bębenek, S. (2006). Geotourist management of abiotic nature objects as a chance for rural areas development. *Proceeding of International Conference GEOTOUR 2006* (pp. 186-187), 5-7 October, 2006, Košice, Slovakia.

Beeton, S., Bergin-Seers, S & Lee, C. (2007). Environmentally sustainable practices of victorian tourism enterprises, A Sustainable Tourism CRC. Retrieved February 02, 2009, from http://www.linkbc. ca/torc/downs1/BeetonPracticesVictorianTourism.pdf?PHPSESSID=0cf903e84ceaabf7cd37ce53ee1939ef .

Bimonte, S. & Faralla, V. (2012).Tourist types and happiness a comparative study in Maremma, Italy. *Annals of Tourism Research,* 39(4), pp. 1929-1950.

Buckley, R. (2003). Environmental Inputs and Outputs in Ecotourism: Geotourism with a Positive Triple Bottom Line?. *Journal of Ecotourism*, 2(1), 76-82.

Cato, M.S. (2009). *Green Economics: an introduction to theory, policy and practice,* Earthscan, London.

Cooper, C., Fletcher, J., Gilbert, D. and Wanhill, S. (1998). *Tourism: principles and practice.* London: Financial Times Prentice Hall.

DeLacy, T., & Lipman, G. (2010). GreenEarth.travel: Moving to carbon clean destinations. In C. Schott (Ed.), *Tourism and the implications of climate change: Issues and actions - Bridging Tourism Theory and Practice*, 3, pp. 299–312. Bingley: Emerald Group Publishing.

Dolnicar, S., & Matus, K. (2008). Are green tourists a managerially useful target segment? *Journal of Hospitality Marketing & Management,* 17(3/4), pp. 314–334.

Dowling, R. (2009). Geotourism's contribution to local and regional development. In: C., Neto de Carvalho & J.C. Rodrigues, (Eds.), *Geotourism and local development* (pp. 15-37) , Geopark Naturtejo, Idanha-a-Nova, Portugal.

Dowling, R., & Newsome, D. (Eds.). (2006). *Geotourism, sustainability, impacts and management*. Elsevier, Butterworth Heinemann, Oxford.

Escher, H. (2010). Enjoying Geopark by E-bike, a new infrastructure for TERRA.vita. Proceeding of 9th European geoparks conference (pp.91), 1-5 October, Lesvos Island-Greece.

Eusébio, M. C. (2006). *Avaliação do impacte económico do turismo a nível regional – O caso da Região Centro de Portugal*, PhD Thesis, Universidade de Aveiro, Aveiro.

Eusébio, C., & Lima, J. (2012). Economic Benefits of Geotourism: Theoretical Reflections and Management Implications. In N.T., Farsani, C., Coelho, C., Costa, &C., Carvalho (Eds.), *Geoparks and Geotourism: New Approaches to Sustainability for the 21st Century* (pp. 137-149), Brownwalker Press Publisher, Florida, USA.

Farsani, N.T., Coelho, C., Costa, C., & Carvalho, C. (2012a). *Geoparks and Geotourism: New Approaches to Sustainability for the 21st Century*. Brownwalker Press Publisher, Florida, USA.

Farsani, N.T., Coelho, C., & Costa, C., (2012b). Tourism Crisis Management in Geoparks through Geotourism Development. *Journal of Tourism and Development*, 17/18, 1627-1638.

Farsani, T. N., Coelho. C. & Costa, C. (2011). Geotourism and geoparks as novel strategies for socio-economic development in rural areas. *International Journal* of *Tourism Research*, 13 (1), 68-81.

Fennell, D.A. (2003). *Ecotourism. An introduction*. London and New York: Routledge.

Godfrey, K. and Clarke, J. (2000).*The tourism development handbook: a practical approach to planning and marketing*, New York: Continuum.

Gorman, C.E. (2007). Landscape and Geotourism: market typologies and visitor needs. Paper presented at European Tourism and the Environment Conference: Promotion and Protection, Achieving the Balance, Dublin, Ireland, October 2007.

Hartling, J. and Meier, I. (2010).Economic Effects of Geotourism in GeoparkTERRA.vita, Northern Germany, *The George Wright Forum*, 27 (1), pp.29-39.

Hughes, H. L. (1994). Tourism multiplier studies: a more judicious approach. *Tourism Management*, 15(6), pp. 403-406.

Inskeep, E. (1991). *Tourism Planning. An integrated and sustainable development approach*. New York, Van Nostrand Reinhold.

James, h. C. L. and Hose, T. A. (2008). Are We in Danger of Losing the "Geo" in Geotourism? An assessment of the geological potential of selected sites in southern-central Britain. In: Dowling, R. K. andNewsome, D. (Eds.). *Geotourism. Proceedings of the Inaugural Global Geotourism Conference, 'Discover the Earth Beneath our Feet',* Fremantle, Western Australia, 17-20 August. Promaco Conventions, pp. 199-208.

Kadt, E. De (1979).*Tourism: Passport to Development?* Oxford University Press, Oxford.

Law, A., DeLacy, T., McGrath, G. M., Whitelaw, P. A., Lipman, G. & Buckley, G. (2012).Towards a green economy decision support system for tourism destinations. *Journal of Sustainable Tourism*, 20(6), pp. 823-843.

Lima, J. (2008). *Turismo e Desenvolvimento Económico – Segmentos de maior valor económico para a Covilhã*, Master Thesis, Universidade de Aveiro, Aveiro.

Mathieson, A. and Wall, G. (1990).*Tourism: economic, Physical and social impacts*, Longman Scientific and Technical (1st publication 1982).

Mill, R., and A. Morrison (2006).*The Tourism System*. Kendall/Hunt Publishing Company. Fifth Edition.

Naturtejo Geopark Authorities. (2010). 30-31st October –Dinosaurs invade the Geopark and Travelling Dinos come to an end. *Geopark Naturtejo da Meseta Meridional-European and Global Geopark Monthly Report*, Number 50, pp: 8.

Naturtejo Geopark Authorities. (2011). 28th to 31st July – XV Borderland Fair of Idanha-a-Nova: presenting Earth products. *Geopark Naturtejo da Meseta Meridional-European and Global Geopark Monthly Report*, Number 59, pp: 11.

Nickerson, N. (2010). Geotourism as a Basis for Infrastructure Planning. Paper presented at the Conference for the Montana Association of Planners September 22, 2010, Institute for Tourism and Recreation Research University of Montana, U.S.A.

Nijaki, L. K. & Worrel, G. (2012). Procurement for sustainable local economic development, International Journal of Public Sector Management, Vol. 25 Iss: 2 pp. 133 – 153.

Patzak M. & Eder, W. (1998). UNESCO Geopark. A new programme-A new UNESCO label. *GeologicaBalkanica*, 28, (3-4), 33-34.

Reynard E, and Coratza, P. (2007).Geomorphosites and geodiversity: a new domain of research. *Geographica Helvetica, the Swiss Journal of Geography*, 62(3), 138-139.

Sharpley, R. and Telfer, D. J. (Eds). (2002). *Tourism and Development: Concepts and Issues*, Channel View Publications, UK.

Stokes, A., Cook, S., & Drew, D. (2003).*Geotourism: The New Trend in Travel*. Travel Industry Association of America (TIA) and National Geographic Traveler.

Tourtellot, J.B. (2000). Geotourism for your community. *National Geographic drafts*, Washington DC, pp. 2.

UNCSD Secretariat. (2012). A Green Economy Knowledge-Sharing Platform: Exploring Options. Rio+20, United Nations Conference on Sustainable Development. Retrieved May 24, 2012, from http://www.uncsd2012.org/rio20/content/documents/290A%20Green%20Economy%20Knowledge%20Platform_ExecSum%20only.pdf.

UNESCO. (2006). Guidelines and Criteria for National Geoparks seeking UNESCO's assistance to join the Global Geoparks Network, Paris, January 2006.Internal document, 10.

United Nation.(1992). Green economy in the context of sustainable development and poverty eradication. Retrieved May 24, 2012, from http://www.uncsd2012.org/rio20/index.php?menu=62.

United Nations Environment Programme (UNEP) (2011). *Towards a green economy: Pathways to sustainable development and poverty eradication*. Retrieved from www.unep.org/greeneconomy/

Williams, A.M. and Shaw, G. (Eds.) (1991). *Tourism and Economic Development – Western European Experiences*. London.

World Economic Forum. (2010). *Green Investing 2010 – Policy Mechanisms to Bridge the Finance Gap*. Available at: http://www3.weforum.org/docs/ WEF_IV_GreenInvesting_Report_2010.pdf

Xun Z. & Ting, Z. (2003). The socio-economic benefits of establishing National Geoparks in China. *Episodes*, 26 (4), 302-309.

Zouros, N., & McKeever, P. (2009).Tools for Earth Heritage Protection and Sustainable Local Development, European Geoparks. In: N., Zouros, T., Ramsay, P., McKeever, & M., Patzak, (Eds.), *European Geoparks, earth heritage protection and sustainable local development* (pp. 15-30), The Natural History Museum of the Lesvos Petrified on behalf of the European Geoparks Network, SIGRI LESVOS, Greece GR-81103.

In: Economic Developments and Emerging Markets ... ISBN: 978-1-62417-986-0
Editors: Guy Boscarino and Derek Notte

Chapter 5

PERSPECTIVES ON POVERTY ALLEVIATION: COMMUNITY-BASED ENTERPRISE, DEVELOPMENT OF INNOVATION CAPABILITY, AND THE VERTICAL INTEGRATION MINDSET

Roland Gau,[*] Edward Ramirez, Ricardo Gonzalez[†] and Ray Perez
University of Texas at El Paso, Texas, US

ABSTRACT

Recent movements in business research have placed greater emphasis on the "Base of the Pyramid," the billions of people living in extreme poverty. These movements include Transformative Consumer Research, research on Subsistence Marketplaces, and the BoP Protocol, all of which have subtly different goals, but all focus on the improved conditions of consumers living in poverty through a deeper understanding of the individuals and marketplaces. A review of the research in this area reveals a number of articles that have examined these consumers and marketplaces from the bottom-up. On the other hand much of the traditional work in poverty alleviation addresses issues from the national-level, down. We propose that a missing, but vital link is the deeper understanding of the marketplaces and systems at the community-level. We argue that poverty alleviation efforts focused on empowering individual entrepreneurs, while potentially life-changing for these individuals and their families, tends to result in limited benefits for the community at-large. We then discuss the importance of innovation in lifting communities out of from what we describe as "Stage 0," (where communities are exploited for their labor and raw materials). We propose a 4-stage model, ranging from "Stage 0" to "Stage 4" (as communities progress from being exploited for labor/raw materials, to being able to manufacture components/products with older technologies, to being able to manufacture finished systems/products with modern innovations, to finally, being able to innovate independently). Finally, we discuss the importance of vertical

---

[*] Corresponding author: Roland Gau. 500 University Ave. College of Business Administration. El Paso, TX 79968, (915) 747-6036, E-mail: rgau@utep.edu.

[†] Ricardo Gonzalez is a 2nd year MBA student at UTEP. Ray Perez is an undergraduate research assistant for the Department of Marketing and Management at UTEP.

integration, through a community-based perspective as an enabler of progress, from Stage 0 through the stages. This perspective is unique in that it brings the experience of bottom-up perspectives on poverty alleviation, with the raw potential of innovation as an agent for transformation at the community level.

INTRODUCTION

"Millions of people in the world's poorest countries remain enslaved by the chains of poverty. It is time to set them free."

Nelson Mandela

Despite the development of technological advances that have markedly improved the quality of life for many of the world's inhabitants, millions remain shackled by poverty (Shah 2010). In developed economies poverty is characterized as much by relative wealth, as it is by absolute wealth. For example, the threshold for an individual in poverty in the US is just over $11,170, and represents the purchasing power that satisfies the everyday comforts that a typical American takes for granted (Schwartz 2005). On the other hand, an examination of subsistence marketplaces in less developed countries reveals a stark contrast. Whereas extreme poverty is relatively uncommon in developed countries, it is a relatively common characteristic of subsistence marketplaces. Earning less than $1.25 USD a day (World Bank 2012b), those living in extreme poverty face a number of unmet basic needs that represent challenges that are likely unfamiliar to residents in developed economies (e.g., malnutrition, lack of clean water, increased rates of disease), and often think and behave quite differently than individuals in the developed world (Gau, Jae, and Viswanathan forthcoming; Viswanathan et al. 2009). Moreover, the effects of poverty hamper subjective well-being, as life satisfaction and positive emotions correlate with the absolute level of wealth. In short, poverty represents a substantial and pervasive problem with serious implications that need to be addressed. Future generations will judge our sense of humanity based on our response to this issue.

This chapter examines the idea of business as a facilitator of poverty alleviation, and proposes the use of community-based enterprises as a catalyst for poverty alleviation. We focus on two elements of business strategy that are proposed to positively impact community-based enterprises: the development of innovative capability for community-based enterprises, and a presence of a vertical integration mindset to facilitate the goal of innovation.

The rest of the chapter is organized, as follows. First, a brief review of poverty alleviation efforts is presented. Next, a case is built for importance of a community-based approach, as opposed to one centering on the individual or nation-state, which employs a business-oriented approach to poverty alleviation that develops local capacity in subsistence marketplaces. This approach is couched in sustainability in the sense that it requires focuses on indigenous economic growth efforts fueled by self-reliance and autonomy, where patronage from wealthy outsiders becomes unnecessary. Concerning this approach to poverty alleviation, the characteristics of subsistence marketplaces are suggested to furnish the needed power to spur community development. A 4-stage model is developed that demonstrates how communities can transition from a subsistence-level marketplace to a thriving local economy, with the development of innovation capability. Within this vein, innovation at the community-level is cited as a key goal in poverty alleviation. Finally, the conceptualization of a vertical integration mindset is suggested to facilitate the community's transition through the 4 stages of development.

POVERTY ALLEVIATION AND THE ROLE OF BUSINESS

In this section, a case is offered for *community-based enterprises* (i.e., collectively-owned businesses that seek to jointly maximize community welfare and economic growth) as a critical element in poverty alleviation in subsistence marketplaces. In doing so, we begin with a brief discussion of firms' and economic organizations' poverty alleviation efforts, noting a distinction between macro- and micro-level efforts. Finally, the need for a community-level focus centering on sustainability is presented as a viable solution for poverty alleviation, concluding with a call for the use of community-based enterprises.

Two Perspectives on Poverty Alleviation

Whereas both firms and global economic organizations represent two change agents that seek to eliminate poverty, this discussion first addresses organizations which have received wide-spread attention for their multifaceted efforts. Consider that, since its inception in 1945, the International Monetary Fund (IMF), which includes human welfare and poverty alleviation amongst its top priorities, has endeavored to eradicate poverty through issuing developmental loans to nation-states (International Monetary Fund 2012). In this regard, the IMF has not only served to finance major infrastructure projects, but it also has taken on the responsibility of advising governments on macro-economic policies that are suggested to engender economic growth. Similarly, the World Bank seeks the eradication of extreme poverty and hunger through providing low cost loans to developing countries (World Bank 2012a). Both the IMF and the World Bank, however, has also come under fire for making its loans conditional on the implementation of sweeping economic reforms by recipient countries, which are often tied to free-market economic policies, such as the promotion of foreign direct investment, adherence to current multilateral trade agreements, and conformance with internationally recognized environmental protection standards (Bretton Woods Project 2005). Despite the well-intended efforts towards poverty alleviation, some nations have grown reluctant to accept funding from the IMF and the World Bank as it features contingencies which could prove to have negative long-term repercussions

A third major, international organization of note is the United Nations (UN), an international organization whose principal aim is to facilitate cooperation amongst member nation-states through the creation and implementation of international law, economic development, world security, and social and humanitarian advancements. The UN administers a number of efforts which target poverty alleviation in a manner more directly focused on the poor population, versus on the development of national infrastructure. More specifically, the UN has established a number of prominent agencies including the World Health Organization, the World Food Programme and United Nations Children's Fund whose overarching missions center on social justice and whose operational activities involve local interventions designed to raise the living standards of the world's poor (United Nations 2012b). Furthermore, the UN, in partnership with the World Bank and a number of other international organizations, has crafted a set of Millennium Development Goals, which call for multilateral agreement on poverty eradication, universal education, and pro-environmental initiatives (United Nations 2012a).

The business community has also looked to address poverty alleviation in a variety of ways. An exemplar is the breakthrough work of Nobel Laureate Muhammad Yunus, who created the Grameen Bank, a microfinance lender in Bangledesh. By providing very small value, low-interest loans that are collateralized through social norms, Yunus' banking concept has facilitated economic development in subsistence marketplaces (Yunus 2004). The efforts of Yunus and Grameen Bank have led to improved well-being for countless entrepreneurs who have created businesses from these micro-loans. Additionally, a number of businesses have engrained the notion of implementing pro-social elements as a central part of business objectives, vis-à-vis profit maximization.

Apart from organizational and firms' efforts at poverty alleviation, a body of literature catalogues how such efforts can be implemented. C.K. Prahalad's book *The Fortune at the Bottom of the Pyramid*, has been influential in opening the business community's eyes to the potential for business activities that serve poor consumers (2004). In this regard, Prahalad suggests that those living in poverty represent an underserved marketplace with a burgeoning demand for basic goods and services. While marketing mix elements (product, price, promotion, and points of distribution) must be modified in order to serve this market, firms with the flexibility and ingenuity to do so will benefit greatly on multiple dimensions (such as satisfying currently unfulfilled needs and the prospect of capitalizing on a "first mover advantage"). Together, these efforts have led to a number of movements in business research, such as the Subsistence Marketplaces Initiative, founded by Madhu Viswanathan, and the creation of Transformative Consumer Research, currently chaired by David Glenn Mick. Taken collectively, these efforts imply that the key to poverty alleviation lies in leveraging the actions of the individual.

Although efforts have been made towards poverty alleviation, the focus has primarily been limited to nation-state and individual-level interventions. With regards to the former, the success of organizations such as the World Bank and the IMF in reducing the level of global poverty is contingent on recipient nation-states' effective utilization of funds and enactment of economic development policies. While efforts by these organizations contribute towards the goal of poverty alleviation, such broad based programs tend to overlook implementation issues that may emerge on the frontlines. As such, the specific mechanisms that undergird successful implementation have yet to be exploited. Furthermore, programs initiated at the nation-state level fail to develop a blueprint for direct engagement with the poor, implying that such efforts may be suboptimal.

In terms of the latter, whereas microfinance and marketplace literacy programs have the potential to radically transform individuals' lives, successful strides towards poverty alleviation are limited to individuals' efforts. In this regard, although poverty alleviation based on the actions of individuals makes for heart-warming stories, questions remain regarding the scalability and operationalization of such efforts. Critics of this approach might argue that even though the actions of a single individual can reach a sizable number of people, the prospect that such efforts will truly impact a community's overall welfare are minimal. Furthermore, the individual-level route to poverty alleviation could be suggested to overemphasize the improved outcomes of microfinance loan and marketplace literacy education recipients, while overlooking the continued impact of poverty on the non-beneficiaries. As such, individual-level efforts may only serve to redistribute the small amount of wealth currently held by the community to the intervention beneficiaries, instead of actually generating new wealth.

Subsistence Marketplaces and the Emergence of the Community-Based Firm

Subsistence marketplaces are characterized by ubiquitous poverty. In such marketplaces, which have also been described as those residing at the bottom of the pyramid, buyers and sellers represent "two sides of the same coin," as both parties to an exchange live in poverty and must cope with its deleterious effects in a similar fashion (Viswanathan and Rosa 2007; Viswanathan et al. 2009). In addition, the subsistence marketplace is plagued by a lack of physical and social mobility, resulting in largely self-contained, closed-system marketplaces, featuring a rigid social structure. As a consequence, the negative effects of the individual-level poverty alleviation schemes alluded to above are exacerbated in this context. More specifically, this situation increases the likelihood that any of the benefits derived will primarily accrue to the intervention recipient, as the associated positive outcomes merely represent a redistribution of resources encapsulated by a closed-system. It should finally be noted that this same effect is found to impact the distribution of quality of life improvements, where an individual-level poverty alleviation approach has a minimal overall net effect on those living in a subsistence economy.

One of the powerful, positive characteristics of subsistence marketplaces, however, is found in its potential to harness social capital. In this regard, such marketplaces can benefit across multiple dimensions by leveraging the heightened level of cooperation that exists between individuals in close-knit, rural communities. As such, subsistence marketplaces stand to gain from community-based actions that systematically overcome marketplace barriers that would otherwise be insurmountable for either an individual buyer or seller. More specifically, through the pooling of economic resources, which are relatively scarce in a subsistence marketplace, communities would be empowered to circumvent financial barriers to entry. For example, while the cost of a used plastic injection-molding setup, perhaps $1,000 USD, is likely to represent a prohibitive expense for an individual entrepreneur, this investment could be realized through a concerted, community-based effort. Similarly, the cost of raw materials used in the production of a finished good would dramatically decrease through community-based collective purchasing efforts that reap scale economies. It is noteworthy to also mention that self-help groups in many subsistence markets have already instituted such practices, as collective purchasing efforts enable group members to avoid paying retail markup prices (Deininger and Liu 2009).

In short, a community-based solution to poverty alleviation not only represents the most empowering of the aforementioned approaches, but it may also be the most sustainable. This statement is predicated on the notion that such an approach could infuse a closed-system, subsistence economy with a source of newly generated wealth. The surplus acquired through scale economies would allow for increased buying power, and hence an increase in the collective standard of living. In other words, by harnessing the social capital embedded in the community, economic activity would not be restricted to wealth redistribution occurring as a by-product of an individual-level poverty alleviation approach.

Community-Based Enterprises and Sustainability

In order to ensure the successful implementation of a community-based enterprise approach to poverty alleviation, such efforts must be conducted in a sustainable manner. In this regard, community-based enterprises should develop sensitivity to the impact of their organizations on the economy, environment, and social welfare (Roseland 2000). The converse, however, also applies, as community-based enterprises should recognize how these factors can and will impact their longevity. As a result, community-based enterprises should adopt a long-term perspective with regards to their operations. For instance, approaches to poverty eradication that require patronage, or a subsidy from a wealthy buyer of a given product offering, may not represent an economically-sustainable strategy. More specifically, although consumers have been found to place value on innovative, social business models, such as the "buy-one, give-one" (e.g., Toms Shoes, One Laptop Per Child) or the fair-trade model (e.g., Ten Thousand Villages; various coffee and chocolate manufacturers), their willingness to pay a price premium or to make a charitable contribution for these offerings may hinge on economic conditions, disposable income levels, or individual difference variables (Mohan and Wilding 2009). In fact, the recent economic downturn has resulted in a precipitous decline in charitable donations, as well as in the sales volumes of firms providing pro-social goods and services (Barton and Hall 2010), lending credence to the idea that reliance on donors' continuing kind-heartedness in light of economic volatility may not make for a sound business strategy.

In addition to economic sustainability, the idea of environmental and cultural sustainability is also crucial in subsistence marketplaces. From an environmental perspective, subsistence populations often live more closely-connected to the environment than those in developed economies (International Institute for Sustainable Development 2012), as these consumers spend a relatively large portion of their time engaging in activities related to their basic survival, such as drawing water and gathering cooking fuel. Thus, business ventures in subsistence communities must take care to understand the environmental impact of their operations, as maintaining a healthy relationship with the environment will ensure the longevity of the communities that they serve. This relationship with the environment also represents a deeply-embedded social norm whose violation could result in animosity towards the firm. Likewise, from a cultural perspective, care must be taken to maintain the fundamental social elements of a community, as these factors are perhaps the most valuable resources that can be leveraged. Notions such as inclusive capitalism, which focus on social justice for those who typically have the least power, are critical in building poverty alleviation solutions in subsistence marketplaces (The Henry Jackson Initiative for Inclusive Capitalism 2012). The maintenance of the cultural fabric of these marketplaces allows community-based solutions to leverage social capital and to facilitate community-based learning. Thus, whereas firms' adherence to a more sustainable business approach in developed economies may simply represent a new marketing ploy, an attempt to reduce costs by streamlining operational inefficiencies, or efforts at strategically paying "lip service" to a pro-social cause, enacting a sustainability-oriented approach is a vital element for business' survival in subsistence marketplaces (Ramirez forthcoming).

In summary, the development of community-based enterprises represents an effective approach to be instituted in poverty alleviation efforts. By implementing the governance and profit-sharing systems that have proven to be instrumental in the success of co-ops (Reiss 2010), the actions of such enterprises would resonate with a community-centric perspective, in addition to supporting the community's overall health and welfare. Nevertheless, the overarching goals of community-based enterprises closely align with those of other firms taking part in free-market capitalism. As such, community-based enterprises' need to operate in sustainable fashion is of paramount importance. This is particularly the case with regards to such enterprises' evaluation of their economic sustainability, as many pro-social businesses tend to rely heavily on the generosity of wealthy patrons. Furthermore, the true test of the long-term viability of the community-based enterprise approach to poverty alleviation efforts will lie in its ability to compete and survive in the turbulent market-driven environment.

SUSTAINABLE DEVELOPMENT THROUGH INNOVATION

Although the use of community-based enterprises should overcome many of the weaknesses inherent in nation-state and individual-level interventions for poverty alleviation, the continuing viability of such an approach is contingent on enterprises' ability to innovate. Widely identified as a key component in business success in general, and broadly construed of as "a firm's tendency to engage in and support new ideas […] for the development of new processes," innovation allows firms to keep, and win new, customers (Lumpkin and Dess 1996, p. 142). Within this vein, community-based enterprises' innovations that create value-adding goods and services will cement long-term business relationships with community members, in addition to enabling these firms to export their wares. As a result, innovative community-based businesses that endeavor to continuously improve their offerings versus merely focusing on exploiting competitive advantages derived from their core competencies will avoid product obsolescence and thus declining sales. In essence, the purpose behind community-based enterprises' need to innovate is that doing so will facilitate their ability to compete in the highly-competitive free-market economy, as well as raising the community's collective standard of living. For this reason, the following section uses a 4-stage model to describe how innovation fosters community development, while strengthening community-based enterprises' competitive positions for the onslaught of challengers found in the free-market environment (See Figure 1).

A Stage 0 community is one that is characterized by extreme poverty and the presence of unmet basic needs, such as sanitation, clean water, and reliable food sources. As a result, Stage 0 communities face a looming threat from hunger and disease. Furthermore, in this stage, a substantial portion of a community's financial resources and economic activity is driven by external sources, such as foreign direct investment, foreign aid, and charity.

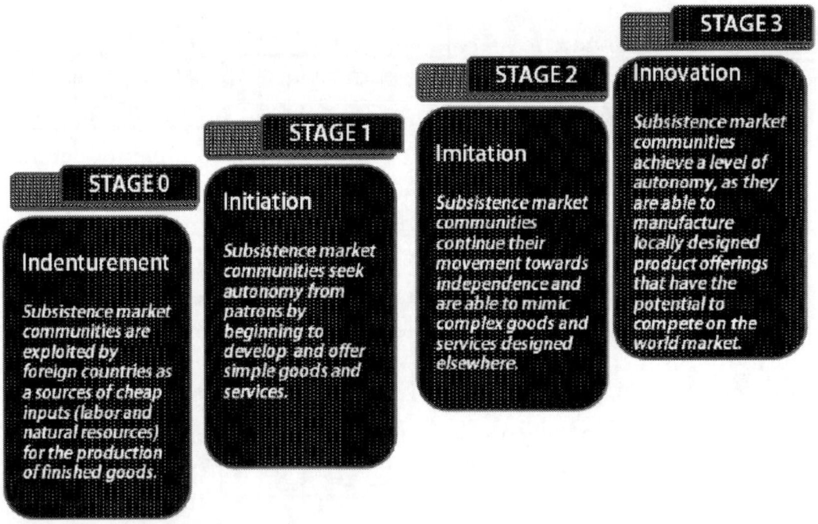

Note: This figure depicts and describes the evolution of Subsistence market communities. As indicated by the figure, communities progress through four discrete stages of development. Community dependency on generosity from wealthy benefactors declines and local capacity and autonomy increases as communities evolve.

Figure 1. Stages of Susbsistence Market Community Development.

Thus, the economic activities of Stage 0 communities are largely restricted to those that address basic needs, such as the development of agriculture or the provision of medical care. In addition, some might argue that these communities suffer from exploitation by their benefactors. Stage 0 communities have been historically viewed by imperialistic nation-states as cost-effective sources of labor or raw materials. This situation creates a dependency that negatively impacts the collective psychology of members of such communities. Finally, it should be noted that examples of Stage 0 communities include those already suffering from poverty and which might be recovering from disasters, whether natural (i.e., post-earthquake Haiti) or man-made (i.e., various locales immediately following wars).

Stage 1 communities represent those that have progressed beyond the struggle for mere survival, but are still impacted by poverty. In Stage 1, fundamental physical needs, such as water and food, are met, so that community members are not at risk of starvation or dying of thirst. Despite these monumental advancements, the availability of food and water in such communities would best be characterized as unreliable and relatively unsafe when compared with that in the world's developed economies. As a result, Stage 1 communities are particularly susceptible to illness associated with unclean food and water sources. Furthermore, the possibility of regressing into extreme poverty (or Stage 0) is an ever-present threat. Many of the subsistence marketplaces described in prior research could be categorized as Stage 1 communities. These communities have functioning, largely informal, economies that produce basic goods and services. In this regard, hand-made products, such as textiles and crafts, more sophisticated agricultural business models, and basic products or components for more complex products are produced by these communities. Finally, a number of Stage 1 communities are present throughout the world, including the subsistence marketplaces that comprise the colonias throughout Latin America and the poor villages found in rural India.

Stage 2 communities represent those that have evolved from a point in which poverty is the norm to one with an emergent middle-class. At this stage, the loss of productivity caused by illnesses is reduced, and increases in absolute income result in substantial quality of life improvements. In addition, infrastructure (i.e., power, running water, communications, and roads) is created, and becomes relatively reliable. Stage 2 communities also feature specialized firms, which develop the competencies to engage in large-scale production processes and to reverse-engineer their competitors' complex products. As a result of mounting industrialization efforts and increased specialization, Stage 2 communities are able to play a more significant role in the production of finished goods than are less evolved communities, which have been relegated to producing raw materials. By capturing more downstream functions in the value chain, Stage 2 communities develop immunity to the indenturing effects associated with declining terms of trade for commodities in international markets (cf., Cardoso and Faletto 1979), and enjoy dramatic quality of life improvements. Communities that could be described as belonging to this stage include those found in Asia (i.e., Taiwan, South Korea, Japan) that developed manufacturing-based economies prior to their ability to innovate.

Ultimately, the transition from a Stage 2 to a Stage 3 economy is a more subtle one. Whereas individuals' quality of life may not noticeably change, advancements are made to improve infrastructure reliability, and investments in institutions to develop education and culture are present. From a business perspective, a community's evolution to Stage 3 represents an important leap for its economic sustainability. Whereas Stage 2 communities are limited in manufacturing capability, and thus can only manufacture products based on imported design and engineering skills, Stage 3 communities have developed the intellectual capital needed to develop new products from the ideation to the finished goods phase. This ability to generate knowledge and to fortify intellectual capital paves the way for autonomy. As a consequence, in addition to making hefty investments in research and development, a strong emphasis is placed on providing Stage 3 community members with a broad-based primary education, as well as highly-specialized skills in secondary and post-secondary schooling.

In summary, subsistence markets transition through four distinct stages in an evolutionary process, with each transition calling for different interventions. For instance, the move from Stage 0 to Stage 1 is akin to triage, where the focus is on the immediate availability of basic provisions required for survival. In this vein, external sources of funding play a central role in the transition. The transition from Stage 1 to 2, however, leverages the most important concepts found in this chapter, and given that the transition from Stage 2 to Stage 3 is beyond its scope, the balance of the discussion is organized around this transition.

The transition from Stage 1 to Stage 2 community requires two distinct elements. First, it requires that the community is able to effectively leverage social capital (Bourdieu 1979; Granovetter 1973). As previously described, while subsistence marketplaces lack financial resources, they contain a potential wellspring of social capital. The reasoning behind this stems from the fact that individual community members share similar life experiences, as well as a cultural orientation toward interdependence and collectivism, thus solidifying the prospects of accepting a shared fate. In addition, a community-wide sense of empathy develops, where individual members stand united and embrace the idea of a common objective. As a result, higher levels of trust and a greater willingness to cooperate for the common good follow. As a final outcome, the community, as well as the potential community-based enterprise, is more willing to accept change, which is a precursor to innovation within organizations.

Although social capital can be leveraged to lead to a community's readiness to accept innovation, this alone is inadequate for its development. In addition to each individual member of the community being psychologically primed to unite and, thus, to accept change, a second element that is instrumental revolves around the formation and exploitation of a collective mind (March 1991). Based on writings in organizational studies, a collective mind is defined as when a group of individuals contributes, and subordinates itself, to a codified knowledge base (Weick and Roberts 1993). As such, the interrelationships among members represent the storehouse or locus of the organization's collective mind, solidifying a strong bond between group members. A group's collective mind, furthermore, has the capacity to learn, as individuals compare and contrast their maps of codified knowledge to the realities found in the real world, making adjustments where the mental representation does not fit actuality (Walsh 1995). Taken together, the presence of a collective mind results in not only the development of knowledge in the organization (community) effective for community learning, but it also opens the doorway to receptiveness to innovation, and hence an evolution from a Stage 1 to a Stage 2 community.

Downstream Integration Mindset as a Catalyst

As community-based enterprises increase their capabilities, through familiarity and experience with their commonly used business practices, they will inevitably need to grow. How this growth occurs can be described through the process of integration, which occurs either vertically or horizontally (e. g., Balakrishnan and Wernerfelt 1986). Vertical integration represents the expansion of a firm through the acquisition, typically via a purchase or merger, of another firm that performs a different function in the value chain, producing a different product or service than that produced by the original firm. Vertical integration can occur in one of two ways. The first is forward, or downstream, integration, where firms control more of the processes that are closer to the end-user. The second is backward, or upstream integration, where firms control more of the processes that lead back to the procurement of the raw materials used in finished product. On the other hand, firms can also expand through horizontal integration, accomplished by acquiring firms that perform the same function (i.e., offer the same good or service), resulting in the control of a larger market share. A stark reality is that community-based enterprises may never be able to acquire or merge with other firms, so the discussion is not restricted to the actual transaction between firms. Instead, the focus is on the mindset that accompanies the desire to use vertical or horizontal integration strategies. As a community-based enterprise grows organically, the organization will benefit from guiding that growth.

The example offered by Delphi in Juarez, Mexico suggests that growth through a vertical integration mindset is desirable for community-based enterprises, particularly with regards to poverty alleviation. Much of the value produced by the goods and services offered by Stage 1 communities results from access to cheap labor or raw materials. As community-based enterprises gain familiarity and experience in producing relatively simple goods and services, there will be an increase in organizational capabilities, as well as an increase in the abilities of the individual employees. A seemingly natural outcome is the shift from the creation of simple goods and services to the creation of more complex goods and services. However, the expertise that organizations develop may create cognitive entrenchment, resulting in an overreliance on familiarity with a given process (Dane 2010). For community-based enterprises in Stage 1 communities, this can become a trap that prevents the community from progressing into Stage 2 and beyond. In many ways, this sort of growth would reflect a horizontal integration mindset, where Stage 1 communities strive for the production of more raw materials, the availability of more cheap labor, or the manufacture of simple components. This is attractive because of the seemingly lower level of risk, as well as requiring more of the same. If the products and services produced by a Stage 1 community are commodities, this process would result in an entrenchment in the production of commodities, and restrict the value of the products and services offered. While this process is distinctly better than remaining in a Stage 0 state, this also restricts the ability to transition into Stage 2. The net result is the continued dependence on more developed economies, a likely unsustainable, long-term path.

On the other hand, a vertical integration perspective allows community-based enterprises to offer increasingly complex products and services, as they increase their organizational capabilities. Specifically, the implementation of forward, or downstream, integration offers community-based enterprises the ability to control the more complex parts in the creation of products and services, which represent a greater proportion of the value chain. From a broad perspective, this is helpful for the goal of increasing the ability to innovate. Gaining familiarity with increasingly complex processes pushes organizational capabilities toward the ability to produce unique, "new-to-the-world" products and services. Ultimately, although this pathway does not represent a guaranteed transition into Stage 2 and Stage 3, it this allows for the possibility, whereas the horizontal integration mindset likely prohibits the transition.

Community-based enterprises in subsistence marketplaces are also uniquely suited to benefit from a vertical integration mindset. In terms of innovation, vertical integration improves the level of trust and collaboration between firms (Nocke and White 2007). In a similar fashion, the presence of trust would allow for the presence of a vertical integration mindset. However, individuals in subsistence often think concretely and are focused on the immediate, which would lead to the desire for the "more of the same" that comes with a horizontal integration mindset. This can be overcome through the increased levels of trust that arise from the leveraging of social capital.

As mentioned previously, subsistence marketplaces have relatively greater levels of social capital, which arise from the tendencies to think and behave interdependently and collectively. Individuals living in subsistence are often willing to sacrifice their own preferences, and even their own well-being, for the good of the community, if they trust the people, processes, and the institutions for which they sacrifice. So the presence of trust can overcome the desire for "more of the same," and facilitate the vertical integration mindset. The leveraging of this social capital would result in greater efficiency in organizational learning, which in turn allows for the facilitation of innovation. Additionally, this increased trust also increases the level of organizational learning, which consequently allows for the facilitation of innovation.

CASE STUDY: MAQUILADORAS IN JUAREX, MEXICO

In 1947, the United States and Mexico entered into a temporary labor agreement in which unskilled laborers from Mexico could temporarily work in the US From 1942 to 1962, the Bracero program sponsored 4.5 million work visas for Mexican laborers (America on the Move 2012). As one of the major border crossings between the two countries, Juarez became the temporary home to those trying to enter the US via the program. In addition to the pull factor for the Bracero program, the agriculture crisis in Mexico outside of the border region served as a push factor to increase the population of the city (Ullerton and Barraza de Anda 2008). By the end of the Bracero program in 1964, the city's population continued to grow, with no local sources of employment for unskilled neighborhoods.

During that time, the upper classes of Mexican society in Ciudad Juarez thrived on the "quickie divorce" industry, in which single-party, no-fault divorces could be granted in 48 hours (Bender 2012). Seeing the growing population, unemployment, and the foreseeable end of the divorce industry with pending changes to the legal system, a group of concerned citizens commissioned a study in 1964 by Arthur Anderson to find future growth opportunities for the region (Gonzalez 2012). The report suggested that a the best use for the overwhelming supply of unskilled labor in the city be put to use assembling components for export of finished products back to the US, as assembly of US produced components was not considered an import for tax purposes (Gonzalez Alvarez 2000).

At the opening of the first assembly plant, called a "maquiladora", the demand for jobs was so high that the local police department was called assist with crowd control (Gonzalez 2012). Five years later, in 1969, over 30,000 people were employed in the maquiladora industry. In 1975, 60,000 positions were filled, doubling again every five years, showing no signs of slowing until it reached approximately 250,000 workers in 1985 (Cruz Piñeiro 1992). Along with the growth in the unskilled labor market, the city also gained jobs from auxiliary services such as transportation and food services that arose from the new market.

Continuing with the community based perspective, the community and industrial leaders were able to anticipate and solve issues in education, housing, and transportation for the benefit of all stakeholders. . By understanding local culture and contexts, the community leaders offered solutions to problems before they arose and quickly establish Ciudad Juarez as the largest trading partner with the US (Gonzalez 2012).

The directed approach that the community leaders took in building the city's industry involved several phases. While originally entirely dependent on US plant managers, engineers, accountants, et cetera. In 1969 the city's major university, Universidad Autonoma de Ciudad Juarez, was founded with the patronage of the industrial sector with the goal of helping produce employees to move beyond unskilled labor into managerial and technical positions and supplant the need to import talent from the US (de Urrutia 1982). Eventually, the region would be home to four universities with a diverse range of foci to serve the local community's need and the needs of the growing middle class.

Finally, Delphi—a automotive design company spun off from GM—had over 72,000 workers in manufacturing workforce in Mexico, approximately 23,000 of them in Ciudad Juarez (Soden et al. 1999). Starting in 1995, Delphi created a technical design center in the city, employing 1,000 Mexican engineers at its outset (Soden et al. 1999). At its peak, the Delphi Mexico Technical Center in Ciudad Juarez employed 20,000 engineers working on designs for Delphi products that were then manufactured locally in the city (Weissert 2011). This brings Ciudad Juarez clearly up to the stage 3 phase of community development.

While this case works on a slightly larger scale than our previous examples, it serves as a working example of a community's development. It also serves to highlight the important role of having community leaders vested in the development of their region, as it deviates from the more traditional route of economic development.

Concluding Thoughts

This chapter discusses a novel business strategy as an approach to poverty alleviation. It calls for a reconceptualization of business strategy, not only in terms of firm ownership, but also in terms of the mindset behind growth in subsistence marketplaces. The use of community-based enterprises, rather than the use of individual entrepreneurs or national-level policy, also represents a departure from the majority of approaches found in the literature. A community-level focus, furthermore, provides a more detailed explanation for operationalizing poverty alleviation efforts. Whereas macro-level approaches are often silent on details, and individual-level approaches may not be impactful in communities where poverty is the norm, it is suggested that this mid-level, community-focused approach is "just right."

Additionally, the approach found in this chapter provides a pathway for the transformation of poor communities that are focused on mere survival into economically thriving communities. Specifically, it describes the evolution of communities from Stage 1 to Stage 2, within a 4-stage model, where poverty was once the norm to conditions in which poverty is less prevalent and a growing middle class emerges. This transition is particularly relevant for individual well-being, as it provides for a sharp increase in well-being that is generally associated with an increase in income and wealth, although preceding the plateau that represents the diminishing returns occurring after basic needs are met.

This chapter also suggests that the community-based enterprises' ability to innovate will play a central role in its long-term sustainability, especially as it transitions from Stage 1 to Stage 2. In essence, innovation represents the key to independence from more developed economies, as knowledge generation becomes a normal part of the activities of the community. In terms of facilitating this growth, the development of a vertical integration mindset is necessary. This mindset allows community organizations to leverage the abundance of social capital that is present in subsistence marketplaces, and to avoid the trap of commodification of its offerings.

This chapter is intended to open up the discussion for how business strategies can be impactful in poverty alleviation. However, it is by no means comprehensive. Specifically, it is silent on a number of potentially crucial steps in, and outcomes, of the process. For example, the discussion does not address the potential outcomes associated with successful implementation. In this vein, while the community-based enterprise approach is suggested to increase the absolute level of wealth for individuals residing within a community, research on poverty illustrates instances where relative income disparities can detract from poverty alleviation efforts. In particular, feelings of jealousy and injustice regarding the distribution of benefits may negatively impact the social capital accrued in subsistence marketplaces, and points to the need for exploration of this issue.

Similarly, formal versus informal governance of community-based enterprises need to be carefully considered. One of the factors that contributes to the abundance of social capital in subsistence marketplaces is the prevalence of the informal economy. Buyers and sellers have more familiar, socially enforced relationships, rather than the arm's length, contractually-enforced relationships typically seen in developed economies, such as the US The impact of this factor on the proposed vertical integration mindset is unclear, and needs to be better understood.

In closing, this chapter presented conceptual arguments for how business strategy can impact poverty. Despite this contribution, a great deal of work remains undone. For instance, effort needs to further develop each argument by creating a series of more specific, detailed conceptual frameworks on their enactment. In addition, laboratory and field studies could be conducted to provide additional insights into the effects of such business strategies. All in all, the results derived from these research extensions would be beneficial for garnering a more lucid understanding of a potentially effective mechanism for poverty alleviation in both developed and lesser developed countries.

REFERENCES

America on the Move (2012), "Opportunity or Exploitation: The Bracero Program," http://americanhistory.si.edu/onthemove/themes/story_51_ 5.html.

Balakrishnan, Srinivasan and Birger Wernerfelt (1986), "Technical Change, Competition and Vertical Integration," *Strategic Management Journal*, 7 (4), 347-59.

Barton, Noelle and Holly Hall (2010), "Donations Dropped 11% at Nation's Biggest Charities Last Year," *The Chronicle of Philanthropy*, Oct. 17, 2010.

Bender, Steven (2012), *Run for the Border: Vice and Virtue in US-Mexico Border Crossings*, New York: New York University Press.

Bourdieu, Pierre (1979), "Public Opinion Does Not Exist (1972)," In: *Communication and Class Struggle*

Cardoso, Fernando Enrique and Enzo Faletto (1979), *Dependency and Development in Latin America*, Berkley, CA: University of California Press.

Cruz Piñeiro, Rodolfo (1992), *La Fuerza De Trabajo En Los Mercados Urbanos De La Frontera Norte*: El collegio de la frontera norte.

Dane, Erik (2010), "Reconsidering the Trade-Off between Expertise and Flexibility: A Cognitive Entrenchment Perspective," *The Academy of Management Review*, 35 (4), 579-603.

De Urrutia, D. C. (1982), *Así Empezó: La Verdadera Historia De La Universidad Femenina De Ciudad Juárez, Chih., Y Sus Transiciones a Universidad Mixta, Universidad De Ciudad Juárez, A.C. Y Universidad Autónoma*: Universidad Autónoma de Ciudad Juárez.

Deininger, Klaus and Yanyan Liu (2009), "Longer-Term Economic Impacts of Self-Help Groups in India," Policy Research Working Paper: The World Bank.

Fund, International Monetary (2012), "The Imf at a Glance," http://www.imf.org/external/np/exr/facts/glance.htm.

Gau, Roland, Haeran Jae and Madhubalan Viswanathan (forthcoming), "Studying Low-Literate Consumers through Experimental Methods: Implications for Subsistence Marketplaces," *Journal of Business Research.*

Gonzalez Alvarez, Lorena (2000), "Evolucion De La Legislacion Applicable a La Industria Maquiladora En La Franja Fronteriza Norte Y El Tlcan," Universidad Panamericana, Guadalajara Mexico.

Gonzalez, Jamie (2012), ed. Ricardo Gonzalez.

Granovetter, Mark S. (1973), "The Strength of Weak Ties," *American Journal of Sociology* 78 (6), 1360-80.

International Institute for Sustainable Development (2012), "Livelihoods and Climate Change," http://www.iisd.org/ecp/es/resilience/climate.asp.

Lumpkin, G. T. and Gregory Dess (1996), "Clarifying the Entrepreneurial Orientation Construct and Linking It to Performance," *Academy of Management Review*, 21 (1), 135-72.

March, James G. (1991), "Exploration and Exploitation in Organizational Learning," *Organization Science*, 2 (1), 71-87.

Mohan, John and Karl Wilding (2009), "Economic Downturns and the Voluntary Sector: What Can We Learn from Historical Evidence?," *History and Policy*, 85.

Nocke, Volker and Lucy White (2007), "Do Vertical Mergers Facilitate Upstream Collusion?," *The American Economic Review*, 97 (4), 1321-39.

Prahalad, C. K. (2004), *The Fortune at the Bottom of the Pyramid: Eradicating Poverty through Profits*: Wharton School Publishing.

Project, Bretton Woods (2005), "What Are the Main Concerns and Criticism About the World Bank and Imf?," http://www.brettonwoodsproject.org/ item.shtml?x=320869.

Ramirez, Edward (forthcoming), "Consumer-Defined Sustainably-Oriented Firms and Factors Influencing Adoption," *Journal of Business Research.*

Reiss, Robert (2010), "Ocean Spray's Secrets of Co-Op Success: An Interview with Randy Papadellis, Chief Executive Officer," *Forbes*, September 15, 2010.

Roseland, Mark (2000), "Sustainable Community Development: Integrating Environmental, Economic, and Social Objectives," *Progress in Planning*, 54, 73-132.

Schwartz, J. E. (2005), *Freedom Reclaimed: Rediscovering the American Vision*, Baltimore: G-University Press.

Shah, Anup (2010), "Poverty Facts and Stats," http://www.globalissues.org/ article/26/poverty-facts-and-stats.

Soden, Dennis L., Fred Cady, Rogelio Carassco, Mathew McElroy, and Lorena Orozco (1999), "The Delphi Decision: A Case Study," In: *IPED Technical Reports*: Institute for Policy and Economic Development.

The Henry Jackson Initiative for Inclusive Capitalism (2012), "Towards a More Inclusive Capitalism."

Ullerton, Thomas M. and Martha Patricia Barraza de Anda (2008), "Borderplex Population Modeling," *Migraciones internacionales*, 4 (3), 91.

United Nations (2012a), "Millennium Development Goals," http://www.un. org/millenniumgoals/.

--- (2012b), "United Nations: Humanitarian Affairs," http://www.un.org/en/ humanitarian/index.shtml.

Viswanathan, Madhubalan and Jose Rosa (2007), "Product and Market Development for Subsistence Marketplaces: Consumption and Entrepreneurship Beyond Literacy and Resource Barriers," In: *Product and Market Development for Subsistence Marketplaces: Consumption and Entrepreneurship Beyond Literacy and Resource Barriers* Oxford, UK: Elsevier, 1-17.

Viswanathan, Madhubalan, Anju Seth, Roland Gau, and Avinish Chaturvedi (2009), "Ingraining Product-Relevant Social Good into Business Processes in Subsistence Marketplaces: The Sustainable Market Orientation," *Journal of Macromarketing*, 29 (4), 406-25.

Walsh, James P. (1995), "Managerial and Organizational Cognition: Notes from a Trip Down Memory Lane," *Organization Science*, 6 (3), 280-321.

Weick, Karl E. and Karlene H. Roberts (1993), "Collective Mind in Organizations: Heedful Interrelating on Flight Decks," *Administrative Science Quarterly*, 38 (3), 357-81.

Weissert, Will (2011), "Juarez Maquiladoras Recovering Despite Bloodshed," http://www.boston.com/business/articles/2011/01/23/juarez_maquiladoras_recovering_despite_bloodshed/.

World Bank (2012a), "What We Do," http://go.worldbank.org/7Q47C9KOZ0.

--- (2012b), "World Bank Sees Progress against Extreme Poverty, but Flags Vulnerabilities," http://www.worldbank.org/en/topic/poverty.

Yunus, Muhammad (2004), "Grameen Bank, Microcredit and Millennium Development Goals," *Economic and Political Weekly*, 39 (36), 4077-80.

In: Economic Developments and Emerging Markets ... ISBN: 978-1-62417-986-0
Editors: Guy Boscarino and Derek Notte

Chapter 6

U-City Development for Economic Competitiveness in an Advanced ICT Era

***Hyung Min Kim*[1] *and Tae Sung Kim*[2]**
[1]The University of Melbourne, Victoria, Australia
[2]Yonsei University, Seoul, Korea

Abstract

The world economy has undertaken economic restructuring. In a post-industrialization period, information-related industries, associated with urban creativeness, are driving economic growth in the world. Information and Communication Technology (ICT) is critical to make cities competitive in the 21st century. ICT helps to generate high levels of productivity in business sectors and to attract highly educated human capital for new knowledge creation. Governments attempt to provide the advanced ICT infrastructure for achieving a leading role in a global economy since innovative technologies play an important role in the competitiveness of cities. Thus, the primary objective of this paper is to explore the new trend in urban development coupled with ICT infrastructure for enhancing urban competitiveness, using South Korea as case study.

By virtue of advancement in ICT, Korea has initiated to create a Ubiquitous City (U-City). U-City aims to enable any people to access various U-services anywhere and anytime through U-technology. These facilities can strengthen urban competitiveness and improve quality of life attracting highly educated human resources. U-City development is a new approach to improve urban competitiveness in an advanced ICT era.

1. Introduction

There is an emerging trend in urban development in concert with advancement in technology. The world economy has undertaken economic restructuring from manufacturing-

[1] Email: hyukim@unimelb.edu.au.
[2] Email: sgtkimts@yonsei.ac.kr.

oriented to information-related industries (Thorns, 2002; Fridmann and Wolff, 1982). In the past, industrialization led economic growth, but in the post-industrialization period, high-value added industries are driving economic growth in the world. Information and Communication Technology (ICT) is critical to make cities competitive in the 21st century. During the industrialization period, governments had prioritized building massive urban infrastructure such as road, sewerage, public transportation, ports and airports to support urban activities. However, in a post-industrialization period, ICT is conducive to high levels of productivity in business sectors. Technology has attracted highly educated human capital for new knowledge creation (Colecchia, 2002, Jalava and Pohjola, 2007). Moreover, ICT has helped a variety of working class to access economic opportunities (Jorge, 2002). In addition, cities have been increasingly wedded to the idea of connection through ICT in the new global economy (Graham and Marvin, 2001). Therefore, contemporary governments attempt to provide the advanced ICT infrastructure in order to achieve a leading role in the global economy.

Government strategies in response to technological development can facilitate flows of information and direct enhancement of economic competitiveness. In this context, this paper explores the new trend in urban development coupled with ICT infrastructure, using South Korea (hereafter Korea) as case study. Korea is an outstanding example of this urban development due to its rapidly growing technologies. Various indicators have demonstrated the highly developed ICT industry in Korea. According to the ICT Development Index, Korea was recognized as the top country in 2010 followed by Sweden, Iceland, and Denmark (ITU, 2011). Economist Intelligent Units (EIU) ranked Korea in the third position in the IT Industry Competitiveness Index 2007 after the United States of America and Japan. In 2012, furthermore, Korea ranked first in the E-Government Development Index representing innovation and development of E-government, followed by the Netherlands and the United Kingdom (UN, 2012).

By virtue of the world leading ICT, Korea has recently initiated to create Ubiquitous Cities (U-City). U-City aims to enable everybody to access various U-services anywhere and anytime through ubiquitous technologies. U-technologies provide digitalized, intelligent and smart urban facilities in favor of urban competitiveness and the quality of life.

The purpose of this paper is to examine U-City development in Korea as a new approach to improve urban competitiveness in an advanced ICT era. This study explains the concepts of smart cities and U-Cities, and, then, illustrates recent Korean practice epitomized by U-Cities of Gangnam and Dongtan.

2. THE CONCEPT OF SMART CITIES AND U-CITY

Current urban planning practice reflects the significance of information technologies. Various terms have been being used across countries to signify similar concepts. Some academics regard 'smart city' as a bigger concept which includes 'digital city', 'intelligent city', 'ubiquitous city', 'creative city', 'knowledge city' and others. Most countries in Europe and North America use 'smart' or 'intelligent' in their practice. A few countries, such as Korea and Finland, employ 'U' to stress unrestricted accessibility.

Emphasis of these terms has been slightly different depending on main foci. For example, knowledge cities and intelligent cities stress knowledge creation for economic growth and development (Komninos, 2006; Yigitcanlar et al., 2008). Komninos (2006) defines an intelligent city as a space with a high capacity for learning and innovation. He emphasizes three dimensions of intelligent cities: creativeness of human capital, institutions for knowledge creation, and digital infrastructure for communication and knowledge management.

Although there are various terms and different foci, the literature on smart cities has stressed common components such as rationales for smart city development, methods to manage information, locations of smart cities and human capital in these cities.

Rationale: Economic Restructuring Process and Advancement in Technologies

As for the rationale, exuberance of this trend in smart cities is due to the impacts of ICT on economy. Past studies have acknowledged positive effects of ICT on economic growth. According to Castells (2000), knowledge and information are critical elements in urban development because they are the base for the process of production. A new economy is organized around information as access to technological know-how is at the roots of productivity and competitiveness (Castells, 2000). Colecchia and Schreyer (2002) also argue that ICT has contributed to economic growth at 0.2 ~ 0.5 percentage point a year in selected OECD countries. Similarly, Jalava and Pohjola's research (2007) shows the increased capital investment in ICT contributes to growth of productivity at 0.46 percentage points in Finland.

Methods to Manage the Means of Production: Ways to Maximize Accessibility to High-Technologies and Benefits of Information Creation

Smart cities generally provide high-tech ICT infrastructure to maximize accessibility to information. Hall (2000) defines a smart city as the city that monitors and integrates conditions of all of its critical infrastructures including roads, bridges, tunnels, water, power and rails, in order to optimize its resources, to plan its preventive maintenance activities and to monitor security aspects while maximizing services for citizens. Partridge (2004) also argues that smart cities, covering all these attempts to integrate information-related technologies into urban development, can be seen as a place where ICT strengthens the freedom of speech and the accessibility to public information and services.

Location: Depending on Spatial Distribution of ICT Industry, the Agglomeration Economy and Government Investment

Governments, in particular, local governments are eager to immerge into the ICT trend. Due to the agglomeration economy, ICT tends to create clusters in a few selected areas. The concentration of similar ICT industries provides momentum for economic growth and improves productivity. These clusters have been often fuelled and guided by government planning and regulation (Firmino, 2005).

Human Capital: Creative Class Willing to Enjoy Urban Amenities in a Safe, Vibrant Environment

Smart cities are closely associated with human capital. Some studies argue that smart cities attract high quality human resources and stimulate to create human capital, social learning and smart communities (Hollands, 2005) as emphasizing the close relationship between smart cities and human capital.

Making U-City is practice of the Korean government in line with this global trend in post-modern society. The Korean government recognizes U-City as society where anyone can use advanced U-services conveniently anytime and anywhere. Since Weiser (1991) prospected that people would be able to access networks anytime anywhere through computers installed invisibly, the concept of ubiquitousness has become a symbol of future society. Weiser uses the term, ubiquitous computing, to emphasize the most profound technologies that can overcome spatial barriers (Weiser, 1991). Derived from Weiser's concept of 'ubiquitous', U-City is a built environment where people can use U-services through U-technologies based on ICT and its spin-off technologies. The typical U-technologies are Radio Frequency Identification (RFID), Ubiquitous Sensor Networks (USNs), Wireless Broadband (WiBro), Broadband Convergence Networks (BcNs), Fiber To The Premises (FTTP) / Fiber To The Home (FTTH), and Digital Multimedia Broadcasting (DMB). U-infrastructure uses sensors like RFID and sensor networks to communicate with buildings, facilities, infrastructure and any other objects in urban space.

3. U-City Development in Korea

ICT industries play a momentous role in the Korean economy. The rate of annual growth in ICT industry has exceeded two digits on average since the 1990s (see Table 1). Also, ICT industry is becoming increasingly significant in GDP. Proportion of ICT industry in GDP has steadily increased from 6.1 percent in 2001 to 11.8 percent in 2011. Numbers of ICT-related firms also increased from 13.5 thousand in 2005 to 18.6 thousand in 2010, displaying a 37.1 percent increase over recent five years, according to IT Statistics of Korea (ITSTAT).

The strength of ICT industry in Korea is attributable to a steady emphasis on manufacturing industry. Since 1961 when a military leadership took over the presidential authority, the Korean economy has emphasized export-oriented manufacturing for economic growth (Kim and Han, 2012). In the 1960s, given limited capital, technology, and resources, the government encouraged export-oriented light manufacturing products (Cumings, 1997). In the 1970s, labor-intensive heavy manufacturing was a main driving force for economic growth (Shin and Timberlake, 2006). As a result of these export-oriented manufacturing initiatives, the amount of export drastically increased from US $100 million in 1964, US $1 billion in 1971, US $10 billion in 1977, and to US $100 billion in 1995 (KITA, 2012).

Table 1. Contribution of ICTto GDP in Korea

	GDP growth (%)	Growth of ICT industry (%)	Contribution of ICT to the GDP growth (%p)	Proportion of ICT in GDP (%)
1996	7.2	16.4	1.0	6.1
1997	5.8	10.8	0.7	5.9
1998	-5.7	22.6	1.3	7.2
1999	10.7	33.2	2.4	7.9
2000	8.8	34.5	2.7	8.7
2001	4.0	8.7	0.8	6.0
2002	7.2	15.9	1.3	6.4
2003	2.8	13.7	1.1	7.1
2004	4.6	17.1	1.4	8.0
2005	4.0	11.7	1.0	8.6
2006	5.2	12.6	1.1	9.2
2007	5.1	8.7	0.7	9.5
2008	2.3	6.8	0.6	9.9
2009	0.3	3.9	0.3	10.3
2010	6.3	17.7	1.4	11.4
2011	3.6	7.9	0.7	11.8

Source: ITSTAT, www.itstat.go.kr.

In Korea, the 1990s witnessed the growth of knowledge-intensive manufacturing such as semi-conductors and computer-related products. In line with export-oriented manufacturing strategies, advancement in ICT industry has resulted in increases in trade volumes of ICT products. Figure 1 depicts trends in export and import of ICT products in Korea. The volume of exports increased 3.8-fold over fifteen years (see Figure 1). Export-oriented economic strategies continue in urban development. In addition to exporting ICT products, the Korean government attempts to export a Korean urban planning model tied with U-City. The advancement in manufacturing technology and industry has provided foundation for ICT development in Korea. Currently, Korean is a leading country in ICT technologies and ICT-based urban development.

Figure 2 describes nationwide distribution of employment in ICT industry. During the industrialization period, the Korean government took a growth pole approach to facilitate export manufacturing products by selectively choosing areas for economic growth. The strengths of manufacturing in these selected areas have continued for ICT development to a large extent. Also, the Capital Region, Seoul and its surrounding two regions (i.e., Incheon and Gyeonggi), has been dominant in providing ICT employment.

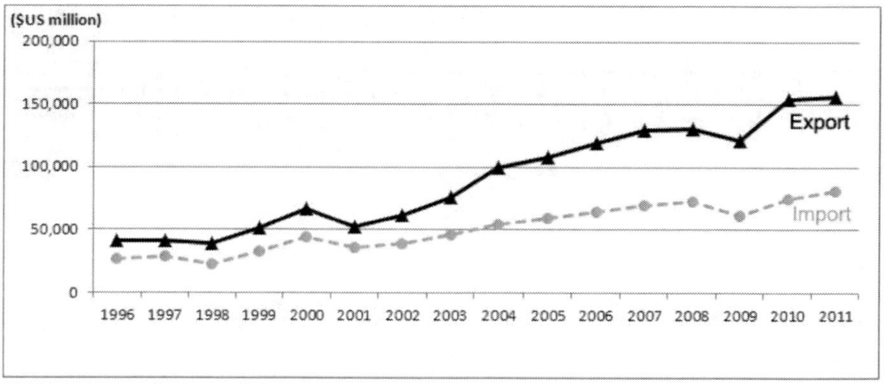

Source: ITSTAT (www.itstat.go.kr).

Figure 1. Amounts of export and import in Korea, 1996 - 2011 (US $ million).

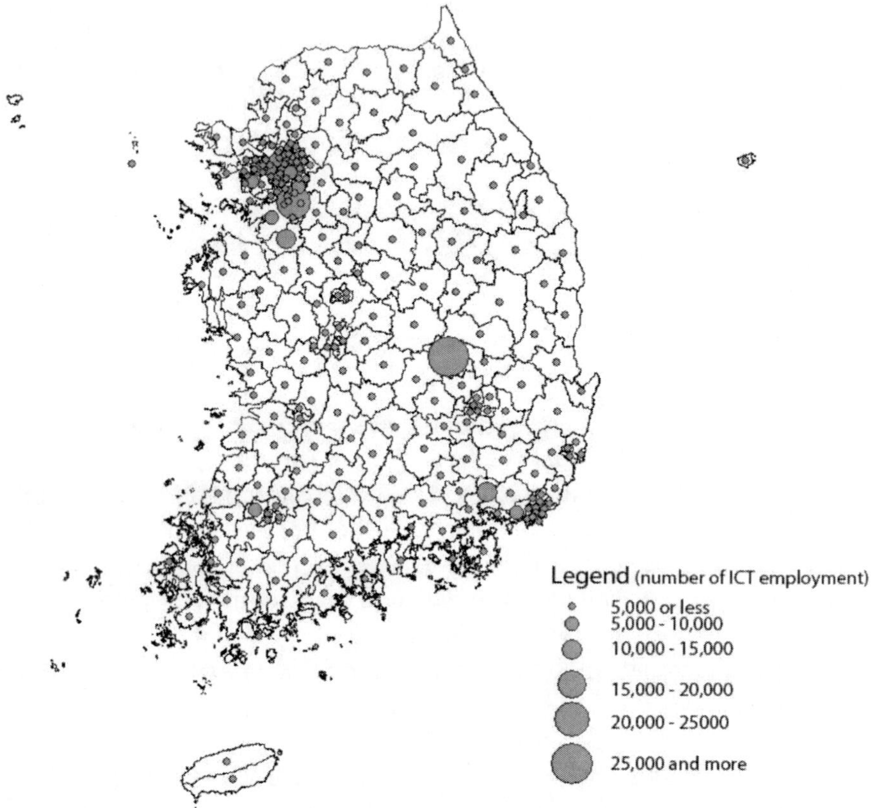

Source: Data regarding ICT employment is adapted from Statistics Korea.

Figure 2. Distribution of ICT employment in Korea, 2010.

The Korean Government has made an effort to take advantage of these growing ICT industries in urban development projects. These urban developments are spatially related to ICT industry. By virtue of advancement in ICT technologies, Korea has continuously developed national strategies for knowledge based urban development through agendas regarding Cyber Korea, E-Korea and U-Korea (Lee, 2008). The government has used U-concepts such as U-City, U-infrastructure, U-technology and U-services in an attempt to establish U-City as a national brand for global smart city markets. In 2008, U-City construction began to be legally supported through the establishment of *Ubiquitous City Construction Act* in order to facilitate the provision of information regarding administration, traffic, welfare, environment, and natural disasters. According to this Act, the Ministry of Land, Transport and Marine Affairs (MLTM) announced detailed guidelines for U-City development in 2009. The guideline identified 228 U-services in eleven categories (see Table 2).

Table 2. Overview of U-services in Korea

Administration	• Administrative field work support • Urban landscape management • Online civil affair services • Neighborhood amenities • Citizen participation
Traffic	• Traffic management optimization • Electronic payments • Traffic information • Extra information for drivers • Public transportation • High-tech road • Taxi call service
Welfare	• Health management service • U-hospital service • Remote medical service • U-health management service • U-health center service • Services for the elderly • Services for the disable • Services for multi-cultural family • Services for childbirth and childcare
Environment	• Pollution management service • Waste management service • Green service • Services for energy efficiency • Alternative energy service
Crime and disaster prevention	• Rescue & First aid • Personal safety • Public safety • Institution safety • Fire management • Natural disaster management • Accident management • Integrated disaster management

Table 2. (Continued)

Facilities management	• Road facilities management • Building management service • River facilities management • Other facilities management • Underground facilities management • Data management and supply
Education	• U-kindergarten services • U-campus services • U-classroom services • Remote education services • U-library service • Education support for the disabled
Culture, Tourism, and Sports	• Cultural facilities management • Cultural spaces experience • Culture information • U-tourism information • U-park • U-playground • U-resort • U-sport
Logistics	• Production records tracking & management • U-distribution center • U-shipping • U-delivery • Delivery records tracking • U-store • U-shopping
Labor and employment	• U-work service • Industrial activity support • Industrial safety management
Others	• Home management service • External connection service • Housing complex management service • U-artifact service • U-theme street service

Source: Ubiquitous City Construction Guidelines, MLTM, 2009.

U-City is proposed to improve the effectiveness of urban space/infrastructure management and to provide urban information to citizens. The purpose of U-City can be fulfilled by providing U-services for better quality of life in various fields such as security, transportation, environment, leisure, education, medical care, employment and governance. Since U-services provide convenience, safety, security, efficiency and entertainment to citizens, U-services are regarded as important factors to strengthen urban competitiveness (Shin et al., 2009).

Strategies to make U-City are dependent on different regional contexts. Some cities have to solve security and safety problems, while some cities need to address economic and/or environmental issues. Some cities have strengths in tourism, whereas others have advantages in commerce. Depending on strengths and weaknesses of cities, different U-services have to be implemented. For example, Incheon is a gateway of Korea due to the presence of an international airport and a seaport. Thus, U-City projects in Incheon have focused on international activities with a free economic zone. Another example is Sejong City. This city is under construction to become the national administrative service center. Relevant U-services will be incorporated with this new urban development to achieve the goal regarding administrative services. Therefore, U-City of each region has to concern about uniqueness of the city to maximize the effects on cities and citizens, rather than following a standardized U-services.

The 1st Ubiquitous City Comprehensive Plan includes fifty-two U-City projects nationwide from 2009 to 2013 (MLTM, 2009). According to this plan, there is one completed U-City project; there are thirty-eight projects under construction and thirteen planned projects. The projects are to be implemented both on green-field and built-in areas.

4. Case Study in Korean Cities

U-City development has been implemented in both existing urban fabric and new development. In the application of U-technology within an existing town, it is difficult to cover entire urban areas with U-infrastructure due to costs and development process such as land acquisition and harmonization of diverse stakeholders who have already lived there. U-City development in a new town, however, can be simultaneously incorporated because U-infrastructure is designed and built with the beginning of the new town project. This paper provides two case studies for gaining an understanding of U-City development. Firstly, Gangnam, a new elite town in Seoul will be discussed as an example in a built-in urban area. Due to the lack of budget, local governments in Korea usually have played a limited role in U-City projects. However, Seoul, in particular, Gangnam, has been transformed into a more intelligent, sustainable community through the efforts of the local government as well as support from the central government. Secondly, Dongtan U-City will be explored as new development because Dongtan U-City is the first case that the initiative has been implemented.

As shown in Figure 3 covering the Capital Region, Gangnam is one of the most apparent ICT clusters. Also, Dongtan is adjacent to a high concentration of ICT industry. Governments intend to make use of current economic strengths regarding ICT industry. Gangnam and Dongtan are globally recognized as listed on the Smart 21 Communities by Intelligent Community Forum in 2008 (ICF, 2008).

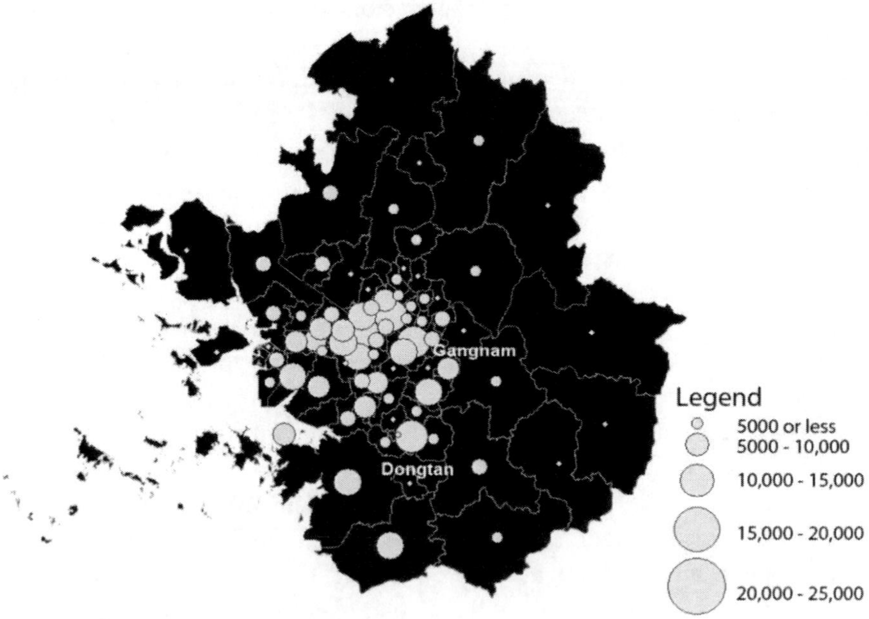

Source: Data regarding ICT employment is adapted from Statistics Korea.

Figure 3. Gangnam and Dongtan with ICT employment in the Capital Region (2010).

1) Gangnam U-City Development

Gangnam, as one of the local governments in Seoul, is located in south-eastern Seoul. Gangnam geographically occupies 6.5 percent of Seoul with a land area of 39.6 km^2 and it contains 2.5 percent of Seoul's total population. However, it provided 14.3 percent of the total employment and 13.0 percent of ICT employment in 2010 according to Statistics Korea. Since Gangnam was newly developed in the late 1960s, it has provided high quality environments for working and living. New establishments of small ICT firms have chosen this area due to favourable environments for business. Gangnam is currently a mecca for business, high level services, shopping, fashion and high-end housing (Kim and Han, 2012). Also, there are passengers in the Gangnam station more than any other stations in Seoul (SM, 2010).

There are three distinctive strengths in Gangnam as U-City. First, the city enhances the quality of administrative services provided by the local government. U-services improve accessibility to the administrative services and widen the opportunities of community participation. According to the Intelligent Community Forum (ICF), ICT in Gangnam has played a crucial role in increasing public participation that helps to build a transparent government. Since the Korean government launched the Electronic Government (E-Government) project in 1995, Gangnam has been eager to utilize ICT. In 1997, Gangnam had a local network to provide services such as tax payments and other administrative services. In 1999, an electronic system of the government was capable of processing citizen affairs such as permits and registrations. In 2002, the local government upgraded this system to the web-based one, thus, the government collected fifteen percent of the total tax payments through online by 2006. Also, this system issued two million documents to citizens through the

Internet and public kiosks (ICF, 2008). The Gangam local government estimates that this system has reduced a quarter of labor costs in the local government since 1995, and this is equivalent to US$30million (ICF, 2008).

Second, ubiquitous security and safety are core aspects in Gangnam. This area is the most affluent place in Korea but a crime rate is the highest among the five richest districts in Seoul. Hence, the U-Gangnam City Control Center was established in 2011 to manage more than one thousands Closed-Circuit Televisions (CCTVs) for neighborhood security, public safety, parking control, and traffic information collection (Gangnam District Official Blog, 2011). In addition, data from sensors and broadband is collected at this control center where operators and police officers are ready to react to sudden events 24/7 (Gangnam District Official Blog, 2011). With a high-resolution of three-dimensional Geographic Information System (GIS), the control center is capable of spotting the location of crime quickly and accurately (Klimowicz, 2012). Furthermore, Gangnam has employed a new method called 'intelligent image-analysis technology' to prevent and respond to crimes involving children (Klimowicz, 2012).

Third, Gangnam has the most futuristic image in Korea and become the test bed for intelligent street furniture. Currently, Gangnam is the most popular meeting place. Along Gangnam Great Road (GangnamDaero in Korean), one of the main streets in Gangnam, twenty-two media poles stand and shape new streetscape (see Figure 4). Redevelopment of the Gangnam Great Road is a project of *Design Seoul* by *the Seoul Metropolitan Government* (SMG). Gangnam Great Road with twelve-meter-tall Media Poles is called 'U-Street'. Media Poles with touch screens provide various services, which are similar to functions in smart phones. Pedestrians can search for maps, read news, check transport information, take photos, send the photos to their e-mail addresses, and play games. Media Poles also function as street lights, free Wi-Fi hotspots and digital billboards. In addition, these poles contain CCTVs to monitor the street for security and safety. Media Poles are not only street furniture with cutting-edge ICT, but also digital street arts. They change their colors to create novel ambience on the street, which can become a digital gallery when the poles show visual arts and images.

U-City development of Gangnam is noteworthy because it supports the place where the agglomeration economy actively takes place. Initiatives of U-City development are currently strengthening creative environments for commercial and residential activities. Producer services tend to be concentrated on Gangnam. Thirty-four percent of Seoul's total employment in producer services is located in Gangnam and surrounding two districts[3] (Kim and Han, 2012). Venture firms are likely to choose Gangnam and tend to cluster within a walking distance (Park, 2008). Despite advancement in communication technology, firms are likely to cluster to access quality information. In particular, a face-to-face contact is considered efficient as it can enhance relationship and provide quality information (Storper and Venables, 2004). Accordingly, physical infrastructure of cities plays an important role in facilitating these activities by shaping creative environments. Media poles in the Gangnam Great Road as fixed facilities contribute to the enhancement of streetscape and accessibility to information.

In addition, residential areas in Gangnam have been recognized as the most expensive housing prices in Korea (Kim et al, 2009) and are emerging as a bourgeois town. Advanced

[3] Surrounding two districts mean Seocho and Songpa.

administrative services and CCTVs can improve security and livability for residents in Gangnam. The emergence of Gangnam as an elite town has been further encouraged by these advanced technologies combined with residential services.

Source: taken by authors in 2012.

Figure 4. Media poles on Gangnam Great Road.

Partially by virtue of these new infrastructures to improve urban creativeness and amenity, Gangnam plays a leading role in economic growth of Korea. While the number of jobs in Seoul increased eighteen percent between 2003 and 2010, the growth rate of Gangnam far surpassed Seoul's average at twenty-nine percent in Gangnam (SK, 2011). Amounts of export of establishments in Gangnam reached US$14,409 million, exhibiting a quarter of Seoul's total export volumes in 2011 (SMG, 2012). Another example of growth of Gangnam is seen in the relocation of headquarters of Samsung Electronics and other Samsung's sister companies. These companies moved into Gangnam in 2008, signifying Gangnam is increasingly attractive for business environments.

2) Dongtan U-City Development

Dongtan U-City is a new town governed by the local government of Hwaseong in the Gyeonggi province. Dongtan is located in the south area of the Capital Region, 40 km distance from Seoul (see Figure 3) and it is in an important traffic node to link Seoul to the second largest city, Busan. In addition, there is high concentration of ICT industries due to semiconductor factories of Samsung in Hwaseong. The presence of this multinational corporation facilitates activities of relevant small- and medium-size firms linked forward and backward. These locational advantages stimulated to develop Dongtan into U-City.

Construction of Dongtan New Town I began in 2005 and was completed in 2008 as the first attempt to combine new town development with U-technology. Dongtan New Town I with an area of 904 ha has capacity to accommodate 120 thousand residents. Dongtan New Town II will be completed on a land area of 2,160 ha by 2013. The government has invested US$46 million[4] in Dongtan U-City since Dongtan was chosen for U-City development (Choi et al, 2010). Dongtan U-City is designed to provide twelve U-services to improve quality of life, livability, and sustainability (see Table 3). The twelve U-services are classified into four categories: U-safety, U-environment, U-media and U-transportation (Hwaseong City, 2006).

Table 3. U-services in Dongtan and costs for establishment

	U-services	Proportion of costs for establishment (%)
U-safety	License plate detector	4.6
	Anti-crime	28.9
U-environment	Air-pollution information	5.2
	Water leakage management	6.6
U-media	Media board	6.7
	U-placard	6.3
	Internet portal	6.3
Transport	Traffic information	7.4
	Real Time Signal Control	16.9
	Bus Information System	3.6
	U-Parking	3.1
	Detecting Illegal Parking	4.3

Source: Adapted from Choi et al (2010).

U-Safety: Anti-Crime and License Plate Detector Service

U-safety service is critical in Dongtan U-City because the area was notorious for horrible crimes such as serial murders and rapes before the new town development. Thus, one-third of total public expenditure has been spent for safety as shown in Table 3. Anti-crime service is designed to provide safe neighborhoods for residents. 231 CCTVs have been installed in the city and they are connected to the Hwaseong Dongtan U-City Information Center (see Figure 5) that collects urban data and operates 24/7, as U-Gangnam City Control Center does (Hwaseong Dongtan U-City, 2007). Through the CCTVs, operators monitor the city and they contact relevant agencies such as police office when an emergent situation happens (see Figure 5).

License Plate Detector Service oversees vehicles in Dongtan U-City for safety purpose. Using eighteen CCTVs, real-time surveillance is possible and operators can identify stolen vehicles and trace vehicles involved in crime (Hwaseong Dongtan U-City, 2007). Those detectors are installed on nine main roads.

[4] US 1$ = KR 1,000 WON.

Source: Taken by authors in 2012.

Figure5. Hwaseong Dongtan U-City Information Center and a Monitoring Room.

U-Environment: Air-Pollution Information and Water Leakage Management Service

U-environment services attempt to create pleasant living environments for local residents. Air-pollution information service provides environmental information about water quality, atmosphere and waste. This service performs scientific evaluation regarding environmental pollution (Hwaseong City, 2006). In particular, three LED monitors display the degree of air pollution including CO_2, SO_2, O_2, NO_2 and others (Hwaseong Dongtan U-City, 2007). This service includes texting SMS to residents. Citizens are also able to access this information on the internet portal website.[5]

Water leakage management service is to ensure stable tap water supply by detecting leaking pipes and improving water delivery. Twelve flow-meters and forty-seven pressure-gauges have been installed for efficient water management (Hwaseong Dongtan U-City, 2007). This service also provides information regarding water leakage, water stopping, water pipe construction and water quality.

U-Media: Media Board, U-Placard and Internet Portal Service

Media board service provides public information of Hwaseong such as cultural events, government announcements and urgent notices through media boards in public space. Two media boards (size: 8m x 6m) have been installed in the busiest streets in Dongtan (Hwaseong Dongtan U-City, 2007). These electronic boards offer information regarding traffic and neighbourhood (Hwaseong City, 2006).

U-placard functions similar to the media board by providing information regarding cultural events, neighbourhood, and government notices (Hwaseong City, 2006). There are ten U-placards (size: 5.4m x 0.9 m) (Hwaseong Dongtan U-City, 2007). The purpose of U-placard is to improve efficiency on advertisement and administration services. U-placard is eco-friendly because it can reduce the use of disposable banners for advertisement (see Figure

[5] Hwaseong environmental information service, uenv.hscity.net.

6). The difference between media board and U-placard is that information on U-placard targets local residents and their business, while media boards present important information about the city and its local government, and the large media board functions as a local land mark.

Three webpages[6] comprise internet portals in Dongtan: Hwaseong City homepage, U-traffic homepage and U-environment homepage. One portal was in operation in the past, but it was specialised into three webpages to save public expenditure.

Source: taken by authors in 2012.

Figure 6. U-placard in Dongtan.

U-Transport: Traffic Information, Bus Information, U-Parking, Detecting Illegal Parking, and Real Time Signal Control Service

Traffic information service provides residents with information on traffic conditions such as congestion and traffic flows on roads. Citizens are able to access information through twelve Variable Message Signs (VMSs) established on roads and the U-traffic portal website (HwaseongDongtanU-City, 2007).

A bus information system provides information on bus arrival time, routes and operating issues (Hwaseong City, 2006). Because information on bus arrivals is shown on Bus Information Terminals (BITs), the bus patrons can be more relaxed when they are waiting for buses. Dongtan has thirty-six BITs with LCD monitors which display information at bus stops (Hwaseong Dongtan U-City, 2007). This information is also available online so that users can access this information through a smart phone or a personal computer.

U-parking is a service to inform numbers of possible parking spaces through intelligent parking LED systems equipped at the entrance of parking lots. Currently, there are two parking lots in operation of U-parking service. This service offers extra information regarding

---

[6] Respectively, www.hscity.net, utraffic.hscity.net, uenv.hscity.net.

location verification of parked car and real-time available parking location (Hwaseong City, 2006). The information is also available online.

A parking system spots illegal parking vehicles. Eighteen CCTVs are operational to improve the quality of neighborhood as they are installed at busy areas (Hwaseong Dongtan U-City, 2007). Once a car enters crackdown zones (e.g., busy shopping, business, and/or public areas), CCTVs oversee the car (Hwaseong City, 2006). When the car is identified as violating parking rules, the CCTV automatically captures the image of the car as a proof.

Real time signal control supports smooth traffic flows. This control system makes use of eighty-one signal controllers, thirty-nine VDSs and thirteen CCTVs to collect traffic information, control traffic signals depending on congestion, and provide information regarding road traffic conditions (Hwaseong Dongtan U-City, 2007).

Due to the recent operation, it is too early to evaluate Dongtan U-City's performance at this stage. However, there is evidence that supports positive impacts of U-City development. As the first U-City development project in Korea, Dongtan has become a model for other cities. Dongtan U-City reported that 1,944 people visited the Hwaseong Dongtan U-City Information Center in 2011 (Kim 2012). Choi et al (2010) argues Dongtan U-City is economically feasible. As a result of their cost-benefit analysis, social benefits outweigh costs for establishment and operation of U-services. Also, Dongtan U-City has encouraged products to grow up to 2.9-fold larger than the amount of investment (Kim, Cho & Jeong, 2006).

Table 4 demonstrates the growth of Hwaseong city and the Gyeonggi province where Dongtan is located. Hwaseong produces the highest per capita Gross Regional Domestic Product (GRDP) among thirty-one areas in the Gyeonggi province. In addition, the growth rate in Hwaseong exceeds other areas. Per capita GRDP increased 1.5-fold over eight years in Hwaseong: this was higher performance than neighboring areas. Population increased spectacularly from 195 thousand in 2000 to 472 thousand in 2008, displaying a 2.4-fold increase over eight years. This population growth rate is the highest within the province. Of course, a new town development seems to result in the increase in population, but it is also anticipated that Dongtan U-City development has contributed to a massive influx of new residents.

Table 4. Population and GRDP growth, 2000 – 2008

	Population		GRDP (US$ million)		GRDP per capita (US$ thousand)		Per capital GRDP Growth Rate (2000-2008)
	2000	2008	2000	2008	2000	2008	
Hwaseong	195,000	472,000	5,180	18,802	26.5	39.9	150.3%
Gyeonggi (except Hwaseong)	9,085,000	11,077,000	109,816	182,894	12.1	16.5	136.6%

Source: Statistics Korea.

Note: The base year of GRDP is 2005. US$1 = KR 1,000.

Conclusion

Development of technologies has stimulated the contemporary economy to pursue high-value added products. Rationales of the Korean initiative to create U-City are in line with the worldwide economic restructuring from low-end manufacturing to high-end products. Two case studies reveal the strength of Korean technologies. Gangnam is currently transforming into a global business district where high level services are concentrated and Dongtan has been grown by virtue of the concentration of ICT firms.

Information is becoming increasingly significant for administration, production and dwelling. Methods to manage information are to establish visible infrastructure as seen in the examples of U-City. Various tangible facilities such as the U-City information center and digital media facilities have been established in Gangnam and Dongtan to collect and provide information. For the competitiveness of cities, governments go through great lengths to provide the advanced ICT infrastructure using innovative technologies.

However, locations have been deliberately chosen in conjunction with current strengths. At a regional level, public investment in U-facilities has to be selective in location choice. Gangnam is a leading area in terms of the economy, education, entertainment and retail, thus it attracts a great number of new business establishments, passengers and customers. Gangnam functions as a laboratory to test new products and new franchises in Korea. U-facilities in this area can benefit a multitude of people. On the contrary to Gangnam, Dongtan U-City, as green-field development, is based on existing ICT industries. Dongtan can benefit from the concentration of ICT firms. At a local level, the motivation of ubiquitousness is to make everywhere accessible to information, but the embodiment of U-facilities, as a physical infrastructure, has to be bound to certain places. Once these infrastructures are established, it is hard to change locations due to transition costs. Accordingly, in both case studies, physical facilities have been operational in the busiest areas so that more people can access these facilities. CCTVs, media poles, U-placards and media boards are established in main arterial roads and major public places. Taken altogether, therefore, spatial barriers seem to exist in ubiquitous cities.

Current practice of making U-City is to improve liveability, sustainability and accessibility, while infrastructure during the industrialization period stressed basic needs for economic and residential activities. Liveability, sustainability, and accessibility seem to have little relevant to economic growth, but infrastructure regarding these elements plays contributing roles as a basing point for the influx of human capital in a post-industrialization era. The creative class is critical to economic growth and prefers to stay and enjoy in vibrant, safe urban environments (Florida, 2003). U-services contribute to creating more salubrious and more vibrant environments as seen in the examples of Gangnam and Dongtan. U-City development not only increases sales of ICT products, but also has an indirect impact on economic growth by attracting high level human capital.

This paper has examined current initiatives to make U-City in Korea as one of the most developed countries in this field. This planning practice is at a burgeoning phase. Thus, no one can exactly predict how much technologies affect urban forms in the future. However, integration of technologies with urban development seems to be one of the probable planning strategies in the 21st century due to the conviction that advancement in technologies benefits contemporary society.

REFERENCES

Bramwell, A., Nelles, J., & Wolfe, D. A. (2008). Knowledge, innovation and institutions: Global and local dimensions of the ICT cluster in Waterloo, Canada Regional Studies, 42(1), 101-116.

Castells, M. (2000). *The rise of the network society* (2nd edition), Oxford: Basil Blackwell.

Choi, D. C., Kwak, Y. C., & Jung, C. M. (2010). A study of economic analysis of construction project in HwasungDongtan U-City, Journal of the Korea Planners Association, 45(2), 247-259.

Colecchia, A. and P. Schreyer (2002). ICT investment and economic growth in the 1990s: Is the United States a unique case? Review of Economic Dynamics 5, 408-442.

Cumings, B. (1997). *Korea's place in the sun: A modern history*. New York, W.W.Norton&Company.

Gangnam District Official Blog (2011), *u-Gangnam City Control Center: the safety protector of our district*.

EIU 2007. The means to compete benchmarking IT industry competitiveness.Economist Intelligent Units.

Firmino, R. J. (2005). Planning the unplannable: how local authorities integrate urban and ICT policy making. Journal of Urban Technology, 12(2), 49-36.

Florida, R. (2003). *The Rise of the Creative Class: And how it's transforming work, leisure, community and everyday life*. New York: Basic Books.

Friedmann, J. & Wolff, G. 1982. World city formation: an agenda for research and action. International Journal of Urban and Regional Research, 6, 309-344.

Graham, S. and S. Marvin (2001). *Splintering urbanism: networked infrastructures, technological mobilities and the urban condition*. New York, Routledge.

Hall, P. (2000). *Creative cities and economic development*. Urban Studies, 37(4), 633-649.

Hollands, R. G. (2008). Will the real smart city pleasestand up?: Intelligent, progressive or entrepreneurial?, City, 12(3), 303-320.

Hwaseong City (2006). *Hwaseong Dongtan Ubiquitous City*. http://www.udongtan.or.kr

Hwaseong Dongtan U-City (2007). *U-City Story: Hwaseong Dongtan U-City*. Brochure published by Hwaseong City and HwaseongDongtan u-City.

ICF (2008).*The Smart21 Communities of 2008*.New York: Intelligent Community Forum www.intelligentcommunity.org

ITSTAT (2011). Major IT Index. Daejeon, Ministry of Knowledge Economy. www.itstat.go.kr

Jalava, J. and M. Pohjola (2007). ICT as a source of output and productivity growth in Finland. Telecommunications Policy, 31, 463-472.

Jorge, S. N. (2002). The economics of ICT: challenges and practical strategies of ICT use for Women's economic empowerment. Division for the advancement of Women, United Nations.

Kim, H. M. and S. S. Han (2012). City profile: Seoul. Cities, 29(2), 142-154.

Kim, P. R., Cho, B. S., &Jeong, W. S. (2006), The Propagation Effects on the Regional Economy Induced by U-City Construction in Wha-sung and Dong-tan City. The Journal of The Korean Institute of Communication Sciences, 31(12B), 1087-1098.

Kim, S. (2012). *DongTan U-City catches 44 wanted criminals*, Suwon Ilbo Newspaper, 30 Mar 2012.

KITA (2012). Korea Trade Statistics. Seoul, The Korea International Trade Association.

Klimowicz, G. (2012). South Korean District implements integrated security centre, Future GOV, 22 May 2012. http://www.futuregov.asia/articles/2012/may/22/south-korea-establishes-ubiquitous-city-control-ce/.

Komninos, N. (2006).*The Architecture of Intelligent Cities:integrating human, collective and artificial intelligence to enhance knowledge and innovation*, Conference Proceedings Intelligent Environments 06, Institution of Engineering and Technology, 53-61.

Lee, J. W., Baik, S. H., & Lee, C. H. (2011). Building an Integrated Service Management Platform for Ubiquitous Cities. Computer, 44(6), 56-63.

Lee, S. (2008). *Towards Ubiquitous City: Concept, Planning, and Experiences in the Republic of Korea*. In T. Yigitcanlar (Ed.), Knowledge-Based Urban Development: Planning and Applications in the Information Era, (148-170). New York: Information Science Reference.

MLTM (2009). *Ubiquitous City Construction Guidelines*.Gyeonggi, Ministry of Land, Transport and Marine Affairs.

MLTM (2009). *The 1st Ubiquitous City Comprehensive Plan*. Gyeonggi, Ministry of Land, Transport and Marine Affairs.

Nam, T., &Pardo, T. A. (2011). *Conceptualizing smart city with dimensions of technology, people, and institutions*. Paper presented at the Proceedings of the 12th Annual International Digital Government Research Conference: Digital Government Innovation in Challenging Times, College Park, Maryland.

Park, J.-h. (2008). A study on the venture location characteristics considering industrial categories and regions: focusing on Gangnam-gu and Geumcheon-gu in Seoul. Seoul, Hanyang University.Master's degree.

Partridge, H. (2004). *Developing a human perspective to the digital divide in the smart city*.In Proceedings of theBiennial Conference of Australian Library and information Association (Queensland, Australia).

R&D Information Center (2011). *The new growth drivers of industrial classification,* Seoul: Knowledge Industry Information Institute.

Shin, J. D., Lee, J. W., Lee, S. H., & Kim, H. B. (2009). A Study on Using U-Service to Strengthen Urban Competitiveness. Journal of Geographic Information System Association of Korea, 17(4), 417-430.

Shin, K.-H.and M. Timberlake (2006). Korea's global city: structural and political implications of Seoul's ascendance in the global urban hierarchy. International Journal of Comparative Sociology, 47(2), 145-173.

SK (2010). GRDP in Gyeonggi province.Daejeon, Statistics Korea.

SK (2011). Survey of business activities.Daejeon, Statistics Korea.

SM (2010). Transport statistics. Seoul, Seoul Metro.

SMG (2012). Seoul Statistics. Seoul, Seoul Metropolitan Government.

Storper, M. and A. J. Venables (2004). Buzz: face-to-face contact and the urban economy. Journal of Economic Geography, 4, 351-370.

Thorns, D. C. (2002) *The transformation of cities: urban theory and urban life*. New York: Palgrave Macmillan.

UN (2012).*E-Government Survey 2012*.New York: United Nations, http://www2.unpan.org /egovkb/global_reports/12report.htm.

Weiser, M. (1991). *The computer for the 21st century*. Scientific American, 265(3), 94-104.

Yigitcanlar, T., O'Connor, K., Westerman, C. (2008), *The making of knowledge cities: Melbourne's knowledge-based urban development experience*, Cities, 25 (2), 63-72.

In: Economic Developments and Emerging Markets ... ISBN: 978-1-62417-986-0
Editors: Guy Boscarino and Derek Notte

Chapter 7

EU LOW-CARBON REVOLUTION'S INFLUENCE ON CHINA AND ITS COUNTERMEASURES

Anwen Wang Dongmei*

Low-Carbon Economy and Technology Research Institute of Hohai University, Changzhou, China

ABSTRACT

Taking Copenhagen summit on climate change as a symbol, the European countries are focused on the development of low-carbon technologies, and vigorously promote low-carbon revolution to respond to the challenge of climate change, protect secure energy future and capture the economic strategy. The rapid development of low-carbon in Europe, which has a positive impact on China, but at the same time, also constitutes a serious challenge to the status of a new pattern of development in the future world. China should be prepared to develop a low carbon economy strategic planning, strengthen the reserves of long-term carbon strategy and promote the work of the RMB-denominated carbon finance market development, in order to achieve energy conservation and economic development win-win.

Keywords: Europe, low carbon revolution, opportunity and challenge, countermeasure

INTRODUCTION

Since 2003, the energy white paper of UK proposed the concept of low carbon economy for the first time, all the countries of the world gave a lot of attention, with increasing of energy shortage and environmental pollution, the development of low-carbon economy had become the development direction of the world economy and the inevitable choice of human survival.

* Low-carbon Economy and Technology Research Institute of Hohai University, Jinling North Road, No. 200, Xinbei District, Changzhou 213022, China.

Table 1. The distinctions of "low carbon economy" and "high carbon economy"

	Name	Features	Connotation	Growth	energy sources	Consumption	Civilization
Low carbon economy.	Green economy.	Low pollution, low energy consumption, low emission.	Pay attention to the quality of economic growth.	Sustainable growth .	Renewable energy, solar energy.	Pay attention to health, simple, sustainable.	Ecological civilization.
High carbon economy.	Brown economy.	High pollution, high energy consumption, high emission.	Pay attention to the quantity of economic growth.	Cost growth.	Fossil energy, carbon fuel.	Pay attention to enjoy, convenience, luxury.	Industrial civilization.

Development of EU Low-Carbon Economy

From 2006, Europe has published "energy efficiency action plan" and the "green book" of European energy strategy. In 2007 Europe passed the energy security and climate change development strategy, expected that by 2020 the proportion of renewable energy consumption will increased to 20%. By the end of 2008, Europe passed the climate energy package plans, including the European emissions trading scheme amendment, European members supporting measures of task allocation decisions, vehicle carbon dioxide emissions regulations and the fuel quality directive and so on, in order to reach that greenhouse gas emissions in 2020 reduce at least 20% based on 1990. Therefore, European countries have actively developed low carbon economy.

Set up Low Carbon Revolution Concept and Law

In Nineteen nineties, the United Nations Intergovernmental Panel on Climate Change has published 4 copies of global climate assessment report, warning the risk of elevated temperature. European got consensus to reduce carbon dioxide emissions to curb global warming, relatively high "carbon economy" low carbon model became the trend of Europe's economic development, the public began to generally agree with low carbon Revolution (see Table 1) concept, low carbon economy become green tide of European.

Establish Clear Low Carbon Strategy Goal

To develop low carbon economy, first, it must have a clear strategic goal; therefore, European countries have developed in line with the national conditions and low carbon economy development goal. Take UK as an example, according to "energy white paper" published in the UK, the long-term emission reduction plan strategic goal is that, reduce the amount of carbon released 2300~3300 million tons by 2020 and reduce at least 60% in the total emissions by 2050.

Take Measures to Realise Low Carbon

Current European countries generally take three low carbon actions: first, through carbon capture, carbon sequestration, carbon storage to reduce the unit energy carbon consumption which is the essence of low carbon economy. Second, develop alternative energy, low carbon energy and carbon-free energy to control carbon emissions, terminate the situation that economic development depends on the energy emission, which is the core of low carbon economy, third, change people high carbon consumption preference, reduce the energy consumption of fossil, realise low carbon life, which is the root of the development of low carbon economy.

Build Perfect Low Carbon Policy Supporting System

Economic development together with policy formulation can better reach the low carbon economy strategic target. European countries formulated trenchant rewards policy measures to develop low carbon economy, such as collection of carbon tax, giving tax subsidies on carbon emission reduction technology, which is the major feature of the Europe low carbon management mechanism. Like UK, the government direct investment, financial support and private investment in low-carbon areas totaled 50 billion pounds from 2008 - 2011.

Established Diverse Governance Mechanisms

The developed countries in Europe and America basically have 4 kinds of governance mechanisms to develop low carbon economy.

The first one is the public and private sector management, taking Germany as the typical, second one mainly rely on the government regulation to control ecological pollution, taking Holland and Sweden as the tipical, the third one mainly based on market mechanism to solve the low carbon economy development process of various problems, taking Australia and Canada as typical, the last is that senior management of companies develop and implement low-carbon economic development strategies, this is the common moda for European countries.

Increase Innovation of Science and Technology

Technical innovation is an important support for the development of low carbon economy. European countries improve the efficiency of energy by information communication technology. Such as UK, takes low carbon power station technology like wind energy as the key of carbon emission reduction, plans to build the first carbon capture and storage (CCS) and Zero coal Discharge demonstration project, prompting more power stations.

EUROPEAN LOW CARBON REVOLUTION'S INFLUENCE ON CHINA'S ECONOMY DEVELOPMENT

Low carbon economy is the third technological revolution after the industrial revolution and the information revolution, will become the only road to develop economy of developed countries and developing countries. after financial crisis Europe is not only China's largest trading partners, but also the main sources of high technology introduction. Under the background of economic globalization, the European low carbon revolution will have far-reaching impact. on Chinese economic development.

European Low Carbon Revolution Challenges to Chinese Economic Development

Challenges of Economic Structure Locking Effect

"Locked-in Effect" is that once infrastructures and equipments (Service life about 15 years to 50 years) are invested, investors will bear enormous cost when abandoned. British industrialization has already been one or two hundred years, thermal power and nuclear power plants are not new, the Locked-in Effect of developing low carbon energy is much less than China which in the middle stage of industrialization. In China, coal has accounted for about 70% in the energy structure for a long time, if China continues to use the traditional or modified technology, when it is required a commitment to reducing greenhouse gas emissions or obligations, China may be "locked". It is an urgent and realistic challenge to avoid "Locked-in Effect" in the process of economy development.

If developing low carbon economy, China will encounter three kind of "locking" bottleneck: first, constraints of foreign dependency. China's FDI has accounted for more than 40% of GDP, significantly higher than that of the developed countries and some Asian countries and regions, and reduces the control and safety of local economics.

Second, bottom restrict of "Smiling Curve." According to smile curve theory, the R and D in the upstream and branding in the downstream share the most value-added in the value chain, and the less value for manufacturing. Chinese manufacturing industry is in the bottom of smiling curve, has little additional value, but has been blamed for the world's largest polluters by western countries who got huge profits. Chinese manufacturing industry must break up high technology blockade from western countries to get rid of the dilemma to extend to the both ends, which still must be a long and difficult process from both their own strength and Western attitudes.

Third, constraint of the low quality of human resources. Chinese manufacturing industry labor are mainly transfer from rural surplus labor force, lack high level labors who can be engaged in the sides of smiling curve. If China abandons the traditional labor-intensive industries, tens of millions of unemployed workers may cause a chain reaction, which will bring disastrous consequences to social stability.

Challenges of Traditional Route City of Spatial Structure

Some scholars have calculated, an individual's carbon consumption included 65% for housing construction and transportation, only 35% for food, leisure and fashion. At present, China's developed cities almost draw further apart from low carbon city. The "residential energy conservation", "science and technology into residential" become luxury houses, the pursuit of large-sized apartment, luxurious decoration of space, squanders lots of resources and energy. The greening rates of communities are much lower than the request, fewer trees resulted in less carbon absorbed by the nature. Some cities don't have bicycle roads, no position to park bicycles. With increasing number of private cars, traffic jams, resulted in huge amounts of carbon emissions. It is very difficult to reverse the high carbon city tendency since it has existed for longtime.

Challenges of Development Stage Feature

At present, China is in the industrialization stage, the basic task is to achieve modernization, which brings unimaginable pressure to develop low carbon economy. The International Energy Agency forecasted that China's energy consumption and carbon emissions will be more than the United States around in 2010. China speeds up the realization of modernization requires that GDP grows at least 9% every year, however, according to the energy-saving emission reduction targets of the "eleven five", energy intensity decreased in 4% every year, China's carbon emissions in 2010 still increased above 20% than in 2005, indicating the carbon emissions will sustained in growth trend for a long time. The contradiction between accelerating the modernization and low carbonization will always exist. With the continuous growth of energy demand and carbon emissions, international pressure grows with each passing day. The time for China, a developing country, not assume the emission reduction obligations will not exceed a maximum of 2020. Actually, it is not much time to solve this contradiction.

The Opportunities on China Low Carbon Development by European Low Carbon Revolution

Promotes Economic Structural Transformation

From long-term look, on the one hand, European low carbon revolution development will drive Chinese industrial structure transform. Low carbon revolution will curb the high carbon industry development, lead low carbon industry which does not depend on the traditional energy to rise abruptly, low carbon agriculture will reduce the consumption of fossil energy, instead of organic, ecological and efficient development mode. With the Shanghai Disneyland opened, tourism and other low carbon service industries will develop better. On the other hand, European low carbon revolution will change the social life of China. Low carbon city construction will be the trend, the gas penetration rate, city greening rate and waste disposal rate will be improved, the research of energy-saving appliances and residential energy management system and development will be valued, bus and light rail development will be focused on to improve public transport travel rates, private automobile emission standards will be regulated strictly, the social responsibility, emission reduction, of enterprises also will get more attention.

Promote Low Carbon Technology Progress

China low carbon technology level and European countries there is a big gap. As in the aspects of renewable energy technology, large-scale wind power equipment, high cost of solar photovoltaic battery technology, fuel cell technology, hydrogen energy technologies have not little difference; transportation vehicle fuel economy, hybrid electric vehicle related technology is still in the exploratory stage; metallurgical, chemical, building in areas such as energy conservation and energy efficiency improvement room for the larger.

On the other hand, the low carbon technology international cooperation the opportunity to increase. "The United Nations Framework Convention on climate change" provisions of the developed countries have the obligation of technology transfer to developing countries. Central Europe signed "Joint Declaration" on climate change, the United States launched the "Asia-Pacific Partnership on clean development and climate, and so on to the low carbon technology development and cooperation as the foundation. China as large energy consumption and carbon emissions can make use of multiple pathways from European countries experience, narrowing in low carbon technology gap, promote the rapid progress of technology of low carbon.

Stimulates the Carbon Finance Market to Accelerate the Formation and Development

European carbon trading help establish the new carbon finance market with rapid expansion, including direct financing, carbon trading indicators, bank loans, and carbon credit derivatives such as environmental protection, which become new trading varieties that institutional investors interested. China is the largest supplier of international carbon trading market, but CDM project has no advantage, after the financial crisis, the carbon price is even below the guidance price.

Drawing lessons from European financial market and carbon financial derivatives innovation experience, our country is badly in need of strengthening Carbon Asset Management, establishing carbon finance platform, optimizing the carbon financial innovation environment, encouraging financial institutions to set up carbon finance related business, initiating establishment of carbon industry fund which dedicated in carbon management and low carbon technology development, supporting the issue of energy-saving emission reduction bond and improving trade size and the related asset liquidity. In addition, as a big carbon asset status, China should set up a fair, impartial, open competitive carbon trading market mechanism platform to obtain CDM market even the pricing power at CER (Certified Emission Reduction).

CHINA'S COUNTERMEASURES TO EUROPEAN LOW CARBON REVOLUTION

Under the background of economic globalization, European low carbon revolution will have far-reaching impact on China's economic development. Therefore, Chinese "Twelfth Five-Year Plan" must be focused on the system study of how European low carbon revolution challenges the previous pattern of economy growth, to grasp development opportunity.

Formulate and Implement Low Carbon Economy Regulations and Development Strategy

Establish and Implement Laws and Regulations of Low Carbon

From the experience of European countries, government and legal support is the indispensable condition for the development of low carbon economy. The government should formulate "low carbon economy law", "recycling economy law" and other relevant laws, and modify the "renewable energy law" and other related to energy, environmental protection, resource regulations, to provide strong policy support and legal protection. for China's low carbon economy development.

Formulate and Implement National Low Carbon Economy Development Planning as Soon as Possible

Formulate national low-carbon economic development planning by integration resources of government, enterprises and society, coordination policies of science, economy, tax and legislation strategically. Formulate and implement carbon development plan of town, industry and especially corresponding industry. Infiltrate the strategic thoughts development of low carbon economy into science, technology, economic and social development.

Establish and Perfect the Public Participation Mechanism in Carbon Emission Reduction

Learn from European experiences, all levels of Chinese government should give the public a convenient low carbon information channels, develop public evaluation on carbon emissions intensity compared with social economy development, put forward a low carbon economy statistics and evaluation system, guide and lead enterprises and public to operate and live as low carbon way, coordinate all the social forces to promote low-carbon economic demonstration area construction. Proceed with establish the concept of low carbon life, encourage the public to improve behavior, reduce carbon footprint by publicity and education, and establish public participation carbon emission reduction mechanism to expansion the masses foundation of sustainable development strategy.

Take Low Carbon as Center of Structure Adjustment

Optimize Energy Structure and Improve Energy Efficiency

The thematic of the "Twelfth Five-Year Plan" on China economy development is that adjust the economic structure and change the mode of economic growth. In 2006 China energy consumption per unit GDP is 11.5 times more than Japan and 7.7 times more than Germany and France. From 2001 to 2007 Chinese energy intensity increased from 4.21t standard coal per million to 4.77t standard coal per million, energy efficiency is only 33.4%. Therefore, China has great potential to change carbon burden to carbon asset through adjustment of energy structure, technological innovation and energy conservation management and development of renewable energy and other ways of energy saving and emission reduction, reducing carbon emissions intensity.

Adjust the Industrial Structure, Limit High Carbon Industry Development

For the same level about technology and economy, carbon emissions varies significantly due to the different industrial structure. According to UK energy white paper announced in 2003, the GDP of the British economy has increased 1 times, but total energy consumption increased only 10% in nearly 30 years because of the development of modern service industry and the improvement of energy efficiency. Compared China's industrial development to economic growth, the tertiary industry, the low energy consumption, developed slow, economic growth relied too much on the high energy consumption, high pollution and heavy chemical industry. From 2005 to 2008, China's secondary industry proportion of GDP rose from 47.7% to 48.6%, the growth rate of high energy consumption industrial was significantly higher than the average growth rate, high technology industry proportion dropped 2.1%. IT industry and modern service industry has been the most potential low carbon industry in Europe, the modern service industry proportion of GDP has reached as high as 60%~70%, in China this part has huge promotion space, should be the important content of industrial structure adjustment.

Adjust Area Structure to Promote Low Carbon Urbanization Construction

European countries put forward six elements on the ecological city that meet the comfort requirements of human health, increase economic value of natural system, deal with the relationship between protection and development, sustain utilization of land, water, air and mineral resources, keep development of city traffic, construction, municipal infrastructure with the ecological requirements, upgrade the living environment aesthetic value and protect natural systems biology integrity. Chinese low carbon urbanization system should also contain four part construction including sustainable environmental carrying capacity, circulation and effective economic growth, harmonious social development and intensive use of resources.

Strengthen International Cooperation and Independent Innovation Ability of Low Carbon

Strengthen Import and Exchange of Low-Carbon Technology

Advanced technology is the supporting strength of European countries to develop low carbon economy but also the biggest constraint to the Chinese economy turn from the "carbon" into "low carbon." There are two ways to improve Chinese low carbon technologies, one is the direct introduction of foreign advanced technology, the other is independent research and development. Climate change is one of the largest public phenomenon, it cannot rely on a country or region to complete the carbon reduction task, the developed countries has promised developing countries to transfer technology of greenhouse gas emission reduction in the United Nations Framework Convention on climate change, and the United States, "Asia-Pacific Clean Development and Climate Partnership Programme." China should grasp the opportunity that European countries are eager to get rid of financial crisis to strengthen with governments, non-governmental organizations, enterprises, training institutions for technical cooperation and exchanges, using multiple channels to accelerate the introduction, absorption and popularization of low carbon technology.

Improve the Ability of Low Carbon Technology Independent Innovation

Due to intellectual property, commercial interests, and many other reasons, European countries tend to be reserved about technology transfer. Therefore, China should speed up the introduction and cultivation of carbon reduction technology talents, foster the ability of low-carbon technology independent research and development, strengthen storage of low-carbon strategy technology, integrated with market existing low carbon technology while import European technology to gain the initiative at the low carbon economy technology revolution.

Support the Establishment of Common Low Carbon R and D Mechanism, Promote Breakthrough and Transfer of Low-Carbon Technology

With high cost, low short term gain, Chinese enterprises has little initiative so that low carbon technology developed slowly in China. Therefore, it is very important for the current government to support construction of low carbon public R and D mechanism by finance and tax preferential policies.

The government should also actively guide enterprises and private capital to develop renewable energy technology, greenhouse gas emissions control and disposal technology, biological engineering and carbon sequestration technology, advanced coal, nuclear power and other major energy equipment manufacturing technology, carbon capture, utilization and storage technology to make manufacturing extend to the both ends of "smile curve".

Construct Taxation and Financial Environment Which Is Good at Low Carbon Economy Development

Construct Low Carbon Tax System

Chinese governments at all levels should shift from production tax to public finance which is low-carbon oriented with harmonious and sustainable development of natural and social.

Among fiscal budget arrangement, attention should be paid to the low carbon economic investment, in order to form a stable and diversified funds guarantee mechanism. Implementation of rewards and punishment tax policy is that reduce or exempt income tax, refund tax, provide preferential loans and depreciation policy, give fiscal subsidy to the enterprises who work on pollution control, clean production, development and utilization of new energy, comprehensive utilization of waste, and use energy-saving emission reduction equipment.

Take measures to expand government demonstration effect to promote the development of low carbon economy, such as expanding the scope of government green procurement, purchase ecological public products with public finance, including the subsidies of rivers and lakes Environmental management, natural forest protection, farmland returning to forest, herd returning to grass, sandstorm source control, soil desertification control, wetland protection, fossil desertification control, ecological construction, natural protection area enlarge, division of ecological functional area and coastal ecosystem, biodiversity conservation.

Construct Low Carbon Finance Service System

According to the fact that government and market have no efficiency in microcosmic action of energy saving and emission reduction, as well as the carbon emission reduction have a global effect, but the cost is quite different from different countries, "Kyoto Protocol" make carbon emissions rights commercialization to form carbon trading market and established there international market operation mechanism, emissions trading, joint implementation, and clean development. Although the mechanism exists many disadvantages, as a new solution to carbon reduction problem, it is still regarded as a kind of system innovation and provides policy support and systematic guarantee for our country to promote the development of low carbon economy. The cost of carbon emissions per ton is around $15 in China, in contrast to $30 in developed countries, China has a cost advantage. With the growth in energy demand, increased in reduction condition projects and obvious economic scale effect characteristic, carbon emissions trading quite attracts international capital. According to the United Nations and the World Bank forecast, global carbon market size in 2012 would be 150 billion dollars, China would share a lot, but China still lacks of a unified national market. Therefore, China should establish Shanghai carbon trading market to promote the development of low carbon economy.

REFERENCES

Bo Jun. Boulevard invisible: the operational mode of Japanese business[M]. *Nanjing: phoenix publication*; 2009.

Lin Hong. Research on the development of low-carbon economy at home and abroad and the recommendations on the province[J]. *Economic Review*; 2009(5): 17-18.

Low Carbon and Environmental Goods and Services: An Industry Analysis, Innovas Solutions; 2009.

Pan Xi, Zhao qiguo. Our country must walk ecological high value agricultural road [N]. *Science times*; 2010-09-06.

Pan yiyong. More effective low carbon and safe U. K. energy policies[J]. *Views on the global technology economy*; 2007(9): 56-59.

[Spain] Pablo Troy Booth. China's economic prosperity and its international influence[J]. *Foreign policies*; 2007(9-10).

Stern N. *The Economics of Climate Change: The Stern Review*[M]. Cambridge: Cambridge University Press; 2006.

U. K. Energy White Paper 2003. Our energy future-creating a low carbon economy[EB/OL]. http://www.Berr.gov.uk/files/ filel0719.Pdf.

Wang Yang, etc. Research on the linkage mechanism of low-carbon economy based on multi-center governance theory[J]. *Technology Progress and Policy;* 2010(22); 1-3.

Wang Zhilun. Low carbon economy will have a significant impact on the world economy [J]. *Selected reference materials*; 2009(34): 6-10.

Yuan Lei. Low-carbon urban planning should be fit to the local conditions[N]. *Guangdong construction newspaper*; 2010-06-18.

Zhang aijun, Li xiaodan. Policy choice for China to develop low-carbon economy [J]. *Macroeconomic management*; 2010(1): 55-56.

In: Economic Developments and Emerging Markets ... ISBN: 978-1-62417-986-0
Editors: Guy Boscarino and Derek Notte

Chapter 8

THEORY AND PRACTICE ON RISK MANAGEMENT OF FLOODWATER UTILIZATION BASED ON THE WATER UTILITIES IOT: A CASE STUDY OF THE NORTHERN PART OF JIANGSU PROVINCE, CHINA

An Wen and Qi Wenjuan
Low-Carbon Economy and Technology Research Institute of Hohai University, China

ABSTRACT

This paper uses theoretical analysis combined with case studies to introduce the meaning and conditions of the floodwater utilization in northern Jiangsu, and the key technologies of water utilities IOT architecture and remote sensing equipment, transmission network, water utilities management and feedback mechanism. The paper explores the function of water utilities IOT technology to strengthen risk management in space, time and capacity, and puts forward suggestions on the formation and improvement of the water utilities IOT, including the establishment of water management information standardization system, the acceleration of the breakthrough of water information sensor core technology, the removal of the industry barriers, regional barriers and departmental barriers.

Keywords: Water Utilities IOT, the Northern Part of Jiangsu Province, floodwater utilization, risk management

INTRODUCTION

The poor usage of water resources pumping water resources in the drought after floods discharging is very common in the country, especially in the northern part of Jiangsu, is not only a waste of cost but also a waste of water resource especially the floods resource, which makes a strong demand for the floodwater utilization. Floodwater utilization means under the use of water storage engineering such as reservoir, sluice gate structure, natural depression, artificial lakes and underground reservoir to combine the use of system theory, risk management, IT and other modern technology to implement effective flood risk management, in order to increase the supply of water resources and maintain good ecology under the premise of ensuring safety. The main bottleneck of the current risk management of floodwater utilization is technological innovation.

The application of water utilities IOT can solve the problem effectively. Jiangsu province successfully tested the changes in water quality through a wireless network around the clock real-timely online in Taihu Lake in the year of 2009, which inspired us to actively explore the risk management of floodwater utilization based on the water utilities IOT in the northern part of Jiangsu Province.

1. Network and Key Technologies of Water Utilities IOT

1.1. Network of Water Utilities IOT

Water utilities IOT uses the wireless networks and the Internet to link up each sensor node of water information in fields such as water resources development, protection and others to monitor and use the water information of water consumption, water quality, water level, flow ect to form a huge sensor networks, and make a more sophisticated way for the dynamic management of water resources, intelligent management of water management, risk identification and control to improve the efficient use of water resource, promote the harmony between human and water. Water utilities IOT combined by remote sensing equipment, transmission network, water management and applications (application platform) and feedback mechanisms.

1.2. Key Technologies of Subsystems

The organic combination of communication system, computer and the sensor can achieve intelligent characteristics which is similar to the human. Currently used wireless senor in water system can only get some simple information such as the water volume, water level and water quality, it can't depict the rich and colorful water world. How can the scalar information, video, audio and image vector information be collected is main technical innovation problems for the water system, besides, the future wireless sensor network also have the information processing such as compression, identification and reconstruction to produce the information more vividly and lively.

Collection points of Water information mainly distributed in the wild, the transmission of information for wireless network which is based on the cellular network which has center control structure, such as China Mobile GPRS networks. The domestic water industry traditional network mode is in the specific application unit set up branch, internet-facing networks and configuration fixed IP address, wireless sensor network nodes connected directly to the center launched. The transmission way has three disadvantages: Firstly, the need of maintaining node sensor. Secondly, it is needed to maintain the stability between node and center communication. Thirdly, the maintenance is very passive when number of nodes is large and network coverage is not full-time. In recent years M2M platform business and wireless terminal standard launched by telecommunications, mobile launched effectively solve these problems, stable operation, management convenient, can reduce nodes and transmission network maintenance cost, and increase the application service application.

Business application platform is an important component of the whole water content networking which gathers the public or reuse handle business handling logic forms the standard and open software source. Application platform can be divided into such three levels, system resources services, public basic services, application service. The key of the business application platform development is the public share software (especially the data transmission software) development, expanding enrichment sharing software resources set, using the existing software and reducing the cost in the development of new software and at the same time.

Feedback mechanism is made up by quality information collection and processing. Quality information collection includes information recording, quality problems collection, major quality accidents registration and quality problems status inquiring. Quality information processing includes reform inquiring, reform and feedback, problems re-examination and problem archive.

2. The Significances and Conditions of Floodwater Utilization in Northern Jiangsu'

2.1. Necessity

Characteristics like location and climate, as well as the high density in demographic, economic capacity determine the problem of whether or not Northern Jiangsu has enough water and the task of Flood control and drought are existent at the same time, and they are the same grim.

To begin with, the local total water resources are seriously inadequate. The quantity of local water resources is 32.16 billion m^3 per year. In addition, the Huai River regions in Northern Jiangsu accounts for 0.91 billion m^3, becoming the most serious water shortage area in Jiangsu.

Next, it's difficult to transfer the storage water in plains. The Northern Jiangsu is flat, and its natural conditions are not suitable for construction of reservoirs and embankments. The ability to regulate coastal and local earth mound is not enough. At present, the major water storage projects in Northern Jiangsu are Hongze Lake, Luoma Lake and Weishan Lake, as well as the 21 large and medium-sized reservoirs which are distributed in Xuzhou, Lianyungang and Huaian. The normal water storage capacity of "Three Lakes" at non-flood season is 5.87 billion m^3, after deducting the fixed storage capacity, it is 4.31 billion m^3. While the normal water storage capacity of 21 large and medium-sized reservoirs is just 780 million m^3. In addition, as Hongze Lake, Luoma Lake, Weishan Lake and Shiliang River reservoir has the task of flood control of Huai River and Yishusi River basin, they all affect the interests of the neighboring province except Luoma Lake. The water level in flood season is strictly limited with the total capacity of “Three Lakes and One Reservoir” is just 4.86 billion m^3 when compared to the limited water level in flood season and the available capacity is just 3.26 billion m^3. But from June to September, the water demand is maximum because of it is the critical period of rice planting and growing in this region, while the water storage of lakes and reservoirs is strictly limited at this period.

Then, the uneven spatial and temporal distribution of water resources. The rainfall on the allocation of time is very uneven. About 50 percent of the annual rainfall is concentrated in July and August in northern Jiangsu. The local runoff and transit water of upper-middle reaches of Huai River and Yishusi are corresponding highly concentrated in the same period. The inter-annual differences of rainfall are huge.

In the regions of north Huai River, there is 1 to 4 times difference between the serious water shortage years and ampling years. This leads to very uneven distribution of local and transit water resources in inter-annual years and during the years.

The quantity of water resources in Huai River Basin is 19.3 billion m^3. In 1991, it was very abundant in water resources, amounting to 67.1 billion m^3, while in 1978, it was facing very serious water shortage problems, and the water was just 1.25 billion m^3, up to 53 times difference between them. As to the transit water, let’s take Huai River for example, the transit water is 67.52 billion m^3 in 1954 and 1.681 billion m^3 in 1978. There’s 40 times difference about runoff in the flood seasons (from May to September) in the abundant and shortage years.

Finally, the pollution of water threats to water security. Since 1990s, as economic and social rapidly developing in the Northern Jiangsu, the sewage discharge increases significantly and corresponding laggard sewage control make the water pollution very serious. As a result, many rivers cannot be used as a drinking water source.

There are large amounts of sewage from upstream spilled into the Northern Jiangsu territory. The sewage pollutes the main lakes, such as Hongze Lake. It also affects urban and rural life and production water. To solve water shortage problems, some areas exploit underground water excessively, which caused the underground water descent funnel and reduce the flood control standards. With the decline in the quality of surface water, groundwater pollution has gradually increased.

2.2. Efficiency

The above problems need to save water, increase the environmental and ecological water and the water environmental capacity of rivers and reservoirs, with the routine water supply as basis. Try to promote the flow of water to improve the water environment. Scientific and rational flood utilization management can create considerable social, economic and ecological effectiveness. Besides the lakes and reservoirs are filled with water, it can significantly extend the time of the multi-cascade small hydropower to generate electricity. The economic benefit is considerable.

First, it brings economic benefits. The increasing of the water supply, saving much fees used in turned water and reduce the cost of water supply. For example, after flood season, if the water level of Hongze Lake is higher than the normal level 0.3m, it will store 600 million m^3 water more. Second, it brings social benefits. The management of floodwater utilization can promote the flow of water and improve the water quality of water sources, to ensure drinking water security of both urban and rural citizens. The flood resource is intermittent in northern Jiangsu, the management of floodwater utilization also alleviates the water supply and demand contradiction and creates a lot of social benefits. Third, it brings ecological benefits. The relief of flood resource improves the water environment. In late July 2001, the Hongze Lake is nearly exhausted; there is just a small amount of sewage flow into the lake from the Huaihe River. On August 12, Jiangsu seized the chance that the Luoma Lake water level was higher than the flood control level, even higher than the normal level in a long time, do not open the Zhangshan Gate to release water downstream. Jiangsu conducted the project named "Water Transfer from Yi River to Huai River", let the 841 million m^3 clean Luoma flood flow into Hongze Lake and its downstream irrigation area through the Zhongyun River and Xuhong River. Then the sewage was diluted and the vicious water environmental degradation accidents were avoided.

2.3. The Favorable Conditions

The northern Jiangsu implement the floodwater utilization and it has many favorable conditions on transit water, engineering fundamentals and so on.

First, the North Jiangsu is abundant in transit water and it has resource conditions. The flood season of rivers like the Yangtze River, Huai River, Yishusi is from June to September. Each year, on the premise of flood control security, it can almost make sure that the water of Hongze Lake higher than the normal water lever in most years after floods, and the water of Luoma Lake reach or higher than the normal water lever during the half years after flood season.

Second, the corresponding completion of the system of water conservancy project in northern Jiangsu lays a perfect engineering foundation for the utilization of flood. One is a high-standardized system for the control of flood and waterlogging has been established. The premise of floodwater utilization is the insurance of flood-controlling safety. Throughout more than 60 years of constant construction since the foundation of new China, the comprehensive flood-controlling standards of Hongze Lake and the downstream area of Huai River are now approximately once-in-a-century, thus serving as a safety guarantee for the flood utilization in northern Jiangsu. The other is an inter-basin configuration net of water resource has been completed, connecting three water system include Yangtze River, Huai River and Yishusi and realize the mobilization among them. The other is an inter-basin configuration net of water resource has been completed, connecting three water system include Yangtze River, Huai River and Yishusi and realize the mobilization among them. Since 1961, the project of water transfer from the Yangtze River to the northern plains of the Jiangsu Province has begun to take shape after 40 years more construction and improvement.

The project of water transfer from the Yangtze River to the northern plains of the Jiangsu Province series connects Hongze Lake, Luoma Lake, Weishan Lake and Shiliang River reservoir, linked the Yangtze River, Huai River and Yishusi, an inter-basin configuration net of water resource has been completed, connecting three water system include Yangze River, Huai River and Yishusi and realize the mobilization among them. It can not only transfer the water to north, help Huai River and Yi River, but also transfer the water in Huai River to the north and Yi River. In the years when it is abundant in Yi water and shortage of Huai water, the opposite is realizable. Most of the water storage projects in Jiangsu Province was built in the 1950s to 1960s, so the engineering was old and didn't been well repaired for a long time, then the design capabilities were reduced. It not only endangered the safety of flood control, but also affected water storage. It is difficult to solve these problems just through the routine maintenance. Since 1990s, Jiangsu increased the investment in the south dam of Luoma Lake and Shiliang River reservoir, for the project of eliminating risk and reinforcement.

Then the flood controlling capacity and water storage capacity was greatly improved. At the same time, Jiangsu also conducted the project of eliminating risk and reinforcement in allusion to a large number of the small and mid-sized reservoirs. Thus these reservoirs were filled with the engineering condition of recovery and improve the storage level.

Chart 1. The Time Risk Management of Flood Water Utilization in Northern Jiangsu

NO.	NAME	TIME	SITUATION	MEASURES
The first stage	Pre-flood-season	May to June	1. The quantity of Peach flood is not sure. 2. Agricultural irrigation water use is increased significantly in this period. 3. Water supply and demand is the most prominent.	1. The current lake and reservoir water storage is not simple to flood limit water level control, but should the water supply associated control of a little more than normal or the normal water. Amounted to less than the normal water level, such as lakes and reservoirs need to water impoundment, if necessary, adjust the implementation of the fill Lake replenishment measures cited a large number of foreign water to meet the needs of agricultural irrigation water. 2. Pay more attention to weather changes by the use of the water utilities IOT closely, to strengthen the forecasting analysis to ensure flood safety, the vent is to vent.
NO.	NAME	TIME	SITUATION	MEASURES
The second stage	Main flood season	July to the middle of August	There are a lot of heavy rains and flood disasters	1. The water level in lake library is generally controlled by flood limit water level. 2. Using water content networking to get a measures implementation of precipitation. 3. Water level can be driven up when there is a pronounced drought or middle weather forecast occurs the drought is coming.
The third stage	The end of flood season	late August to September	Reducing precipitation and water storage	The current library should consider north lake dynamic water and two impounding need to use water content networking in flood control security manageable tail water use of appropriate super storage, to avoid the lake water level of the normal storage level library just fell, not a long time maintain normal storage level.

3. Technologies of Water Utilities IOT to Strengthen the Risk Management of the Floodwater Utilization in Northern Jiangsu

Taihu Lake experience has shown that the water utilities IOT can technically support intelligent water management of water, water quality, water level, flow, thereby strengthening the risk management of floodwater utilization.

3.1. Supporting the Time Risk Management of Floodwater Utilization with Smart Metering

Flood season in northern Jiangsu was June to September. The water resources management faces two big contradictions: on the one hand, before the arrival of a flood, the agricultural water requirement is large, if lake and reservoir water level are controlled to the flood control level, we cannot guarantee agricultural water requirement.

On the other hand, the local precipitation mainly concentrated in flood season, if we just discharge flood in flood season and store flood until the end of flood season, lake library will face the risk of no water storage.

Water utilities IOT smart meter with RS-485 communication standard, with other sensors and wireless network communications equipment, two-way communication between managers and water units will be realized, it can real-time accurate perceive the situation of water resources, timely eliminate water anomalies, and make full use of ample flow in demand.

According to the particularity of the northern Jiangsu, the flood season of Jiangsu is classified into three stages, and the risk management of floodwater utilization is implemented intelligently.

3.2. Intelligent Scheduling in Support of Floodwater Utilization Space Risk Management

Based on the water utilities IOT inter-basin water environment intelligent monitoring system, on the base of a comprehensive grasp of river in water quality and quantity and other sorts of base material and underground water level real-time data, based on the water point water resources consumption, remote monitoring network of backed real-time data can be used to calculation and analysis, and according to the calculated results of the real time to control the water quality information and supply and demand situation, realize the flood resources control across river basin, improve flood resources utilization ratio of water resources development, utilization, configuration and protection services. For example, when it is abundant in Huai River and short in YiShuSi, it can carry out "transfer the water to YiShuSi ".

In the years when it is abundant in Yi water and short in Huai water, the opposite is realizable. It also can transfer the water to north, when it is dry in the northern Jiangsu.

3.3. Intelligent Management Supporting Flood Resources to Enhance the Ability of Risk Management

In order to improve the risk management of floodwater utilization in northern Jiangsu, water utilities IOT must be used to improve the normal storage capacity of "three-lake and one- reservoir ".Jiangsu did a lot of work to drive up the normal storage level of Hongze lake for the shortage in Huai river downstream from the 1950 s. Finally this level is13.00 m in 1982, but it hasn't reached the first goal of 13.50 m. Construction and improvement of water utilities IOT can help to realize the goal of 13.50 m, ensure the safety of flood control, tackle the contradiction of "water more" and "water less", to strengthen risk management ability of floodwater utilization in northern Jiangsu.

Using water utilities IOT to strengthen supervision of water environment is indispensable to the automatic and real-time monitoring of water quality and quantity and other hydrological index for rivers , lakes and the key water function area of rivers, drinking water reserves, especially the river drainage outlet and those existing the key pollution source of river.

4. Suggestions of Building and Developing of Water Utilities IOT

Although IOT technology has been developed from the concept to water sector practice, and the management of taihu lake made a initial success, but there are many problems need to be solved in order to form a real sense of water utilities IOT.

4.1. Building Unified Water Management Information Standardization System

IOT in water management application is still at the primary stage, there is no unified standard system at present all over the world, it will greatly restricted the development of water utilities IOT as well as the development and application of intelligent measuring facilities, water quality sensor, water level sensor and application software. Jiangsu province has established related standards in water resources management.

In order to build the national unity standardization system of water utilities IOT, the Ministry of Water Resources should take the leader to get all the river administrative agencies participated in and the latest achievements in water utilities IOT should be absorbed.

4.2. Speed Up the Breakthrough of the Core Technology of Sensor of Water Information

With the application and promotion of the networking, Sensor connotation has a rich development process that developed from Dumb Sensor to Smart Sensor, and to Embedded Web Sensor, gradually realize miniaturization, intelligence, and information and networking.

Using these results to perceive the natural water is particularly important. Speeding up the microelectronics, opto-electronic materials, nanotechnology and other areas of science research to form the manufacturing system of water information sensor to get the world's advanced level, and arm water utilities IOT with a more accurate "sense organs".

4.3. Related Departments and Regions to Strengthen the Cooperation

Water management department has overlapping relationships with Construction department, Environmental protection department and many other functional departments.

At present, these departments have set up their wireless network in the water, water taking, sewage discharge, etc. Each region is starting their water resources information network too. Establishing the water utilities IOT with sensing technology as the foundation, must strengthen the coordination and interaction between these departments and regions to share information resources effectively, and perfect water utilities IOT in risk management of floodwater utilization commonly.

EPILOGUE

Water utilities IOT technology is effective to the risk management of floodwater utilization, but is not everything. Because of the floodwater's dualism, the implementation of floodwater utilization must rely on technological, administrative, economic and other kinds of means to provide its security and make sure the flood risk within acceptable limits.

For example, relying on the water utilities IOT and establishing a perfect command system including all the departments to improve the level of risk management, implementing active financial measures to eliminate the negative impact of floodwater and increase the economic compensation to related cities, strengthening the construction and maintenance of water infrastructure to reduce the influence of the flood disaster and ensure that the risks of the floodwater utilization is under control.

REFERENCES

[1] Wen An. Primary Study on SSIS based on MIS and GIS[J], *Telecommunication Information,* 2009 (11) 21-26.

[2] Huiyun Zhai. Primary Study on Application of IOT in Water Resources Management[J]. *Water Resources Informatization,* 2010(4): 6-10.

[3] Yu Song. Exploration and Practice on Floodwater Utilization in Jiangsu Province[J]. *Chain Flood and Drought Management,* 2009(11): 21-26.

In: Economic Developments and Emerging Markets ...
Editors: Guy Boscarino and Derek Notte
ISBN: 978-1-62417-986-0

Chapter 9

POVERTY, INEQUALITY AND SOCIO-ECONOMIC DEVELOPMENT: A DYNAMIC APPROACH FOR COLOMBIA

Alexander Cotte Poveda
Department of Economics, University of Göttingen, Göttingen, Germany
Faculty of Accounting and Administration, University of La Salle, Bogotá, Colombia

ABSTRACT

This study estimates poverty, inequality and socio-economic trends in Colombia's departments. We use the data panel over the period 1993–2007. In this analysis, we suggest a dynamic OLS (DOLS) model to measure poverty, inequality and development trends. Using panel data analysis and cointegration techniques for a dynamic heterogeneous panel the results from DOLS model show variation in the scores across Colombian departments during the sample period. Findings of this analysis should demonstrate that the decrease in poverty and inequality could be achieved through adequate strategies that guarantee development and economic growth with policies concentrated to improve social welfare.

JEL: O18, I32, D63, C14, C23

Keywords: Economic development, poverty, inequality, data panel, Colombia

1. INTRODUCTION

Poverty and inequality are fundamental for socio-economic development especially in developing countries. Different studies have demonstrated that countries and regions with higher socio-economic disparities implies the most critical socio-economic-complications mainly because lower economic growth increases violence and crime, lesser educational achievement, decrease civic or electoral participation or higher mortality rates. In contrast, regions and countries with a minor level of socio-economic disparities achieve a better control in all these aspects (European Commission, 2010).

In regions and countries, it is fundamental to understand the processes that determine poverty and inequality with the aim to improve patterns of social mobility. Profit (2007) proposed a model to recognize the links between the micro process (family) and macro process (state and economic contexts) and how they are fundamentals in the formulation of adequate socio-economic policies that increase social inclusion and development. However, to determine these relationships is necessary to apply several methodology approaches in diverse countries and regions but especially in developing countries to achieve decrease poverty and inequality whereas it promotes development and social inclusion.

In the literature, studies on poverty, inequality and socio economic development have been analysed from in several empirical and theoretical studies. Beginning with Kuznets (1955) who explained that in the early phases of development it increases income disparities, but in the advanced phases of development these disparities decrease. Recently, Mehanna (2004) analysed the causality between economic development, poverty and inequality determining the close relationship between economic development, poverty and crime, and Adams (2004) demonstrating the close relationship between economic growth, poverty and inequality determining that economic growth does reduce poverty specially in developing countries.

Other studies have analysed the relationship between poverty and education or poverty and labour market, for example, Bonal (2007) studied this link in Latin American finding that educational policy can be used to combat poverty and inequality as a strategy of social inclusion and mobility; OECD (2009) demonstrated that employment and the quality of employment is decisive for poverty reduction mainly because output growth generate an increase in productive and remunerative employment by a rise in the incomes of the poor and a reduction in poverty.

These studies have allowed evaluate these relationships in different perspectives. However, the analysis of poverty, inequality and socioeconomic development as an only framework has been little analysed.

In Colombia studies on poverty, inequality and socio-economic development have been worked mainly to determine the link between geographical conditions and development, showing that geography affects income via the productivity of the land, the availability of natural resources, the presence of tropical diseases, and agglomeration (Sanchez and Nuñez, 2000). Other researches have also looked at the interrelationships among development, economic growth, and drug production, which suggests that poverty and the lack of economic growth and development drive both violence and drug production (Holmes, et al., 2002), which indicate the importance of this study as a contribution to the empirical analysis of poverty, inequality and socio-economic development in Colombia through a novel, panel cointegration model that takes into account education, saving and labour market. However, there has been limited empirical analysis on the relationships and causality among variables such as poverty, inequality and socio-economic development in one framework integrated regions of a developing country as it is Colombian case.

This study allows all these variables to enter a system identifying any existence of endogeneity by using a panel cointegration technique for the period 1993–2007. The contribution of this work is explored how poverty, inequality and socio-economic development are related in the Colombian departments. Following this section, Section 2 reviews the literature. Section 3 discusses the methodology. The econometric results are analyzed in Section 4. Finally, Section 5 concludes the study and offers some policy implications.

2. Literature Review

Taking into account that studies on the relationship among poverty, inequality and socio-economic development are limited all together in one structure, this work studies the literature from the double relationship.

2.1. Poverty and Inequality

Much of the studies on the relationship between poverty and inequality have demonstrated that an increase in inequality generates to more poverty and that economic growth and inequality are fundamental in changes in poverty (see e.g. Frank, 2000; Ravallion, 2002, Birdsall 2005, van der Hoeven 2008). In this context, it is key to consider both growth and income disparities simultaneously to decrease poverty and improve income distribution especially in developing countries (UNDP, 2011).

2.2. Poverty and Education

Several studies have argued that an increase in education reduces poverty mainly because more educated people are more likely to get jobs, are more productive, and earn more; quality education is important for socio-economic growth to generate better economic opportunities and incomes; and education offers social benefits that improve the situation of the poor (Barro and Lee, 2001, Raffo et al., 2007, Van der Berg, 2008).

2.3. Poverty and Labour Market

Numerous works on the impact of employment on poverty have indicated that living in poverty generates other problems in the search for employment and contributes to a long-term unemployment. At the same time, unemployment and poverty causes social exclusion (Saunders, 2002, Heady and Room, 2002, Gallie, 2003).

Therefore, unemployment increases the risk of poverty and contributes to inequality where policy makers should give emphasis to employment generation as a good strategy to decrease poverty and increase welfare.

2.4. Poverty and Socio-Economic Development

Much of the studies on the relationship between socio-economic development and poverty described an inverse relationship, and mostly implied that more socio-economic development would reduce poverty (Khan 2000, Nadkarni, 2000, Szirmai, 2005). In this context, it is important incentive socio-economic development within national policies to decrease poverty and increase development especially in developing countries.

3. Model Specification and the Empirical Strategy

In this section, we present econometrics models, the data and variables and empirical results.

3.1. Econometric Models and Estimation Methods

The panel cointegration models estimation involved the following steps: the panel unit root, cointegration tests and panel cointegration estimation.

3.1.1. Panel Unit Root Tests

We need to check that all variables are integrated to the same order. Doing this, we use panel unit root tests according to Im, Pesaran and Shin (2003).

3.1.1.1. Im, Pesaranan Shin (2003)

Im, Pesaran, and Shin (2003) initiate by postulating a separate ADF regression for each cross section:

$$\Delta y_{it} = \alpha y_{it-1} + \sum_{j=1}^{p_i} \beta_{ij} \Delta y_{it-j} + X_{it}' \delta + \epsilon_{it},$$

Once estimating the separate ADF regressions, the average of the *t*-statistics for α_i , from the individual ADF regressions, $t_{iT_i}(p_i)$:

$$t_{NT} = \left(\sum_{i=1}^{N} t_{iT_i}(p_i) \right) / N$$

The Im, Pesaran, and Shin (2003) test statistic requires specification of the number of lags and the specification of the deterministic component for each cross-section ADF equation. The test may choose to include individual constants, or to include individual constant and trend terms.

$$W_{t_{NT}} = \frac{\overline{N}\, t_{NT} - N^{-1} \sum_{i=1}^{N} E\,(t_{iT}\; p_i\;)}{\sqrt{N^{-1} \sum_{i=1}^{N} Var(t_{iT}\; p_i\;)}} \rightarrow N(0,1)$$

3.1.2. Panel Cointegration Tests

Kao (1999) Cointegration Tests describes tests under the null hypothesis of no cointegration for panel data. For estimation of long run parameter when it obtains the estimations of w_{it} and w_{it}' then we get

$$\Sigma \begin{matrix} \sigma_{\mu}^{2}\sigma_{\mu\varepsilon} \\ \sigma_{\mu\varepsilon}\sigma_{\varepsilon}^{2} \end{matrix} = \frac{1}{NT} \sum_{i=1}^{N} \sum_{t=1}^{T} w_{it}\, w_{it}'$$

To assess Σ. Ω can be calculated by

$$\Omega = \begin{matrix} \sigma_{0\mu}^{2}\sigma_{0\mu\varepsilon} \\ \sigma_{0\mu\varepsilon}\sigma_{0\varepsilon}^{2} \end{matrix}$$

$$= \frac{1}{N} \sum_{i=1}^{N} \frac{1}{T} \sum_{t=1}^{T} w_{it} w_{it}' + \frac{1}{T} \sum_{\tau=1}^{l} w_{\tau l} \sum_{t=\tau+1}^{T} (w_{it} w_{it-\tau}' + w_{it-\tau} w_{it}'$$

Pedroni (1999, 2004) proposes two types of cointegration tests. First, the panel tests are based on the within dimension approach which includes four statistics: panel v, panel ρ, panel PP, and panel ADF-statistics. The within tests (*Panel cointegration statistics*), the test proposed by Pedroni (1999) is based on the between-dimension approach, which includes the group ADF-statistic. These statistics are based on averages of the individual autoregressive coefficients related with the unit root tests of the residuals for each Colombian departments.

i. Panel t-statistic: parametric test. Augmented Dickey-Fuller *t* statistic type

$$Z_{tN,T}^{*} \equiv \; s_{N,T}^{*2} \sum_{i=1}^{N} \sum_{t=1}^{T} L_{11i}^{-2} \epsilon_{it-1}^{*2} \;^{-1\,2} \sum_{i=1}^{N} \sum_{t=1}^{T} L_{11i}^{-2}\, \epsilon_{it-1}^{*} \Delta\epsilon_{it}^{*}$$

These statistics basically pool the autoregressive coefficients across different Colombian departments for the unit root tests on the estimated residuals.

These statistics take into account common time factors and heterogeneity across state. Second, the group tests are based on the between dimension approach which includes three statistics: group ρ, group PP, and group ADF-statistics. The between tests calculated (*Group mean panel cointegration statistics*) are as follows:

i. Group t-statistic: non parametric test. Phillips-Perron *t* statistic type,

$$N^{-1/2} Z_{tN,T} \equiv N^{-1/2} \sum_{i=1}^{N} \left(\sigma_i^2 \sum_{t=1}^{T} \epsilon_{it-1}^2 \right)^{-\frac{1}{2}} \sum_{t=1}^{T} \left(\epsilon_{it-1} \Delta\epsilon_{it} - \lambda_i \right)$$

ii. Group t-statistic: parametric test. Augmented Dickey-Fuller *t* statistic type,

$$N^{-1/2} Z_{tN,T}^{*} \equiv N^{-1/2} \sum_{i=1}^{N} \left(\sum_{t=1}^{T} s_i^{*2} \epsilon_{it-1}^{*2} \right)^{-\frac{1}{2}} \sum_{t=1}^{T} \epsilon_{it-1}^{*} \Delta\epsilon_{it}^{*}$$

These statistics are founded on averages of the individual autoregressive coefficients related with the unit root tests of the residuals for each state in the panel. The tests are asymptotically standard normal distributed. For more details about this test, see Pedroni (1999).

3.1.3. Panel Cointegration Estimation

Although the cointegration methodology proposed by Pedroni allows us to test the presence of long run relationships, it is not able to provide an estimation under an error correction model framework.

In order to estimate a long-run relationship among variables under a panel setting in presence of cointegration, several estimators are proposed: Dynamic OLS (DOLS), Pool Mean Group (PMG), OLS, and Fully Modified OLS (FMOLS). In this investigation, we use the Dynamic OLS (DOLS) estimator. Kao and Chiang (2000) exhibited that both OLS and Fully Modified OLS (FMOLS) have a small sample bias and that the DOLS estimator appears to outperform both estimators.

Pedroni (2001) has also created a between-dimension, group-means panel Dynamic OLS (DOLS) estimator that includes adjustments for endogeneity and serial correlation parametrically.

The dynamic OLS (DOLS) Estimation (Kao and Chiang, 2000)

The Dynamic OLS (DOLS) estimator mentioned by Kao and Chiang (2000) is an addition of Stock and Watson's (1993) estimator. In direction to obtain an unbiased estimator of the long-run parameters, the DOLS estimator extends the static regression with the leads, lags, as well as contemporaneous values of the regressors in first differences. We use a panel model with fixed effect:

$$y_{it} = \alpha_i + x_{it}'\beta + \mu_{it} \qquad i = 1,...,N\, ;\, t = 1,....,T$$

Kao and Chiang (2000) assume that $\{y_{it}\}$ are 1x1, β is a k x 1 vector of the slope parameters, $\{\alpha_i\}$ are the intercepts, and is individual effect, μ_{it} is an error term. We assume that $\{x_{it}\}$ are k x 1 vector is autoregressive process of the first order difference:

$$x_{it} = x_{it-1} + \varepsilon_{it}$$

The Dynamic OLS (DOLS) estimator is shown in the following expression.

$$y_{it} = \alpha_i + x'_{it}\beta + \sum_{j=-q_1}^{q} c_{ij}\,\Delta x_{it+j} + v_{it}$$

3.2. Data and Empirical Results

3.2.1. Data

The data set consists of a panel of observations for the 32 Colombian departments; the analysis covers the period 1993 – 2007. In this study, it is necessary to have information on: i. *Inequality* that refers to disparities in the distribution of monetary resources within or between populations in this study GINI index is used as a measurement of differences in inequality in household incomes over time and between different Colombian departments. ii. *Socio-economic inequalities* show differences in both economic and social resources including earnings, income, education that contribute to welfare. Saving, employment rate and high school coverage were used in this study. iii. Poverty measured in this study as Unsatisfied Basic Needs (UBN) of each Colombian department and it is the dependent variable.

The data come from the National Administrative Department of Statistics (DANE), National Planning Department (DNP). The explanatory variables are calculated from information provided by the Colombian institutions that have the statistics.

These variables are indicators of factors such as level of economic development, inequality, save, the labour market and education were chosen to address the research questions that guide this analysis. To analyse the development level of each departments, the model includes inequality (the GINI index), saving, education (high school coverage), the labour market in the states is analysed using the employment rate.

3.2.2. Panel Unit Root Results

Since our data-base is collected of annually data going from 1993 to 2007, we run panel data unit root tests on all variables. In table 1 reports the results of the Im, Pesaran, Shin panel unit root tests which include an intercept and trend term. The panel unit root tests indicate all the variables are integrated of order one.

3.2.3. Cointegration Analysis

The results of cointegration analysis between inequality and economic development are reported in table 2. We use within-group tests, the panel tests are based on the within dimension approach which include the panel ADF-statistics.

Two between-group tests to check whether the panel data are cointegrated, the group tests are based on the between dimension approach which includes two statistics: group PP, and group ADF-statistics. Table 2 illustrates that among the tests of Pedroni (1999), there is at least one that shows that we reject the null hypothesis of no cointegration in the equation. This allows us to estimate the panel data cointegration relationships.

Table 2 reports both the within and between dimension panel cointegration test statistics. All four test statistics reject the null hypothesis of no cointegration at the 1% significance level.

Table 1. Unit Root test (Individual effects and linear trends)

Test	UBN		GINI		Saving		Employment rate		High school Coverage	
	Level	1st difference	Level	1st difference	Level	1st difference	Level	1st difference	Level	1st difference
Im, Pesaran and Shin W-stat	7.536 (1.00)	-3.582 [a] (0.00)	8.927 (1.00)	-4.865 [a] (0.00)	-0.668 (0.25)	-11.00 [a] (0.00)	-0.559 (0.28)	-10.36 [a] (0.00)	-2.719 (0.00)	-16.73 [a] (0.00)

Note: [a]$p<0.01$, [b] $p<0.05$, [c] $p<0.1$.

Table 2. Panel data cointegration tests

Cointegration Test		Test statistics
Kao Panel Cointegration Test ADF t-statistics		-3.710[a] (0.000)
Pedroni Panel Cointegration Test		Test statistics
Panel Cointegration Test	Panel ADF-Statistic	-4.492[a] (0.000)
Group Mean Cointegration Test	Group PP-Statistic	-13.517[a](0.000)
	Group ADF-Statistic	-6.967[a] (0.000)

Note: [a]p<0.01, [b] p<0.05, [c] p<0.1.

3.2.4. Panel Estimation and Empirical Results

The strategy used to estimate a long-run relationship between poverty and economic development is based on one estimator with error correction: Dynamic OLS (DOLS). The DOLS model used is:

$$y_{it} = \alpha_i + \beta_1 F_{it} + \sum_{j=-1}^{1} \beta_{j+3}\,\Delta F_{i,t+j} + \gamma_1 X_{it} + \sum_{j=-1}^{1} \gamma_{j+3}\,\Delta X_{i,t+j} + \mu_{i,t}$$

Table 3 presents the results from Dynamic OLS (DOLS) estimates of the cointegration relationship. The estimated coefficients for several of the variables were statistically significant for all departments of Colombia. The results of Dynamic OLS (DOLS) estimates ratify the existence of a long-run relationship between poverty, inequality and economic development.

In general, the results indicate that higher inequality generates higher poverty whereas higher saving, employment rate and education leads lesser poverty indicating that social and economic variables are important to achieve development and economic growth to the degree that increase welfare of population.

The relationship between unsatisfied basic needs and Gini index is not significant which concurs with Mehanna et al. (2004). However, in Colombian departments is evident that disparities in the incomes have generated higher poverty and lesser development.

Saving and poverty indicate significant and inverse relationship indicating that an increase in the saving lead lesser poverty in Colombian departments. These results concur with Ivanic (2005) that determined that rising of saving tend to decrease poverty. Employment and poverty indicate that an inverse and significant relationship indicating that higher unemployment leads higher poverty. Several studies have demonstrated that adequate policy strategies against unemployment is key to decrease poverty because increase productivity and income of population which mean higher development and welfare of population (OECD, 1996 and 2009, ILO, 2005).

Education is key strategy to decrease poverty; results indicate significant and inverse relationship. Education can enhance the skills and productivity among poor households which should increase their incomes (Oxaal, 1997, Ladd, 2011). All finding of this study are important to develop adequate policies that decrease poverty whereas increase development and socio-economic development of population especially in developing countries.

Table 3. Panel DOLS long-run estimates

DOLS estimates				
$UBN = \alpha\, GINI - \beta\, SAVE - \eta\, EMPL - \varphi\, CVHS$				
UBN =	0.121 GINI	- 0.182 SAVE	- 0.030 EMPL	- 0.059 CVHS
	(13.471)	(0.658) [a]	(0.331) [a]	(0.375) [a]
Obs: 480	Wald Chi^2 (4) = 673.96			
	Prob > chi^2 = 0.000			

Notes: Standard errors are reported in parentheses. [a] p<0.01, [b] p<0.05, [c] p<0.1.

CONCLUSION

Several studies have analysed the relationship of poverty, inequality and socio-economic development from a two dimension system. In this analysis, poverty, inequality and socioeconomic development is worked as an only framework using panel cointegration techniques for Colombian departments over the period 1993-2007 with the purpose to identify the connections between these variables.

The econometric results suggest that an increase in the saving, employment rate and education decrease poverty measured as unsatisfied basic needs. On the other hand, an increase in the inequality measured as Gini index leads a higher poverty in the Colombian departments. Surprisingly, the Gini coefficient has no significant relation with poverty whereas other variables were significant.

From a policy perspective, socio-economic development policies should incentive key strategies to guarantee economic growth and development and decrease poverty through higher equality, better employment conditions and quality education that generate higher income and welfare for the population.

ACKNOWLEDGMENT

The author is grateful for the support provided by Deutscher Akademischer Austausch Dients (DAAD), the University of Göttingen and the University of La Salle. Any remaining errors are the responsibility of the author.

REFERENCES

Adams R., 2004. Economic Growth, Inequality and Poverty: Estimating the Growth Elasticity of Poverty. *World Development* 32, 1989–2014.

Baltagi (ed.), Nonstationary Panels, Panel Cointegration, and Dynamic Panels, Advances in Econometrics, 15, JAI, Amsterdam, 161-178.

Barro, R., Lee, J., 2001. Schooling quality in a crosssectionof countries. Eco nomica, 68(272), 465-488.

Birdsall, N., 2005, 'The World is Not Flat: Inequality and Justice in our Global Economy', WIDER Lecture 9, United Nations University-WIDER, Helsinki.

Bonal X., 2007.On global absences: Reflections on the failings in the educationand poverty relationship in Latin America. *International Journal of Educational Development* 27, 86–100.

Bourguignon, F, 2004, 'The Poverty-Growth-Inequality Triangle', mimeo, World Bank, Washington, DC.

Breitung, J., 2000. *The Local Power of Some Unit Root Tests for Panel Data*. In B.

Cotte, A., 2011a. Socio-Economic Development and Violence: An Empirical Application for Seven Metropolitan Areas in Colombia. *Peace Economics, Peace Science and Public Policy* (In press).

Cotte, A., 2011b. Economic development and growth in Colombia: An Empirical Analysis with Super-efficiency DEA and Panel Data Models. *Socio-Economic Planning Sciences*. DOI doi:10.1016/j.seps.2011.07.0.

Choi, I., 2001. Unit Root Tests for Panel Data, *Journal of International Money and Finance* 20, 249–272.

European Commission, 2010. Why socio-economic inequalities increase?.Facts and policy responses in Europe.Directorate-General for Research Socio-economic Sciences and Humanities EUR 24471.http://ec.europa.eu/research/social-sciences/pdf/policy-review-inequalities_en.pdf

Frank, R.H., 2000. Does growing inequality harm the middle class? *Eastern Economic Journal* 26, 251–264.

Gallie, D., Paugam, S., Jacobs S., 2003.Unemployment, Poverty, and Social Isolation. Is There a Vicious Circle of SocialExclusion? *European Societies* 5, 1-31.

Heady, C., Room G., 2002. Patterns of Social Exclusion: Implications for Policy and Research. In Barnes M. et al., Povertyand Social Exclusion in Europe. Edward Elgar. Cheltenham, United Kingdom, pp. 146-154.

Holmes J., Gutiérrez S., Curtin K., 2002. Drugs, Violence and Development in Colombia: A Department Level Analysis. *University of Texas at Dallas.Latin American Politics and Society*.

ILO, 2005.Decent Work andPoverty Reduction Strategies.National Policy Group Policy Integration Department.http://www.ilo.org/dyn/infoecon/docs/627/F286029 509/Decent Work-PRSstext.pdf

Im, K., H. Pesaran, and Shin Y., 2003. Testing for unit roots in heterogeneous panels. Journal of Econometrics, 115, 53-74.

Ivanic, M., 2005.Implications of household saving on poverty.Department of Agricultural Eonomics.Purdue University.https://www.gtap.agecon.purdue.edu/resources/download/1918.pdf

Kao, C., Chiang, M.H., 2000. On the estimation and inference of a cointegrated regression in panel data. In: Baltagi, B.H. (Ed.), Nonstationary Panels, Panel Cointegration and Dynamic Panels. Advances in Econometrics, 15. *Elsevier*, New York, pp. 179– 222.

Khan, M.H., 2000.Rural poverty in developing countries. *Finance and Development* 37, 26–29.

Kuznets, S., 1955.Economic growth and income inequality. *American Economic Review* 45, 1–28.

Ladd H., 2011. Education and Poverty: Confronting the Evidence. Working Papers Series SAN11-01.http://sanford.duke.edu/research/papers/SAN11-01.pdf

Levin A., Lin, C., Chu, C., 2002. Unit Root Tests in Panel Data: Asymptotic and Finite-Sample Properties. *Journal of Econometrics* 108, 1–24.

Maddala, G. and Wu, S., 1999. A comparative study of unit root tests with panel data and a new simple test. *Oxford Bulletin of Economics and Statistics*, 61, 631-652.

Mehanna R., 2004. Poverty and economic development: notas direct as it may seem. *Journal of Socio-Economics* 33, 217–228.

Nadkarni V., 2000. Poverty, environment, development:a many-patterned nexus. *Economic and Political Weekly,* 1.

OECD, 1996.Reducing poverty while increasing employment:A primer on alternative strategies.And a blueprint.Economic Studies No . 26. http://www.oecd.org/eco/productivityandlongtermgrowth/2502824.pdf

OECD 2009.The Role of Employment and Labour Markets in the Fight against Poverty.http://www.oecd.org/dac/povertyreduction/43280231.pdf

Oxaal Z., 1997. Education and Poverty: A Gender Analysis. *Swedish International Development Cooperation Agency (*Sida). Report 53.

Pedroni, P., 1999. Critical values for cointegration tests in heterogeneous panels with multiple regressors. *Oxford Bulletin of Economics and Statistic*s 61, 653–670.

Pedroni, P., 2000. Fully modified OLS for heterogeneous cointegrated panels.*Advanced in Econometrics* 15, 93–130.

Pedroni, P., 2001. Purchasing power parity tests in cointegrated panels. *Review of Economics and Statistic*s, Vol. 83, 727-731.

Pedroni, P., 2004. Panel cointegration: asymptotic and finite sample properties of pooled time series tests with an application to the PPP hypothesis: new results. *Econometric Theory* 20, 597–627.

Policy Responses Overcoming Factors in the Intergenerational (Profit), 2007.Transmission of Inequalities, *Final Activity Report.*

Ravallion, M., 2002, Are the Poor Protected from Budget Cuts? Evidence for Argentina. *Journal of AppliedEconomics* V, 95–121.

Raffo, C., Dyson, A., Gunter, H., Hall, D., Jones, L., Kalambouka, A., 2007. Education and poverty: a critical review of theory, policy and practice.Manchester: *Joseph Rowntree Foundation andUniversity of Manchester.*

Sanchez F., Nunez J., 2000. Geography and Economic Development in Colombia: A Municipal Approach. Inter-American Development Bank. *Research Network Working paper* #R-408.

Saunders P., 2002. The Direct andIndirect Effects ofUnemployment onPoverty and Inequality. *Social Policy Research Centre. SPRC Discussion Paper* No. 118.

Stock, J. and M. Watson, 1993, A simple estimator of cointegrating vectors in higher order integrated systems, *Econometrica* 61(4), 783-820.

Szirmai A., 2005. The Dynamics of Socio-Economic Development: An Introduction. *Cambridge University Press.*

UNDP, 2011.Towards Human Resilience: Sustaining MDG Progress in an Age of Economic Uncertainty. http://www.undp.org/content/undp/en/home/librarypage/poverty-reduction/inclusive_development/towards_human_resiliencesustainingmdgprogressinanageofeconomicun.html

van der Hoeven, R., 2008.Income Inequality', in Peil and van Staveren (eds.), Handbook of Economics andEthics, Edward Elgar, Cheltenham.

van der Berg, S. 2008. Poverty and education. Education policy series UNESCO. http://www.iiep.unesco.org/fileadmin/user_upload/Info_Services_Publications/pdf/2009/EdPol10.pdf

Index

#

A

B

C

D

E

F

G

H

I

N

O

P

Q

R

T

U

V

W

X

Y

Z